Cell Motility in Cancer Invasion and Metastasis

Cancer Metastasis – Biology and Treatment

VOLUME 8

Cell Motility in Cancer Invasion and Metastasis

Edited by

Alan Wells
Department of Pathology, Pittsburgh VAMC,
University of Pittsburgh, PA, USA

Springer

A C.I.P. Catalogue record for this book is available from the Library of Congress.

ISBN-10 1-4020-4008-3 (HB)
ISBN-13 978-1-4020-4008-5 (HB)
ISBN-10 1-4020-4009-1 (e-book)
ISBN-13 978-1-4020-4009-2 (e-book)

Published by Springer,
P.O. Box 17, 3300 AA Dordrecht, The Netherlands.

www.springer.com

Printed on acid-free paper

Printed in the Netherlands.

TABLE OF CONTENTS

LIST OF CONTRIBUTORS

Carla Boccaccio
Institute for Cancer Research and Treatment, University of Turin Medical School, Candiolo, Italy

Anja K. Bosserhoff
Institute of Pathology, University of Regensburg, Regenburg, Germany

John G. Collard
Department of Cell Biology, The Netherlands Cancer Institute, Amsterdam, The Netherlands

Paolo M. Comoglio
Institute for Cancer Research and Treatment, University of Turin Medical School, Candiolo, Italy

John Condeelis
Department of Anatomy and Structural Biology, Albert Einstein College of Medicine, Bronx, NY, USA

Candece L. Gladson
Department of Pathology, University of Alabama at Birmingham, Birmingham, AL, USA

Sumanta Goswami
Department of Anatomy and Structural Biology, Albert Einstein College of Medicine, Bronx, NY, USA

Jennifer Rubin Grandis
Departments of Otolaryngology and Pharmacology, University of Pittsburgh, Pittsburgh, PA, USA

Linda Griffith
Biological Engineering Division and Biotechnology Process Engineering Center, Massachusetts Institute of Technology, Cambridge, MA, USA

Amra Hajdo-Milašinović
Department of Cell Biology, The Netherlands Cancer Institute, Amsterdam, The Netherlands

Irene H. L. Hamelers
Department of Cell Biology, The Netherlands Cancer Institute, Amsterdam, The Netherlands

Wendy Ingman
Departments of Developmental and Molecular Biology and Obstetrics and Gynecology and Women's Health, Albert Einstein College of Medicine, Bronx, NY, USA

Elizabeth J. Joslin
Biological Engineering Division, Massachusetts Institute of Technology, Cambridge, MA, USA

Sourabh Kharait
Department of Pathology, University of Pittsburgh, Pittsburgh, PA, USA

Douglas A. Lauffenburger
Biological Engineering Division and Center for Cancer Research, Massachusetts Institute of Technology, Cambridge, MA, USA

Elaine Y. Lin
Departments of Developmental and Molecular Biology and Obstetrics and Gynecology and Women's Health, Albert Einstein College of Medicine, Bronx, NY, USA

Arthur M. Mercurio
Department of Cancer Biology, University of Massachusetts Medical Center, Worcester, MA, USA

Alexander E. Mertens
Department of Cell Biology, The Netherlands Cancer Institute, Amsterdam, The Netherlands

Meera Natarajan
Department of Pathology, University of Alabama at Birmingham, Birmingham, AL, USA

Jeffrey W. Pollard
Departments of Developmental and Molecular Biology and Obstetrics and Gynecology and Women's Health, Albert Einstein College of Medicine, Bronx, NY, USA

Isaac Rabinovitz
Department of Pathology, Beth Israel Deaconess Medical Center, Boston, MA, USA

Gabriel Rezonzew
Department of Pathology, University of Alabama at Birmingham, Birmingham, AL, USA

Jeffrey E. Segall
Department of Anatomy and Structural Biology, Albert Einstein College of Medicine, Bronx, NY, USA

Dong-Wan Seo
Cell & Cancer Biology Branch, Center for Cancer Research, National Cancer Institute, National Institutes of Health, Bethesda, MD, USA

Gene P. Siegal
Departments of Pathology, Surgery, and Cell Biology, University of Alabama at Birmingham, Birmingham, AL, USA

Raj K Singh
Diversified Scientific Inc., Birmingham, AL, USA

William G. Stetler-Stevenson
Cell & Cancer Biology Branch, Center for Cancer Research, National Cancer Institute, National Institutes of Health, Bethesda, MD, USA

Michelle R. Stettner
Department of Pathology, University of Alabama at Birmingham, Birmingham, AL, USA

Donna Beer Stolz
Center for Biological Imaging, University of Pittsburgh, Pittsburgh, PA, USA

Sufi Mary Thomas
Department of Otolaryngology, University of Pittsburgh, Pittsburgh, PA, USA

Kien Tran
Department of Pathology, University of Pittsburgh, Pittsburgh, PA, USA

Weigang Wang
Department of Anatomy and Structural Biology, Albert Einstein College of Medicine, Bronx, NY, USA

Alan Wells
Department of Pathology, Pittsburgh VAMC & University of Pittsburgh, Pittsburgh, PA, USA

Jeffrey Wyckoff
Department of Anatomy and Structural Biology, Albert Einstein College of Medicine, Bronx, NY, USA

Chengsen Xue
Department of Anatomy and Structural Biology, Albert Einstein College of Medicine, Bronx, NY, USA

Clayton Yates
Department of Pathology, University of Pittsburgh, Pittsburgh, PA, USA

Kun Yuan
Department of Pathology, University of Alabama at Birmingham, Birmingham, AL, USA

PREFACE

This book, "Cell Motility in Cancer Invasion and Metastasis", is the fifth contribution in the Series "Cancer Metastasis – Biology and Treatment". Here, we focus neither on a molecule or set of molecules nor on a tissue, but on a cell behavior – motility. As the research community as a whole investigates organismal biology at more and more complex levels, the dynamic nature of the underlying processes come to the fore. Cell movement and migration is an integration of such dynamic multifaceted processes, involving not only the moving cell itself and its supporting matrix, but also adjacent and distant cells. While cell motility has been appreciated and studied for decades, only more recently has its critical role in the pathophysiology of tumor dissemination been probed extensively. This book was commissioned to note the current state of understanding in this arena, and highlight new avenues and key questions for future investigation.

The chapters that comprise this book are presented in three themes. Prior to these themes, I present an introductory chapter that overviews cell motility and presents evidence that suggests that tumor invasiveness may be considered a disease of dysregulated motility. This chapter provides a common framework for the ensuing contributions. The first theme of three chapters focuses on technological breakthroughs that will enable novel modes of study. Yuan and colleagues describe matrices available for study and newer ones being developed. As motility is a biophysical process that depends on the substratum not only for dynamic tractive and contractile forces but also for active signals, the choice of model system determines the range of responses noted. It has been realized that complex matrices are required to provide in vitro models relevance to the human condition. The next chapter mainly describes innovative work from the laboratories of Segall and Condeelis utilizing a novel intravital imaging capability. These studies take the investigation of migration during tumor progression directly into the tumor itself in a living animal. This retains the relevant host environment. Yates, Stolz and Griffith describe novel intermediary modes of study in which organ environments are recreated *ex vivo* to enable higher throughput and more direct manipulations in a system that retains much of the in vivo complexity.

Five chapters comprise the second theme of molecular regulators of tumor cell motility. These select five key classes of molecules that are amenable for intervention. While not comprehensive, the chapters focus on arguably the molecular classes best examined and justified in the context of tumor progression. Starting from outside the cell, Joslin and Lauffenburger explore the concept that peptide growth factors often drive motility in an autocrine modality as a form of cellular sonar; this is in addition to the well-described paracrine chemotactic signaling. As a wide variety of tumors, and practically all carcinomas present autocrine signaling loops involving the motogenic erbB family of receptors, these are highlighted. The next chapter by Comoglio and Boccaccio focuses on peptide growth factors that are defined by their exquisite ability to loosen the bonds between cells and enable an transition to the mesenchymal state. This sets the stage for invasive growth as individual cells break from the primary tumor mass and attain the ability to survival and proliferate as individual cells. At the interface of the cell and surrounding barrier stroma are the proteinases that remodel both the matrix and the cell surface allowing for new cellular interpretations of the external milieu; new advances in these studies are described by Stetler-Stevenson and Seo. Rabinovitz and Mercurio detail the molecules that most ubiquitously provide both traction and interpretation of that supporting matrix, the integrins. In particular, one integrin appears to redirect its activities from maintaining sessile hemidesmosomes in normal epithelia to driving invasiveness in a variety of carcinomas. In the last chapter in this theme, Collard and colleagues describe regulators of the actual motive force, the actin cytoskeleton. In response to signals from growth factor and traction receptors, among others, a family of small GTPases control cytoskeleton assembly, disassembly and connectivity to the substratum.

The last theme aims to integrate these molecules and processes in tumors that exemplify the issues of motility in tumor dissemination. Four neoplasias were selected based on clear human correlative and experimental causative implications of increased motility leading to both localized invasiveness and distant metastases. In all these cases, even the local invasiveness results in significant morbidity and mortality. The first tumor discussed by Stettner, Natarajan and Gladson is highly invasive, with motility being strongly linked to growth factor signaling, adhesiveness, and cell-matrix interactions. A second tumor with significant local invasiveness driven by autocrine growth factor signaling is squamous carcinoma of the oral cavity, as related by Thomas and Grandis. Bosserhoff presents a tumor, melanoma, in which invasiveness is a first step to distant metastases. Lastly, prostate cancer represents a complex invasion/metastasis situation in which motility plays a central, and rate-limiting role.

The compilation of these chapters is a compendium that relates current concepts and techniques, and highlights key questions in the role of motility in tumor dissemination. We hope that these chapters stimulate further research in this emerging field. As editor of this book, I must apologize to the readers and investigators looking for chapters addressing other aspects of tumor dissemination or other tumor types. Particular to the subject matter, this book focuses on the invasion and metastasis of solid tumor, as motility of hematopoietic cells and their neoplasias present unique issues and challenges. In addition, only select key regulators are described at length, not because others are not involved, but due to the strong linkage of these to both motility and tumor dissemination. No doubt future studies will uncover additional key effectors that can be targeted. It is only by focusing on these select topics and representative tumors that we can provide a cohesive and internally comprehensive view that may serve as a starting point for future investigations.

ALAN WELLS
Department of Pathology
Pittsburgh VAMC and
University of Pittsburgh
Pittsburgh, Pennsylvania
USA

1 April 2005

Chapter 1

MOTILITY IN TUMOR INVASION AND METASTASIS – AN OVERVIEW

Alan Wells
Department of Pathology, Pittsburgh VA Medical Center and University of Pittsburgh, Pittsburgh, PA, USA

Abstract: Most cancer morbidity and mortality derives from the dissemination of the tumor cells and subsequent growth in ectopic locations. Local invasion of adjacent normal tissue and adnexia compromise homeostatic functions and prevent surgical or radiological extirpation of the tumor. Recent investigations highlight the central role that induced motility plays in promoting this spread. Metastatic seeding and growth, that dramatically increase tumor burden and relegate local therapies, require additional complex acquired behaviors including resistance to anoikis and shear stresses. Still, tumor cell motility and associated biophysical processes such as adhesion modulation, are central to this terminal progression. Our growing understanding of the function of motility, and key molecular switches that govern this behavior, provide new targets for rationale therapeutics aimed at arresting tumors in their tracks.

Key words: tumor progression, invasion, metastasis, adhesion, protrusion, biophysical processes, migration

1. INTRODUCTION

Most cancer morbidity and mortality occurs upon spread of the tumor from an original localized site. Primary tumors are, by and large, accessible for surgical removal or radiological elimination. However, once the tumor extends beyond the physiological borders, these approaches are either rendered ineffective or engender significant morbidity and mortality in themselves. Subsequent therapeutics suffer from the need to be globally or broadly applied with the concomitant toxicities. Limiting this dissemination

A. Wells (ed.), Cell Motility in Cancer Invasion and Metastasis, 1-23.

would provide palliative benefit, improve physical targeting of cytotoxic therapies, or just stabilize the tumor in a manageable state. Therefore, such limitation of tumor dissemination per se is an appropriate goal for new therapies. However, to realize this in a rational and targeted manner, tumor cell behaviors that enable dissemination must be defined and their key molecular controls identified.

Tumor dissemination takes two forms, invasion and metastasis (Table 1). Localized invasion of unaffected tissue and adnexia compromises normal homeostatic functions of the organs. This invasion is distinct from non-invasive but otherwise growing localized tumors in that the invasive tumor cells breach physiologic boundries and/or organ capsules to interdigitate with normal cells. That this ability is not simply related to cell proliferation was evident when it was noted that the growth rate of invasive tumors was not appreciably greater than encapsulated counterparts (1). A number of cell properties were examined as the basis for this contiguous spread. Induced cell motility has come to be recognized as the dominant regulator of tumor invasion (2), particularly as matrix degradation and remodeling is now understood to be limited rather than extensive, and a part of or partner to motility (3, 4) (see Chapter 7).

Table 1. Properties needed for invasion and metastasis.

Property acquired	Invasion	Metatasis
Dissociate from tumor mass	partial	full
Reorganize/remodel matrix	at invasion	at metastasis
Migrate	major	limited-to-major
Survive in conduit transit	N/A	major
Recognize endothelial cells	N/A	limited-to-major
Proliferate within ectopic stroma	TBD	major

N/A – not necessary, TBD – to be determined

The changes the tumor cell must undergo to survive and proliferate in the invasive state are not well documented. Often the invasion, as in the case of glioblastomas and astrocytomas, remains mainly within the original organ and thus does not require adaptation to a new environment. Even when invasive tumor cells reach into ectopic sites, as for many carcinomas, the contiguity to the primary organ enables wholesale transfer of orthopic stromal cells. This would alleviate the barriers to proliferation normally encountered in an ectopic environment. Thus, the driver of invasion becomes the common rate-limiting step to be targeted, and that is the cell behavior of induced motility.

Metastasis requires multiple new tumor cell properties. Not only must the tumor cell reach a conduit by intravasation or capture in a poorly formed vascularity, but must survive the de-adherent state, extravasate from the conduit, and finally survive and proliferate in an ectopic site dissociated from the normal stromal support elements (5, 6). Most of these behaviors distinguish metastasis from invasion, and define these as two separate developments that may occur in tumors, rather than invasion being simply a stepping stone to metastasis. Clinically, it has long been recognized that some small and localized tumors have a propensity to metastasize well before any evidence of invasion is noted; this has been bolstered by the finding of copious tumor cells in circulation even in non-invasive tumors (7). One cell behavior that they share, however, is induced motility (8, 9); this is involved at both the intravasation and extravasation steps and possibly to find a conducive site in the ectopic organ.

Induced motility of tumor cells and its role in tumor progression is the focus of this compendium. By induced cell motility, or simply motility, the authors are referring to upregulated motility caused by changes noted during carcinogenesis and progression. This often results from the establishment of autocrine stimulatory loops, but may be secondary to altered adhesion receptor profiles or changes in the motility machinery. The chapters will either present the tools by which tumor progression is studied, highlight the molecular cell biology underlying cancer dissemination, or describe individual tumor types that are illustrative of these behaviors and control points. This first chapter will set a common groundwork.

2. EPITHELIAL-MESENCHYMAL TRANSITIONS UNDERLIE TUMOR PROGRESSION

Physiological tissue architecture prevents the migration required for tumor cell invasion and metastasis. This is particularly evident in epithelial organs and their derived carcinomas. Cell-cell adhesion dictates the cell asymmetry that belies the organ function. In many epithelial organs, such as mammary and prostate glands, this cell asymmetry is mirrored by locale-specific extracellular communications (10). Basal stromal cells produce growth factors, in particular TGFα and KGF/FGF7 that act on basolateral cognate receptors in a paracrine fashion. $\alpha 6\beta 4$ Integrins anchor the cells to the underlying basement membrane that is produced by both the stromal and epithelial cells. E-cadherin-dominated adherens junction establish the cellular asymmetry as various growth factors, including those for the EGFR and other expressed receptors, are secreted from the apical membrane domains. An autocrine stimulatoray loop is prevented by the E-cadherin-mediated tight junctions. Such asymmetric cells are non-motile and proliferation inhibited by the cadherin-mediated junctions and other cell-cell communications.

Carcinogenic transformation releases these constraints. Cell undergo a epithelial-mesenchymal transition reminiscent in many aspects to that noted during wound repair (11-13) (see Chapter 9). During the wound healing response, the injury breaks the integrity of the epithelial layer allowing for autocrine signaling by EGFR and the met receptor for scatter factors that drives the transition further (14, 15). However the cells redifferentiate upon cell-cell contact as the wound is re-epithelialized, and the proliferative and motility responses are abrogated. In the case of carcinogenic transformation, a cell-intrinsic change enables the epithelial-mesenchymal transition to occur and continue despite cell-cell contact (16).

The initiating carcinogenic change for the epithelial-mesenchymal transition is still undetermined and likely can be subsumed by a number of gene alterations. Ras is a prime candidate for triggering the diverse proteomic alterations (14, 17), with loss of E-cadherin, changes in integrin profiles to move $\alpha6\beta4$ to the cell front, and an altered cytoskeletal architecture to enable motility (see Chapter 8). In addition, the basement membrane composition is altered to include the pro-motogenic molecules tenascin-C (18-20) and laminin-5 (21). These components, noted during ontogeny, wound healing, and tumor invasion, are produced by both the mesenchymally-transitioned epithelial cells and the reactive stromal cells.

Other initiators must exist in addition to oncogenic ras mutations. In a number of carcinoma cells (see Chapter 13), the epithelial-mesenchymal transition occurs in the absence of mutated ras or p53, and is reversible upon inhibition of growth factor signaling (22, 23) or autonomous expression of E-cadherin (24). In fact, we have found that suppression of EGFR signaling in DU145 prostate carcinoma cells leads to upregulation of the E-cadherin/catenin complex and formation of cell-cell adhesion (23). The same is noted when inhibiting autocrine signaling through c-met (15, 25) (see Chapter 6). This is not unexpected as the scatter factors and many other growth factors, including the carcinoma-ubiquitous EGFR ligands, downregulate E-cadherin-mediated cell-cell adhesions. Thus, we are left with a reinforcing cascade of events in which loss of the E-cadherin-mediated tight junctions leads to autocrine signaling through receptors which further suppress cell-cell adhesions (Figure 1).

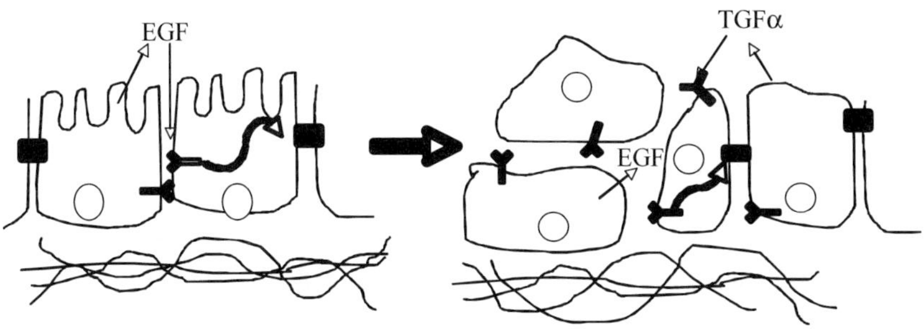

Figure 1. Schematic of reinforced cell-cell de-adhesion. Disruption of the cell-cell tight junctions allows apically-secreted growth factors access to their basolaterally-presented receptors. As growth factor autocrine loops are established, signaling from the growth factor receptors further downregulate E-cadherin-mediated junctions. Shown here is the situation for the EGFR signaling loop in many glandular epithelium and adenocarcinomas.

The implications for tumor motility are immense. In the differentiated state, the cells are essentially non-motile with the exception of lateral (or vertical in the case of keratincytes) shuffling related to normal cell turnover. The cytoskeletal architecture creates a rigid scaffold and the asymmetry of the cell receptors limits motogenic signaling. With the transition to the mesenchymal state, the cell attains a motogenic phenotype (26, 27). The cytoskeleton is reorganized to enable a protrusive actin meshwork at the invadipodia and stress fibers to retract the tail (see Chapters 9 and 3). The now-realized autocrine stimulatory loops are motogenic (see Chapter 4). In the face of underlying genetic alterations that prevent reversion of this phenotype as during physiological wound repair and ontogeny, these events serve to further the fibroblastoid phenotype and motility noted in carcinoma and sarcoma cells (28).

3. BIOPHYSICAL MODEL OF CELL MOTILITY

Cell locomotion consists of moving the cell body from one locale to a distant one, while being attached to a substratum; this involves active motility from the cell itself. This latter point distinguishes locomotion from a passive dissemination noted when tumor cells achieve vascular conduits. The term motility is often used as synonymous with locomotion but also can denote the cell undergoing the biophysical cycle described below without actually translocating as persistence approaches zero. The motility cycle for adherent cells has been best described in fibroblasts (hematopoietic cells and yeast locomote under related principles some of which are nontransferable likely due to differing size, speed and adhesion regimens). These principles can be transferred to disseminating cancer cells as these have undergone a mesenchymal transition to approach the biophysical and biochemical state of fibroblasts. Furthermore, when carcinoma and glioblastoma cells in particular, have been queried for motility, the biophysical and biochemical events have mirrored those in fibroblasts; though there are some cancer-specific events due to presence of autocrine stimulatory loops and altered adhesion and signaling profiles. Thus, this discussion will provide an overall foundational model incorporating information from both fibroblasts and cancer cells.

The foundational model of cell motility flows from a biophysical dissection of the process presented earlier (29) (Figure 2). Prior to actual motility, the cellular skeleton must be labilized to enable the structural changes the cell undergoes. Locomotion then involves lamellipodial extensions at the leading edge, new attachments to stabilize the extension, transcellular contractility, and detachment at the rear. Motility is usually conceived as multiple cycles of these steps, with such a sequential nature often appearing so when locomoting cells are viewed. However, many of the underlying biochemical events occur simultaneously, particularly in rapidly moving cells. This creates a need for temporo-spatial control of the events. For example if cell contractility extended into lamellipod prior to being stabilized by adhesions at the tip, the nascent extension would be retracted abrogating locomotion. Overexpression or overactivation of motility-associated signaling pathways actually leads to failed locomotion as the various biophysical processes lack coordination (30, 31). Thus, locomotion results from a highly orchestrated and segregated control machinery.

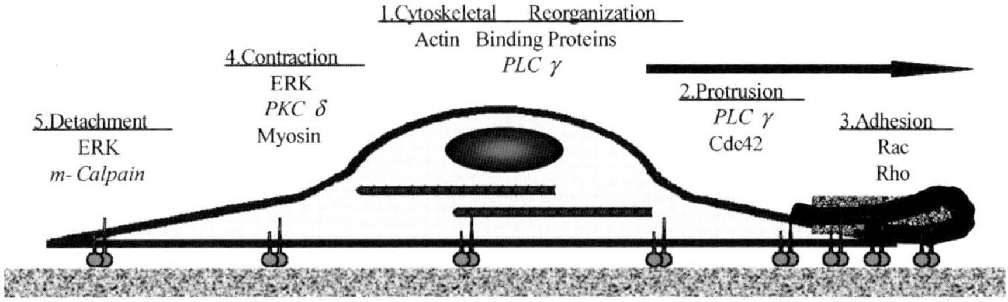

Figure 2. Transition to a motile phenotype involves multiple coordinated changes. Shown are key steps in locomotion with the stylized order superimposed. Underneath the biophysical step are key molecular switches implicated in transmitting the cues to actions for both adhesion and growth factor receptor-initiated motility (plain text) and selectively activated for growth factor receptor-initiated motility (italics).

Locomotion can be triggered by a number of external signals. Adhesion receptors not only provide the adhesive and traction forces, but can generate motility signals from binding to substratum ligands. This mode of motility is referred to as haptokinesis or haptotaxis depending on the absent or presence of a substratum gradient, respectively. Other receptors, those for peptide growth factors/cytokines are most relevant to tumor progression, drive motility by triggering biochemical regulatory cascades; the tractile forces are still provided by adhesion receptors. This mode is referred to as chemokinesis or chemotaxis. What distinguishes the two modes of motility in addition to the inducing receptor are both quantitative, growth factor motility is generally faster and recruits a higher fraction of the cells, and qualitative, a few key regulatory switches appear differentially activated. However, the clear cut distinctions are blurring. First, classic growth factor receptors can signal from matrix-bound and even matrix-intrinsic ligands (19, 21, 32-34), and this can drive motility and provide for adhesion (35). Integrins can be triggered to induce motility from matrix-liberated fragments, as demonstrated by the situation of the α6β4 integrin in carcinoma progression (36, 37) (see chapters 6 and 8). What appears to be key is not the mode of triggering motility but the strength and duration of signaling, and the downstream signaling pathways activated.

In the following chapters there will be exposition of key regulatory biochemical signaling pathways that control motility. Adhesion receptor-mediated (see Chapter 8) and growth factor receptor-mediated (see Chapter 6) induction of lcoomotion will be explored as to initiating steps. Other chapters will focus on key intracellular molecular switches (see Chapters 6 and 9) and extracellar determinants (see Chapter 7). The focus of these are not to describe the operative cytoskeletal and membrane events that occur to produce motility; these alone would fill a book (38, 39). Rather, we aim to highlight key points in the regulatory machinery that can be, and are in many cases actually being attacked to limit tumor progression.

4. BIOCHEMICAL CONTROL POINTS OF CELL MOTILITY

Recent investigations have shed light on many of the critical regulatory biochemical cascades that underlie the foundational model of locomotion. These will be described briefly here, with their cognate biophysical processes. The initial event of cytoskeleton labilization to enable cell shape change is linked to the weakening of cell-cell and cell-substratum adhesions (40, 41). The cell-cell adhesion may be lost by cell intrinsic changes related to carcinogenic mutations (42), or driven by scatter factor signaling (15, 41).

While the signaling pathways between receptor signaling and loss of adhesion are still subject to investigative extirpolation, it is clear that catenin signaling and localization is key to the mesenchymal transition of the cancer cells and the resultant cytoskeletal reorganization. Growth factor receptor signaling most evidently affects cytoskeleton structure when cells are in the mesenchymal state. This is driven in part by PLCγ hydrolysis of phosphoinositide (4,5) bisphospate (PIP_2). Removal of this important membrane docking motif not only releases bio-active signals (diacylglycerol (DAG) for protein kinase C species, and inositol trisphosphate (IP_3) for intracellularly stored calcium release), but removes cytoskeleton regulating proteins from the membrane. Release of gelsolin, cofilin, profilin, and capG enables them to perform their actin filament severing and monomeric actin nucleating functions to initiate new actin networks that can provide protrusive forces (43-47). This lability of the actin cytoskeleton allows for the cell shape changes necessary for a tumor cell to migrate through a dense matrix and extravasate through a functional endothelium.

The cytoskeletal reorganization is linked to changes in cell-substratum adhesion in two major ways. First, the existence of stable, strong focal adhesions is interrelated with the existence of tensed actin cytoskeleton and stress fibers (48, 49). The loss of these structures, and appearance of more labile close contacts and focal contacts, is coincident with changes in the actin cytoskeleton (40). PIP_2 hydrolysis might play a role in lossening focal adhesions. As a key linker between the integrin adhesion receptors and the cytoskeleton, PIP_2 is required to open vinculin to enable its linking function (50). In addition, many enzymes that regulate focal adhesions are activated by both growth factor and adhesion receptors (51-53). This loosening of adhesiveness not only moves the cell into a motility-permissive regimen, but also allows membrane distortions such as lamellipodia. Strong focal adhesions and actin stress filaments are largely absent from the protrusive structures during both adhesion and growth factor receptor motility. The lessened cell-substratum adhesiveness in these areas may also be due to activation of the limited intracellular protease calpain under direction of the integrins at the cell front (54-56).

Lamellipod protrusion is governed at many levels, but the underlying mechanism is of actin polymerization under the control of the Wasp-Arp2/3 complex (for reviews see 38, 47, 57). This machinery for generating a branched actin network is controlled by a number of upstream effectors triggered by both integrin and growth factor receptors. Primary roles have been designated for small GTPases, particularly cdc42 (58), and actin binding proteins, including cofilin (59, 60). As such, phosphoinositide metabolism likely plays a critical role in initiating and reinforcing protrusion. Phosphoinositide 3'(OH) kinase (PI3k) simultaneously generates

PIP_3 and removes PIP_2 from the membrane, altering the docking sites for plekstrin-homology (PH)- and C2 domian-containing proteins. This enzyme is found to be activated in lamellipodia (61, 62), and is responsible at least in part for determining the cell front of *Dictyostelium* (63). Thus, the nature of the molecules localized to and activated in the lamellipod is altered, with PIP_2-binding molecules such as PTEN being largely absent (64). This selective accumulation and activation of protrusion-promoting effectors in lamellipodia further reinforce the extension.

Upon contact with the substratum, new cell-substratum adhesions are formed at the front of a lamellipod, which now becomes the dominant one dictating the cell's direction. The molecular decision tree that leads to the formation and stabilization of such attachments is not fully understood, since many lamellipodia are retracted, particularly during the membrane-ruffling phase that proceeds productive locomotion (65). Whether these failures are ones of establishment of attachments (66) or excessive disassembly and retractive forces (67) needs to be determined by further investigations. The formation of stabilized attachments at the lamellipod front appears to involve the same players as those throughout the rest of the cell body. The small GTPases rac and rho both appear involved in adhesion formation during motility (68), though this is likely linked as adhesions dissipate in the absence of cytoskeleton linkage and mechanical stress. Interestingly, the main traction forces emanate from the base of lamellipod (36), a site of well-formed adhesions. This suggests that the contacts at the lamellipod tip serve mainly as guidance points.

Transcellular contractility is required to both move the cell body forward and detach the tail for productive motility, locomotion, to occur. This occurs mainly via the acto-myosin contractile machinery. These forces have been visualized by displacement of pliant surfaces (69-71). These studies have found that during motility a number of forces are generated dependent on the cellular geometry and cell locale. These forces are directed vectorally forward from the centromer towards the dominant lamellipod and from trailing tail inward towards the cell body. Thus, there appear to be two distinct phases of contractile forces to accomplish motility (70). That there are distinct contractile regions is supported by finding intracellular spatial segregation of myosin isoforms; myosin IIb appears to be preferentially localized to the front of the cell nucleus (72-74) while IIa is required to rip up the tail (75), though both myosin IIa and myosin IIb also might be involved in lamellipodia in migrating fibroblasts (76, 77).

The signals that generate these actin-myosin contractions are likely dependent on the nature of the motive signal (78). During haptokinesis, it is possible that stretch-activated calcium channels lead to calcium-induced contraction near the cell tail (79), providing for localized contractility.

However, it is likely that active signaling from integrins also contributes via ERK phosphorylation and activation of myosin light chain kinase (80). Growth factors acutely initiate cell contractility, that when unopposed by sufficient adhesiveness leads to cell rounding and stasis (81). EGF-induced contractility appears to be triggered by a third signaling cascade, one that does not require ERK or calcium signaling (82), but occurs via PKCδ (83). Other important growth factor receptor-activated pathways could include rho (84), Pak1 (85), or ZIP kinase (86). The outcome of increased transcellular contractility is eminently linked to localized adhesiveness as prevention of m-calpain activation leads to lamellipod retraction (87) and transmittal of the force to the matrix rather than productive motility (82).

Failure to release the trailing edge aborts locomotion (87) and transmits the contractile forces to the substratum (82). The control of this release likely depends on the induction of the motive force. During fibroblastoid motility, the tail adhesion components are shed and left on the underlying substratum (88); this appears to be mimicked by tumor cells moving through three dimensional matrices (89). While this shedding might result from mechanical stress (79, 80), it is likely that the detachment is assisted by enzymatic partial disassembly of adhesion (51, 90, 91) or proteolytic cleavage of adhesion components (92, 93). The limited intracellular protease, calpain, is required for integrin-mediated tail detachment from moderately to highly adhesive substratum (94); while the isoform is not fully determined, though there might be a role for calpain II (95). Growth factor receptor triggered locomotion activates calpain II by direct phosphorylation by ERK MAPkinase (96). The target of the calpain-mediated proteolysis appears to be adhesion complex components including talin, FAK, and paxillin (95, 97, 98).

Cell locomotion is an epigenetic phenomenon. While some tumor cells might contain genetic mutations in genes whose products enhance motility, the need for such mutations is not a prerequisite as various extracellular signals can induce locomotion. Still, for productive motility to be evinced, the expression of various proteins must be controlled. For instance, the (m)ena/vasp proteins prevent forward protrusion and thus must be downregulated (99, 100). Additionally, as various motility-associated proteins are shed or proteolyzed, production must be upregulated to compensate, such as for calpain II (101). It is not surprising to find that transcription factors play a role in motility (13, 102-104), suggesting that motility-associated genes may be coordinately regulated. Not only would transcription increase, but the mRNA translation might be directed to the cell locale that requires protein addition or replacement. It has been found that the β–actin mRNA is directed to the based of lamellipodia (105). This aspect of locomotion control has not been extensively mapped, but likely holds key insights into the molecular bases of tumor cell progression.

5. CELLULAR ASYMMETRY

Cell locomotion inherently requires cellular asymmetry (106, 107). From the above, it is evident that the biophysical processes and the biochemical signaling pathways are distinct in different parts of the motile cell. While we can describe the actual differences operative in each cell locale, an unanswered question remains as to whether there exists a unifying event or process that underlies the spatial segregation. The gradient of motility-inducing ligand was thought to be operative during chemotaxis and haptotaxis, but mathematical modeling suggests that ligand-binding receptor differences are too shallow across the dimensions of a cell; this is borne out by the finding that receptor-ligand complexes are not concentrated at the cell front (108-111). In retrospect this was predictable as active receptor signaling also must occur at the cell rear for tail detachment, at least for adherent fibroblasts and carcinoma cells. Rather, focus has shifted to asymmetry in the signaling mechanisms.

Recent work has supported the postulate that phosphoinositide profiles dictate the cellular asymmetry (Figure 3). This would either release/capture PI-binding proteins or serve as segregated docking sites for various motility-associated proteins. In *Dictyostelium*, PIP_3 is accumulated in the cell front due to PI3kinase enrichment there and exclusion of PTEN to the cell rear as PTEN requires PIP_2 binding for activation (64, 112). In EGFR-mediated carcinoma cell chemotaxis, PI3kinase appears to reinforce the cell front but (113) but only after initiation of the protrusion by PLCγ (114, 115). EGFR-induced motility of fibroblasts and keratinocytes appears not to be dependent on PI3kinase (116, 117), but does utilize PLCγ to create a gradient of phosphoinositide species (116, 118, 119). How PLCγ and PI3kinase are directed to the front of the cell is still an open question, but it appears that cdc42, either directly or indirectly, traffics activated PLCγ to the front (31). This finding is consistent with cdc42 being the master switch for directional motility (68, 120). There is as yet no report of a similar interaction of PI3kinase and cdc42. Still, this begs the question of how cdc42 recognizes the cell direction and does so within seconds of a chemotactic gradient being established.

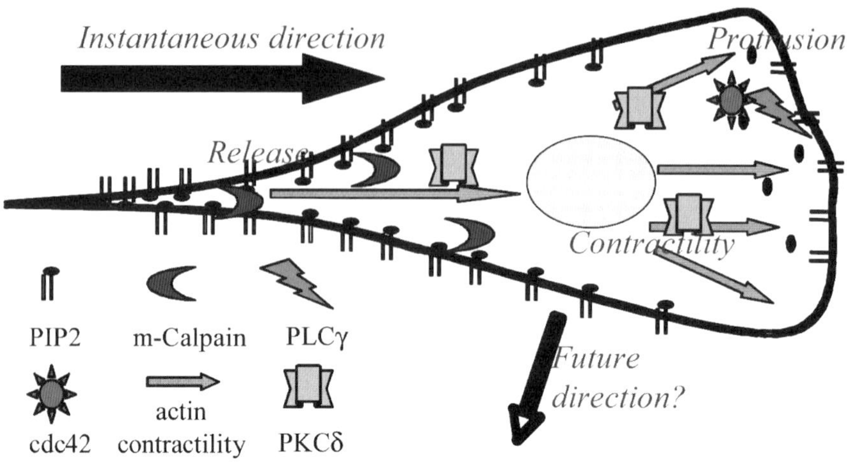

Figure 3. Phospho-insoitides establish a cellular asymmetry during the motile phase. Shown are key molecular switches for growth factor receptor-initiated motility schematized in a locomoting mesenchymally-transitioned cell. In the front, PIP2 is hydrolyzed by PLCγ to release PIP2-bound actin binding proteins (not shown for space reasons) for cytoskeletal reorganization, while at the rear the presence of PIP2 allows binding and co-activation of m-calpain and subsequent rear detachment and retraction. The role of PIP2 in PKCδ activation is only speculative at present.

During chemotactic and haptotactic locomotion, the cells move more or less directly up the gradient. Under such a situation, speed changes with induction of locomotion but persistence, the time a cell continues to move in the same direction, continues until the cell reaches a point of maximal motility in the gradient. Chemokinetic and haptokinetic locomotion are fundamentally different in that no external directional cues exist. The cells appear to locomote in a stochastic manner. Interestingly, upon stimulation of growth factors, cells become actively motile within seconds, with numerous ruffles and aborted protrusions, but only begin to locomote productively after a lag phase that can last hours (121, 122). What determines when and which protrusion becomes established leading to locomotion is unknown, but tight regulation of cdc42 activity plays a role (31). While growth factors increase the translational speed of cells, they also decrease the persistence (121). The cell changes direction often, sending out secondary protrusions that then dominate to send the cell in a new direction. This results in greater dispersion. Such a mode of locomotion enables the cell to sample a greater area. During development and tissue regeneration, this might allow cells to find matrix-embedded cues (123, 124). For cancer dissemination, this would be the perfect strategy to find more permissive regions to penetrate a barrier matrix or an ectopic site that possesses the necessary factors for metastatic seeding and growth. Thus, while chemotaxis might be important for intravasation, though carcinoma cells appear to migrate along collagen cables suggesting a mixed chemokinetic and haptotactic control of directionality (125), during extravasation, a non-persistent search mode would be beneficial.

6. EPIGENETIC AND GENETIC CONTROL OF TUMOR MOTILITY

From the above discussion, it is clear that much of tumor cell motility can be driven in absence of genetic mutations. Over the short term, the signals coming from the external cues can work on the existing proteome. For naïve fibroblasts, there is an extended period of nonproductive motility prior to actual locomotion (122). However, it appears that tumor cells can locomote almost immediately upon expose to growth factors (126). These findings suggest that the entire migration machinery is present. Over the longer term, protein translation and message transcription are required to sustain cell locomotion (127). Specific transcription factors are activated and upregulated during tumor cell motility (13, 102, 103). AP-1 was the first transcription factor to be specifically implicated in tumor motility and invasion (102). Subsequently, others have similarly been shown to function, including STAT3 (128), ERK MAPkinase (103), and, potentially, β-catenin (129).

This transcription upregulation may function to simply replace proteins either shed during tail retraction or proteolyzed during signaling (e.g. as by calpains [93]) or signaling attenuation. An alternate explanation is that transcription and translation are altered during locomotion to enhance and reinforce the motile phenotype. A hint that the latter is at play was the finding that upregulated EGFR signaling in prostate carcinoma cells specifically increase levels of uPAR and HGF (130, 131). As both uPAR and HGF further enhance tumor cell motility and progression (15, 132, 133), this would establish a cascade of autocrine, motility-promoting signaling loops (Figure 4). Additionally, increased levels of thymosins β15 (134) and β4 (129) induce actin polymerization to promote motility.

Genetic changes are the hallmark of cancers and cancer progression. Changes in the cell-cell adhesion machinery lead to the mesenchymal transition and enables autocrine stimulatory loops in carcinomas, thereby promoting invasion and metastasis. Still, most tumor cell locomotion is considered to derive from external cues, either from the surrounding tissue or autocrine secretion, even if the proteomic keyboard upon which they play is altered by genetic mutations (2). Thus, the genetic alterations found in cancers only partly provide the impetus for tumor invasion and metastasis.

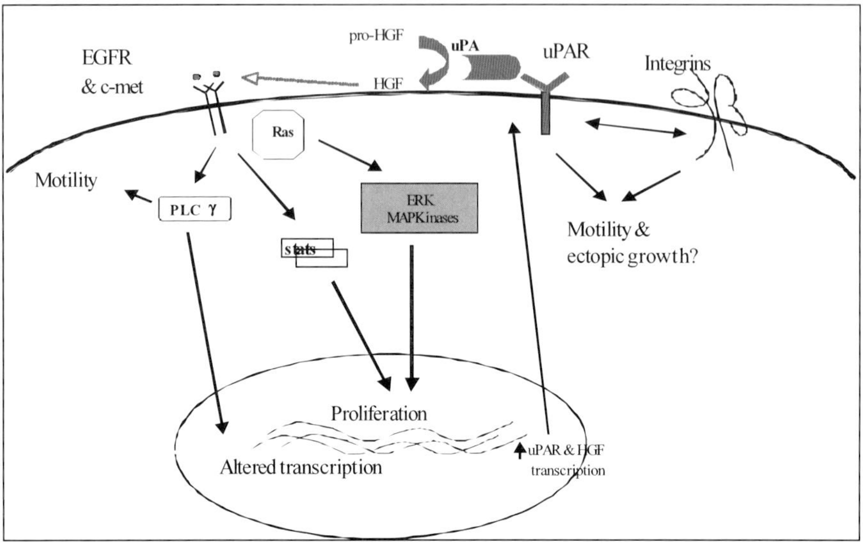

Figure 4. Downstream signaling re-inforces the motility phenotype. Epigenetic signaling of motility occurs not only within the peri-membrane space, but also drives new transcription that generates new paracrine and autocrine signaling that is also motogenic.

7. SUMMARY

The general outlines of the locomotive machinery rapidly are being defined. The commonalities between cell and tumor types and across species and even kingdoms are striking. It is tempting to target common elements in this machinery to limit or stop tumor progression. Abrogating signaling through key molecular switches (Figure 2) or eliminating required proteins, stops cell motility and, when assessed in experimental models, tumor invasion and metastasis. A problem lies in the fact that such cell locomotion is central to many homeostatic and repair processes, including the functioning of the immune response. While many animal-based therapeutic interventions find surprisingly little systemic toxicity, these do not faithfully replicate the stressors and injurious contemporaneously present in human patients. Thus, it is likely that such a global inhibitor will present limiting or even unacceptable toxicities, regardless of how specific the agent is.

Many cancers and subsets therein employ unique variations to these themes in their upregulation or dysregulation of motility. Such specifics are described in the chapters devoted to individual cancers (see Chapters 10, 11, 12, and 13). These changes are more likely to be exploited with less impact on homeostatic and repair-associated locomotion, and thus less toxicity. For instance, if specific thymosin isoforms are upregulated particularly in colon (129) or prostate (134) carcinomas, then reduction to physiological levels should provide some diminution in invasive and metastatic capabilities. A prefered interventional strategy, may therefore, be to develop a multitude of cancer type-specific agents.

Even this limited approach might provide only limited benefit. Inhibitors of EGFR family signaling are only partially effective or effective in only a subset of cancers. Higher doses of agents that reduce signaling to near zero are not tolerated; this is likely due to adverse effects on homeostatic functions of these receptors. Rather, a more successful strategy may be to simultaneously target multiple control points.

The delineation of the key molecular switches that control motility provides for a rational interventional strategy. However, the multitude of inputs and activated pathways suggest a redundancy that may obscure any one particular target. As such, model systems for investigating locomotion and testing agents need to recapitulate the complexity and multicellularity of the human condition. Fortunately, in parallel with hypothesis-driven investigations, newer and more relevant model systems are being developed (see Chapters 2, 3 and 4). These provide a span from *in vitro* to *in vivo*, enabling multiple levels of intervention and assesssment. It is expected that these will enable quantal advances in both our understanding and approach to tumor progression.

Advances in blocking locomotion and thereby tumor progression will require a paradigm shift in clinical testing. Currently, the vast majority of chemotherapy is initially tested on patients with metastatic (advanced) disease. As the locomotion-targeted agents are not likely to be tumoricidal or tumoristatic, little benefit would be predicted and clinical testing must shift to earlier stage cancers. This is possible, as these agents would surely be adjuvants to be administered along with tumoricidal and tumoristatic therapies. Thus, other than the real issue of toxicity, there would be little concern about 'doing harm'. The recent success of an anti-angiogenic agent in such a clinical trials provides hope for more support of adjuvant therapy testing (135). However, the current endpoint measures may not be relevant or realistic to test for locomotion inhibitors. The key intermediary markers of tumor burden, either as response or time to 'clinical progression', would not likely be affected. The real parameter would be patient survival, but in most earlier stage cancers this is both an extended target and confounded by the multitude of therapies utilized during the many recurrences/relapses. New intermediary clinical markers have to be established that measure the key killers of tumor invasion and metastasis. The present body imaging is not sufficient to measure these changes. New modalities need to be developed. One hope might be the in the recently developed approach of finding shed tumor cells in the circulation (7). In sum, while the leaps in understanding tumor cell locomotion and its role in tumor progression have provided us with rational targets and even lead compounds, realization of the promise will await new thinking, and even new technologies in clinical trial design and endpoints. The good news is that such changes are already occurring.

REFERENCES

1. Engebraaten O, Bjerkvig R, Pedersen PH, OD Laerum. Effects of EGF, bFGF, NGF and PDGF(bb) on cell proliferative, migratory and invasive capacities of human brain-tumour biopsies *in vitro*. Int. J. Cancer 1993, 53: 209-214.
2. Wells A. Tumor invasion: role of growth factor-induced cell motility. Adv. Cancer Res. 2000, 78: 31-101.
3. Stetler-Stevenson WG, Yu A. Proteases in invasion: matrix metalloproteinases. Sem. Cancer Biol. 2001, 11: 143-152.
4. Chang C, Werb Z. The many faces of metalloproteases: cell growth, invasion, angiogenesis and metastasis. Trends Cell Biol. 2001, 11: S37-43.
5. Chambers AF, MacDonald IC, Schmidt EE, Morris VL, Groom AC. Clinical targets for anti-metastasis therapy. Adv. Cancer Res. 2000, 79: 91-121.
6. Fidler IJ. The pathogenesis of cancer metastasis: the 'seed and soil' hypothesis revisited. Nat. Rev. Cancer 2003, 3: 453-458.
7. Cristofanilli M, et al. Circulating tumor cells, disease progression, and survival in metastatic breast cancer. N. Engl. J.Med. 2004, 351: 781-791.
8. Wyckoff JB, Segall JE, Condeelis JS. The collection of the motile population of cells from a living tumor. Cancer Res. 2000, 60: 5401-5404.
9. Condeelis J, Segall JE. Intravital imaging of cell movement in tumours. Nat. Rev. Cancer 2003, 3: 921-930.
10. Kim H, Turner T, Kassis J, Souto J, Wells A. EGF receptor signaling in prostate development. Histol. Histopathol. 1999, 14: 1175-1182.
11. Kalluri R, Neilson EG. Epithelial-mesenchymal transition and its implications for fibrosis. J. Clin. Invest. 2003, 112: 1776-1784.
12. Kang Y, Massague J. Epithelial-mesenchymal transitions: twist in development and metastasis. Cell 2004, 118: 277-279.
13. Yang J, et al. Twist, a master regulator of morphogenesis, plays an essential role in tumor metastasis. Cell 2004, 117: 927-939.
14. Edme N, Downward J, Thiery J-P, Boyer B. Ras induces NBT-II epithelial cell scattering through the coordinate activaties of rac and MAPK pathways. J. Cell Sci. 2002, 115: 2591-2601.
15. Birchmeier C, Birchmeier W, Gherardi E, VandeWoude GF. Met, metastasis, motility and more. Nat. Rev. Mol. Cell Biol. 2003, 4: 915-925.

16. Grunert S, Jechlinger M, Beug H. Diverse cellular and molecular mechanisms contribute to epithelial plasticity and metastasis. Nat. Rev. Mol. Cell Biol. 2003, 4: 657-665.
17. Hordijk PL, tenKlooster JP, vanderKammen RA, Michiels F, Oomen LCJM, Collard JG. Inhibition of invasion of epithelial cells by Tiam1-rac signaling. Science 1997, 278: 1464-1466.
18. Jones FS, Jones PL. The tenascin family of ECM glycoproteins: structure, function and regulation during embryonic development and tissue remodeling. Dev. Dyn. 2000, 218: 235-259.
19. Swindle CS, et al. Epidermal growth factor (EGF)-like repeats of human tenascin-C as ligands for EGF receptor. J. Cell Biol. 2001, 154: 459-468.
20. Herold-Mende C, Mueller MM, Bonsanto MM, Schmitt HP, Kunze S, Steiner H-H. Clinical impact and functional aspects of tenascin-C expression during glioma progression. Int. J. Cancer 2002, 98: 362-369.
21. Schenk S, et al. Binding to EGF receptor of a laminin-5 EGF-like fragment liberated during MMP-dependent mammary gland involution. J. Cell Biol. 2003, 161: 197-209.
22. Alper O, DeSantis ML, Stromberg K, Hacker NF, Cho-Chung YS, Salomon DS. Anti-sense suppression of epidermal growth factor receptor expression alters cellular proliferation, cell-adhesion and tumorigenicity in ovarian cancer cells. Int. J. Cancer 2000, 88: 566-574.
23. Yates C, Wells A, Turner T. Luteinizing hormone releasing hormone (LHRH) analog reverses the cell adhesion profile of DU-145 human prostate carcinoma. Br. J. Cancer 2005, 92: 366-375.
24. Wong AST, Gumbiner BM. Adhesion-independent mechanism for suppression of tumor cell invasion by E-cadherin. J. Cell Biol. 2003, 161: 1191-1203.
25. Comoglio PM, Trusolino L. Invasive growth: from development to metastasis. J. Clin. Invest. 2002, 109: 857-862.
26. Friedl P, Zanker KS, Brocker E-B. Cell migration strategies in 3-D extracellular matrix: differences in morphology, cell matrix interactions and integrin function. Microsc. Res.Tech. 1998, 43: 369-378.
27. Friedl P, Wolf K. Tumour-cell invasion and migration: diversity and escape mechanisms. Nat. Rev. Cancer 2003, 3: 362-374.
28. Wolf K, et al. Compensation mechanism in tumor cell migration: mesenchymal-amoeboid transition after blocking of pericellular proteolysis. J. Cell Biol. 2003, 160: 267-277.
29. Lauffenburger DA, Horwitz AF. Cell migration: a physically integrated molecular process. Cell 1996, 84: 359-369.
30. Kundra V, Anand-Apte B, Feig LA, Zetter BR. The chemotactic response to PDGF-BB: evidence of a role for ras. J. Cell Biol. 1995, 130: 725-731.
31. Chou J, Burke NA, Iwabu A, Watkins SC, Wells A. Directional motility induced by EGF requires cdc42. Exp. Cell Res. 2003, 287: 47-56.
32. Shrivastava A, et al. An orphan receptor tyrosine kinase family whose members serve as nonintegrin collagen receptors. Mol. Cell 1997, 1: 25-34.
33. Vogel W, Gish GD, Alves F, Pawson T. The discoidin domain receptor tyrosine kinases are activated by collagen. Mol. Cell 1997, 1: 13-23.
34. Iozzo RV, Moscatello DK, McQuillan DJ, Eichstetter I (1999). Decorin is a biological ligand for the epidermal growth factor receptor. J. Biol. Chem. 274: 4489-4492.
35. Tran KT, Griffith LG, Wells A (2004). Extracellular matrix signaling through growth factor receptors during wound healing. Wound Repair Regen. 12: 262-268.

36. Rabinovitz I, Gipson IK, Mercurio AM (2001). Traction forces mediated by α6β4 integrin: implications for basement membrane organization and tumor invasion. Mol. Biol. Cell 12: 4030-4043.
37. Trusolino L, Bertotti A, Comoglio PM (2001). A signaling adaptor function for α6β4 integrin in the control of HGF-dependent invasive growth. Cell 107: 643-654.
38. Pollard TD (2003). The cytoskeleton, cellular motility, and the reductionist agenda. Nature 422: 741-745.
39. Hall A, ed. *GTPases*. 2000, Oxford University Press: Oxford.
40. Xie H, et al. (1998). EGF receptor regulation of cell motility: EGF induces disassembly of focal adhesions independently of the motility-associated PLCγ signaling pathway. J. Cell Sci. 111: 615-624.
41. Comoglio PM, Boccaccio C (2001). Scatter factors and invasive growth. Semin. Cancer Biol. 11: 153-162.
42. Conacci-Sorrrell M, Zhurinsky J, Ben-Ze'ev A (2002). The cadherin-catenin adhesion system in signaling and cancer. J. Clin. Invest. 109: 987-991.
43. Moriyama K, Yonezawa N, Sakai H, Yahara I, Nishida E (1992). Mutational analysis of an actin-binding site of cofilin and characterization of chimeric proteins between cofilin and destrin. J. Biol. Chem. 267: 7240-7244.
44. Goldschmidt-Clermont PJ, Kim JW, Machesky LM, Rhee SG, Pollard TD (1991). Regulation of phospholipase C-γl by profilin and tyrosine phosphorylation. Science 251: 1231-1233.
45. Chen P, Murphy-Ullrich J, Wells A (1996). A role for gelsolin in actuating EGF receptor-mediated cell motility. J. Cell Biol. 134: 689-698.
46. Schafer DA, Jennings PB, Cooper JA (1996). Dynamics of capping protein and actin assembly in vitro: Uncapping barbed ends by polyphosphoinositides. J. Cell Biol. 135: 169-179.
47. Condeelis J (2001). How is actin polymerization nucleated *in vivo*? Trends Cell Biol. 11: 288-293.
48. Miyamoto S, et al. (1995). Integrin function: molecular hierarchies of cytoskeletal and signaling molecules. J. Cell Biol. 131: 791-805.
49. Zamir E, Katz BZ, Aota S, Yamada KM, Geiger B, Kam Z (1999). Molecular diversity of cell-matrix adhesions. J. Cell Sci.112: 1655-1669.
50. Gilmore AP, Burridge K (1996). Regulation of vinculin binding to talin and actin by phosphatidyl-inositol-4-5-bisphoshate. Nature 381: 531-535.
51. Larsen M, Tremblay ML, Yamada KM (2003). Phosphatases in cell-matrix adhesion and migration. Nat. Rev. Mol. Cell Biol. 4: 700-711.
52. Carragher NO, Frame MC (2004). Focal adhesion and actin dynamics: a place where kinases and proteases meet to promote invasion. Trends Cell Biol.14: 241-249.
53. Wu C (1999). Integrin-linked kinase and PINCH: partners in regulation of cell-extracellular matrix interaction and signal transduction. J. Cell Sci.112: 4485-4489.
54. Bhatt A, Kaverina I, Otey C, Huttenlocher A (2002). Regulation of focal complex composition and disassembly by the calcium-dependent protease calpain. J. Cell Sci.115: 3415-3425.
55. Robles E, Huttenlocher A, Gomez TM (2003). Filopodial calcium transients regulate growth cone motility and guidance through local activation of calpain. Neuron 38: 597-609.
56. Carragher NO, Westhoff MA, Fincham VJ, Schaller MD, Frame MC (2003). A novel role for FAK as a protease-targeting adaptor protein: regulation by p42 ERK and Src. Curr. Biol. 13: 1442-1450.

57. Suetsugu S, Takenawa T (2003). Regulation of the cortical actin network in cell migration. Int. Rev. Cytol. 229: 245-286.
58. Prehoda KE, Scott JA, Mullins RD, Lim WA (2000). Integration of multiple signals through cooperative regulation of the N-WASP-Arp2/3 complex. Science 290: 801-806.
59. Loisel TP, Boujemaa R, Pantaloni D, Carlier M-F (1999). Reconstitution of actin-based motility of *listeria* and *shigella* using pure proteins. Nature 401: 613-616.
60. Kempiak DJ, Yip SC, Backer JM, Segall JE (2003). Local signaling by the EGF receptor. J. Cell Biol. 162: 781-787.
61. Hill K, et al. (2000). Specific requirement for the p85-p110α phosphatidylinositol 3-kinase during epidermal growth factor-stimulated actin nucleation in breast cancer cells. J. Biol. Chem. 275: 3741-3744.
62. Gomez-Mouton C, et al. (2004). Dynamic redistribution of raft domains as an organizing platform for signaling during cell chemotaxis. J. Cell Biol. 164: 759-768.
63. Dormann D, Weijer G, Parent CA, Devreotes PN, Weijer CJ (2002). Visualizing PI3 kinase-mediated cell-cell signaling during *Dictyostelium* development. Curr. Biol. 12: 1178-1188.
64. Iijima M, Huang YE, Luo HR, Vazquez F, Devreotes PN (2004). Novel mechanism of PTEN regulation by its phosphophatidylinositol (4,5)-binding motif is critical for chemotaxis. J. Biol. Chem. 279: 16606-16613.
65. Lee J, Ishihara K, Jacobson K (1993). How do cells move along surfaces. Trends Cell Biol. 3: 366-370.
66. Choma DP, Pumiglia K, DiPersio CM (2004). Integrin α3β1 directs the stabilization of a polarized lamellipodium in epithelial cells through activation of Rac1. J. Cell Sci.117: 3947-3959.
67. Brahmbhatt AA, Klemke RL (2003). ERK and RhoA differentially regulate pseudopodia growth and retraction during chemotaxis. J. Biol. Chem. 278: 13016-13025.
68. Nobes CD, Hall A (1999). Rho GTPases control polarity, protrusion and adhesion during cell movement. J. Cell Biol. 144: 1235-1244.
69. Dembo M, Oliver T, Ishihara A, Jacobson K (1996). Imaging the traction stresses exerted by locomoting cells with the elastic substratum method. Biophys. J. 70: 2008-2022.
70. Burton K, Park JH, Taylor DL (1999). Keratocytes generate traction forces in two phases. Mol. Biol. Cell 10: 3745-3769.
71. Tan JL, Tien J, Pironte DM, Gray DS, Bhadriraju K, Chen CS (2003). Cells lying on a bed of microneedles: an approach to isolate mechanical force. Proc. Natl. Acad. Sci. USA 100: 1484-1489.
72. Cheng TP, Murakami N, Elzinga M (1992). Localization of myosin IIB at the leading edge of growth cones from rat dorsal root ganglionic cells. FEBS Letters 311: 91-94.
73. Conrad AH, Jaffredo T, Conrad GW (1995). Differential localization of cytoplasmic myosin II isoforms A and B in avian interphase and dividing embryonic and immortalized cardiomyocytes and other cell types in vitro. Cell Mot. Cytoskeleton 31: 93-112.
74. Verkhovsky AB, Svitkina TM, Borisy GG (1995). Myosin II filament assemblies in the active lamella of fibroblasts: their morphogenesis and role in the formation of actin filament bundles. J. Cell Biol. 131: 989-1002.
75. Jay PY, Pham PA, Wong SA, Elson EL (1995). A mechanical function of myosin II in cell motility. J. Cell Sci.108: 387-393.

76. Saitoh T, et al. (2001). Differential localization of non-muscle myosin II isoforms and phosphorylated regulatory light chains in human MRC-5 fibroblasts. FEBS Letters 509: 365-369.
77. Ben-Ya'acov A, Ravid S (2003). Epidermal grwoth factor-mediated transient phosphorylation and membrane localization of myosin IIB is required for efficient chemotaxis. J. Biol. Chem. 278: 40032-40040.
78. Mitchison TJ, Cramer LP (1996). Actin-based cell motility and cell locomotion. Cell 84: 371-379.
79. Lee J, Ishihara A, Oxford G, Johnson B, Jacobson K (1999). Regulation of cell movement is mediated by stretch-activated calcium channels. Nature 400: 382-386.
80. Klemke RL, Cai S, Giannini AL, Gallagher PJ, deLanerolle P, Cheresh DA (1997). Regulation of cell motility by mitogen-activated protein kinase. J. Cell Biol. 137: 481-492.
81. Welsh JB, Gill GN, Rosenfeld MG, Wells A (1991). A negative feedback loop attenuates EGF-induced morphological changes. J. Cell Biol. 114: 533-543.
82. Allen FD, Asnes CF, Chang P, Elson EL, Lauffenburger DA, Wells A (2002). EGF-induced matrix contraction is modulated by calpain. Wound Repair Regen. 10: 67-76.
83. Iwabu A, Smith K, Allen FD, Lauffenburger DA, Wells A (2004). EGF induces fibroblast contractility and motility via a PKCδ-dependent pathway. J. Biol. Chem. 279: 14551-14560.
84. Yoshioka K, Matsumura F, Akedo H, Itoh K (1998). Small GTP-binding protein rho stimulates the actomyosin system, leading to invasion of tumor cells. J. Biol. Chem. 273: 5146-5154.
85. Sells MA, Boyd J, Chernoff J (1999). p21-activated kinase (Pak1) regulates cell motility in mammalian fibroblasts. J. Cell Biol. 145: 837-849.
86. Komatsu S, Ikebe M (2004). ZIP kinase is responsible for the phosphorylation of myosin II and necessary for cell motility in mammalian fibroblasts. J. Cell Biol. 165: 243-254.
87. Shiraha H, Glading A, Chou J, Jia Z, Wells A (2002). Activation of m-calpain (calpain II) by epidermal growth factor is limited by PKA phosphorylation of m-calpain. Mol. Cellular Biol. 22: 2716-2727.
88. Regen CM, Horwitz AF (1992). Dynamics of β1 integrin-mediated adhesive contacts in motile fibroblasts. J. Cell Biol. 119: 1347-1359.
89. Friedl P, Maaser K, Klein CE, Niggemann B, Krohne G, Zanker KS (1997). Migration of highly aggressive MV3 melanoma cells in 3-D collagen lattices results in local matrix reorganization and shedding of α2 and β1 integrins and CD44. Cancer Res. 57: 2061-2070.
90. Herman B, Pledger WJ (1985). Platelet-derived growth factor-induced alterations in vinculin and actin distribution in BALB/c-3T3 cells. J. Cell Biol. 100: 1031-1040.
91. Angers-Loustau A, et al. (1999). Protein tyrosine phosphatase-PEST regulates focal adhesion disassembly, migration and cytokinesis in fibroblasts. J. Cell Biol. 144: 1019-1031.
92. Huttenlocher A, et al. (1997). Regulation of cell migration by the calcium-dependent protease calpain. J. Biol. Chem. 272: 32719-32722.
93. Glading A, Lauffenburger DA, Wells A (2002). Cutting to the chase: calpain proteases in cell migration. Trends Cell Biol. 12: 46-54.
94. Palecek S, Huttenlocher A, Horwitz AF, Lauffenburger DA (1998). Physical and biochemical regulation of integrin release during rear detachment of migrating cells. J. Cell Sci.111: 929-940.

95. Franco S, Perrin B, Huttenlocher A (2004). Isoform specific function of calpain 2 in regulating membrane protrusion. Exp. Cell Res. 299: 179-187.
96. Glading A, et al. (2004). Epidermal growth factor activates m-calpain (calpain 2), at least in part, by ERK-mediated phosphorylation. Mol. Cellular Biol. 24: 2499-2512.
97. Tranqui L, Block MR (1995). Intracellular processing of talin occurs within focal adhesions. Exp. Cell Res. 217: 149-156.
98. Carragher NO, Fincham VJ, Riley D, Frame MC (2001). Cleavage of focal adhesion kinase by different proteases during SRC-regulated transformation and apoptosis. Distinct roles for calpain and caspases. J. Biol. Chem. 276: 4270-4275.
99. Bear JE, Loureiro JJ, Libova I, Fassler R, Wehland J, Gertler FB (2000). Negative regulation of fibroblast motility by ena/vasp proteins. Cell 101: 717-728.
100. Kwiatkowski AV, Gertler FB, Loureiro JJ (2003). Function and regulation of Ena/VASP proteins. Trends Cell Biol. 13: 386-392.
101. Glading A, Chang P, Lauffenburger DA, Wells A (2000). Epidermal growth factor receptor activation of calpain is required for fibroblast motility and occurs via an ERK/MAP kinase signaling pathway. J. Biol. Chem. 275: 2390-2398.
102. Malliri A, et al. (1998). The transcriptional factor AP-1 is required for EGF-induced activation of rho-like GTPases, cytoskeletal rearrangements, motility, and in vitro invasion of A431 cells. J. Cell Biol. 143: 1087-1099.
103. Tanimura S, Nomura K, Ozaki K, Tsujimoto M, Kondo T, Kohno M (2002). Prolonged nuclear retention of activated extracellular signal-regulated kinase 1/2 is required for hepatocyte growth factor-induced cell motility. J. Biol. Chem. 277: 28256-28264.
104. Vial E, Sahai E, Marshall CJ (2003). Erk-MAPK signaling coordinately regulates activity of rac1 and rhoA for tumor cell motility. Cancer Cell 4: 67-79.
105. Shestakova EA, Singer RH, Condeelis J (2001). The physiological significance of β-actin mRNA localization in determining cell polarity and directional motility. Proc. Natl. Acad. Sci. USA 98: 7045-7050.
106. Devreotes PN, Zigmond SH (1988). Chemotaxis in eukaryotic cells: a focus on leukocytes and *Dictyostelium*. Annu. Rev. Cell Dev. Biol. 4: 649-686.
107. Nabi IR (1999). The polarization of the motile cell. J. Cell Sci.112: 1803-1811.
108. Xiao Z, Zhang N, Murphy DB, Devreotes PN (1997). Dynamic distribution of chemoattractant receptors in living cells during chemotaxis and persistent stimulation. J. Cell Biol. 139: 365-374.
109. Foxman EF, Campbell JJ, Butcher EC (1997). Multistep navigation and the combinatorial control of leukocyte chemotaxis. J. Cell Biol. 139: 1349-1360.
110. Narang A, Subramanian KK, Lauffenburger DA (2001). A mathematical model for chemoattractant gradient sensing based on receptor-regulated membrane phospholipid signaling dynamics. Ann. Biomed. Eng. 29: 677-691.
111. Servant G, Weiner OD, Herzmark P, Balla T, Sedat JW, Bourne HR (2000). Polarization of chemoattractant receptor signaling during neutrophil chemotaxis. Science 287: 1037-1040.
112. Iijima I, Devreotes P (2002). Tumor suppressor PTEN mediates sensing of chemoattractant gradients. Cell 109: 599-610.
113. Condeelis JS, et al. (2001). Lamellipodia in invasion. Semin. Cancer Biol. 11: 119-128.
114. Mouneimne G, et al. (2004). Phospholipase C and cofilin are required for carcinoma cell directionality in response to EGF stimulation. J. Cell Biol. 166: 697-708.
115. Piccolo E, Innominato PF, Mariggio MA, Maffucci T, Iacobelli S, Falasca M (2002). The mechanism involved in the regulation of phospholipase Cγ1 activity in cell migration. Oncogene 21: 6520-6529.

116. Chen P, Xie H, Sekar MC, Gupta KB, Wells A (1994). Epidermal growth factor receptor-mediated cell motility: phospholipase C activity is required, but MAP kinase activity is not sufficient for induced cell movement. J. Cell Biol. 127: 847-857.
117. Haase I, Evans R, Pofahl R, Watt FM (2003). Regulation of keratinocyte shape, migration and wound epithelialization by IGF-1- and EGF-dependent signalling pathways. J. Cell Sci.116: 3227-3238.
118. Chou J, Beer-Stolz D, Burke N, Watkins SC, Wells A (2002). Distribution of gelsolin and phosphoinositol 4,5-bisphosphate in lamellipodia during EGF-induced motility. Int. J. Biochem. Cell Biol. 34: 776-790.
119. Satish L, Blair HC, Glading A, Wells A (2005). IP-9 (CXCL11) induced cell motility in keratinocytes requires calcium flux-dependent activation of μ-calpain. Mol. Cellular Biol. 25: 1922-1941.
120. Allen WE, Zicha D, Ridley AJ, Jones GE (1998). A role for cdc42 in macrophage chemotaxis. J. Cell Biol. 141: 1147-1157.
121. Ware MF, Wells A, Lauffenburger DA (1998). Epidermal growth factor alters fibroblast migration speed and directional persistence reciprocally and in matrix-dependent manner. J. Cell Sci.111: 2423-2432.
122. Maheshwari G, Wells A, Griffith LG, Lauffenburger DA (1999). Biophysical integration of effects of epidermal growth factor and fibronectin on fibroblast migration. Biophys. J. 76: 2814-2823.
123. Gallio M, Englund C, Kylsten P, Samakovlis C (2004). Rhomboid 3 orchestrates slit-independent repulsion of tracheal branches at the CNS midline. Development 131: 3605-3614.
124. Wells A, Lillien L (2004). Attraction or repulsion - a matter of individual tast. Science STKE pe47.
125. Wyckoff JB, Jones JG, Condeelis JS, Segall JE (2000). A critical step in metastasis: in vivo analysis of intravasation at the primary tumor. Cancer Res. 60: 2504-2511.
126. Segall JE, et al. (1996). EGF stimulates lamellipod extension in metastatic mammary adenocarcinoma cells by an actin-dependent mechanism. Clin. Exp. Metastasis 14: 61-72.
127. Chen P, Gupta K, Wells A (1994). Cell movement elicited by epidermal growth factor receptor requires kinase and autophosphorylation but is separable from mitogenesis. J. Cell Biol. 124: 547-555.
128. Rivat C, DeWever O, Bruynel E, Mareel M, Gespach C, Attoub S (2004). Disruption of STAT3 signaling leads to tumor cell invasion through alterations of homotypic cell-cell adhesion complexes. Oncogene 23: 3317-3327.
129. Chang WS, Chen PM, Hsaio HL, Want HS, Liang WY, Su Y (2004). Overexpression of the thymosin beta-4 gene is associated with increased invasion of SW480 colon carcinoma cells and the distant metastasis of human colorectal carcinoma. Oncogene 23: 6666-6671.
130. Manos EJ, Kim M, Kassis J, Chang B, Wells A, Jones DA (2001). Prostin-1, a novel phospholipase C-γ regulated gene negatively associated with prostate tumor invasion. Oncogene 20: 2781-2790.
131. Mamoune A, et al. (2004). DU145 human prostate carcinoma invasiveness is modulated by urokinase receptor (uPAR) downstream of epidermal growth factor receptor (EGFR) signaling. Exp. Cell Res. 299: 91-100.
132. Blasi F, Carmeliet P (2002). uPAR: a versatile signalling orchestrator. Nat. Rev. Mol. Cell Biol. 3: 932-943.

133. Wang Y (2001). The role and regulation of urokinase-type plasminogen activator receptor gene expression in cancer invasion and metastasis. Med. Res. Rev. 21: 146-170.
134. Bao L, Loda M, Stewart TR, Anand-Apte B, Zetter BR (1996). Thymosin β15: a novel regulator of tumor cell motility upregulated in metastatic prostate cancer. Nat. Med. 2: 1322-1328.
135. Hurwitz H, et al. (2004). Bevacizumab plus irinotecan, fluorouracil, and leucovorin for metastatic colorectal cancer. N. Eng. J. Med. 350: 2335-2342.

Chapter 2

IN VITRO MATRICES FOR STUDYING TUMOR CELL INVASION

Kun Yuan*[1], Raj K. Singh*[3], Gabriel Rezonzew[1] and Gene P. Siegal[1,2]
Departments of Pathology[1], Cell Biology[2] and Surgery[2], University of Alabama at Birmingham, Birmingham,AL, USA and Vivo Biosciences Inc.[3], Birmingham, AL, USA
** These authors contributed equally to this work*

Abstract: Metastasis is the major cause of death in patients with cancer and thus attempts have been made for more than a century to experimentally dissect its component parts. Multiple arbitrary divisions have been offered that often begin with immortalized and transformed cells demonstrating altered cell proliferation, cell cycle and apoptotic pathways initiating their journey of dissemination by detaching from the primary tumor mass, initiating angiogenesis, degrading and penetrating the basement membranes and the surrounding connective tissue boundaries, intravasating, circulating and extravasating through the blood or lymphatic circulation and after evading the immune system eventually reaching one or more distant metastatic sites where they must undergo a similar process in reverse order. Tumor cell invasion is one of the key steps in this complex process. As the metastatic potential of tumor cells is largely dependent on their ability to degrade and migrate through extracellular matrix (ECM) barriers, inhibition of its subroutines (proteolysis, ECM degradation, chemotaxis, haptotaxis etc.) become logical targets for experimental cancer therapy. To explore potential approaches to inhibit tumor invasion, various *in vitro* invasion assays have been devised and become widely-accepted surrogate endpoints to mimic the *in vivo* condition. Most invasion assay systems are based on measuring the ability of cells to invade across purified ECM components such as collagen or complex artificial or reconstituted ECMs. Each has certain strengths, quantitative abilities, ease of manipulation of treatments etc. and like the tumor cells themselves, each is counterbalanced by inherent weaknesses. These are reviewed in this chapter.

Key words: metastasis, extracellular matrix, collagen, amniotic membrane matrix

A. Wells (ed.), Cell Motility in Cancer Invasion and Metastasis, 25-54.

1. INTRODUCTION

Tumorigenesis is a multi-step process involving genetic and epigenetic cellular changes recognized as phenotypic events including uncontrolled proliferation, immortality, dysregulation of signal transduction and cell cycle, altered expression of proteases and their inhibitors, active degradation of host matrix barriers and tumor cell dissemination ending in metastasis (1-11).

The neoplastic progression cascade is also much more than a series of on-off binary events. Rather, it involves a series of coordinated cell-cell and cell-matrix interactions in response to locally produced soluble factors and insoluble matrix signals. Tumor invasion as a discrete step in this cascade has also been studied in depth and can be viewed itself as a series of events including, but not limited to adhesion/de-adhesion states, transmigration of activated cells and degradation of tissue matrix by tumor-secreted and matrix containing proteases (12, 13).

However, the discrete nature of the molecular events driving these cellular processes remains less than fully understood in human tumors due, in large part, to lack of defined functional models.

To that end, and as outlined in this chapter, a number of *in vitro* pure matrix and more complex cell-matrix models of tumor invasion have been developed over the past few decades which employ both animal derived purified matrix e.g., rat fibrin, collagen I gels, rodent tumor biomatrix, pure normal human biomatrix and complex animal-human chimeric matrices with cells.

2. THE EXTRACELLULAR MATRIX

Since the earliest days of microscopy, it has been evident that cells are embedded in an intricate extracellular matrix (the ground substance), which not only provides mechanical support and conglutination, but also influences cell growth, adhesion and motility. During tumor invasion, cancer cells must degrade ECM to penetrate into deeper tissues and vasculature. Individual components as they were recognized and characterized by advances in separation technology, biochemistry and molecular biology became valuable as coating layers for the testing of cell populations for phenotypic changes during neoplastic progression. These cells were later shown to contain a myriad of cell surface matrix receptors, thereby driving these cell-matrix interactions.

2.1 Major Components of ECM (14)

The ECM contains tens of thousands of bioactive molecules but is composed primarily of fibrous proteins and polysaccharides, which are secreted locally and assembled into a gel-like organized meshwork. Fiber-forming proteins have been divided into mainly two functional types: structural (collagens and elastin) and adhesive (fibronectin and laminin).

Collagen is the major scaffold of the ECM, its major functions including strengthening, organizing and stabilizing other matrix components. The main types of collagen in connective tissues are types I, II, III, V, and XI; type I collagen being the most abundant. The other groups of collagens are fibril-associated collagens, including types IX and XII collagen. They attach to and intertwine collagen fibrils to modulate the interactions of the fibrils with other matrix components. Type IV collagen is the major collagenous constituent of the basement membrane (basal laminae). It forms a sheet-like meshwork with other basement membrane components to support usually epithelial and endothelial cells. Each of the two dozen or so collagen family members has a characteristic role and distribution which varies among the different tissues and organs. For example, collagen type VII assembles into anchoring fibrils thereby contributing to the attachment of the basal lamina of epithelium to the underlying connective tissue.

Elastin molecules form a highly cross-linked network, providing elasticity and resilience to the ECM. The core of elastin is covered by microfibrils, whose major component is fibrillin. Fibrillin, in turn, contribute sto the integrity and assembly of the elastic fiber.

Glycosaminoglycan (GAGs) chains and other proteoglycans form a hydrated gel meshwork interwoven by various protein fibers. The GAGs are proteoglycans with negatively charged polysaccharide chains covalently linked to protein. GAGs provide mechanical support, resist compressive forces and selectively sieve the trafficking of molecules and cells. There are mainly four main groups of GAGs: hyaluronan, chondroitin sulfate and dermatan sulfate, heparan sulfate and heparin, and keratan sulfate. Hyaluronan is the simplest type of the GAGs. The other GAGs covalently attach to proteins in the form of proteoglycans. The functions of GAGs are mediated by both core proteins and GAG side chains. Large glycoproteins and proteoglycans in the ECM also bind growth factors and latent proteases, participate in chemical signaling and regulate the activities of proteolytic enzymes.

Fibronectin is a large, multi-domain, adhesive glycoprotein widely distributed in the matrix. Fibronectin binds both collagen and GAGs through multiple binding domains. Its major functions include organizing the ECM, guiding cell migration and facilitating cell adhesion through interaction with integrins. The integrin family members bind to specific domains of collagen, fibronectin, and laminin as well as heparan sulfate, and entactin.

Laminin, an adhesive glycoprotein mainly found in basement membranes, promotes the attachment of epithelial cells to the basal lamina. Laminin molecules can self-assemble into mesh-like sheet polymers and tightly associate with entactin, which also binds type IV collagen. Thus, **Entactin** serves as a structural link between type IV collagen and laminin networks in the basal laminae. Laminin is also the ligand for alpha6/beta4 integrin; the latter interacts with intermediate filaments of the hemidesmosomes, anchoring epithelial cells to the basal lamina. These complicated interactions weave the cross-linked networks in the basal lamina.

2.2 Basal Lamina

Basement membranes are distinguished from other connective tissue elements by their 3-dimensional sheet-like structural complexity containing specific proportions of type IV collagen, laminin and other minor constituents include tenascin, perlecan, nidogen/entactin, fibronectin, agrin, fibulins, BM-40, and calcium-binding proteins (15) The basal lamina is predominantly synthesized by the cells that rest on it. Most basal laminae consist of two distinct layers as viewed by the electron microscope, the lamina lucida or rara and lamina densa just below. The penetration of the basement membrane is the hallmark for malignant tumor progression beyond carcinoma in-situ for epithelial malignancies. Tumor cells metastasizing across tissue and vascular spaces often have to breech multiple membranes to be successful in surviving and proliferating at a site distant from the primary disease.

As laminin and type IV collagen assemble into a flexible sheet-like network, this serves as the scaffolding for the other components including heparan sulfate proteoglycans to bind. This intricate network provides mechanical support, forms selectively permeable barriers for tissue compartments, and provides immobilized ligands that interact with cell surfaces by binding to specific cellular receptors, which eventually affects cell differentiation, polarity, metabolism and migration (Figure 1 [16]).

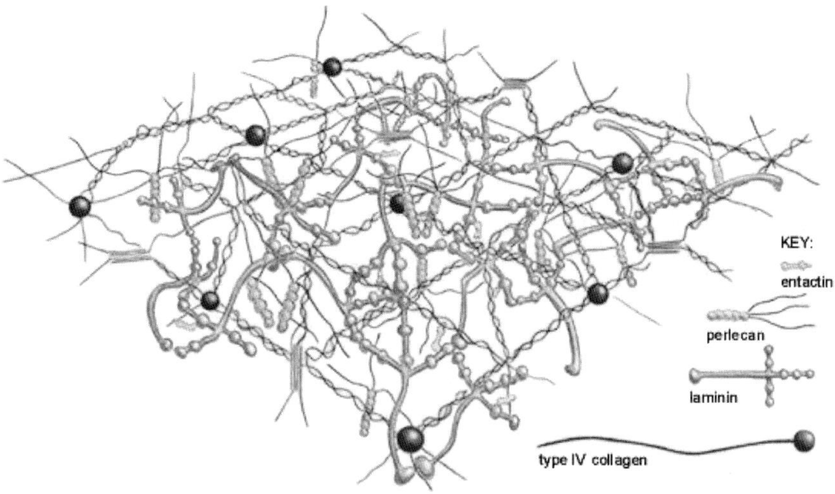

Figure 1. The complicated inter-relationships of the major components of the basement membrane are shown in a diagram to scale based on detailed analyses of molecular interactions by Dr. Peter D. Yurchenco. In the diagram, Collagen Type IV (Col-IV) forms one network. Laminin (Lm) forms another. Entactin (En) and Perlecan (Perl) interact with the two networks. HS is the heparan sulfate glycosaminoglycans linked to the perlecan core protein. (16 adapted with permission). See also Color Plate Section page 339.

3. THE ECM AND CANCER

The dissemination of cancer from the site of origin to non-contiguous locales depends on numerous interactions between the cancer cell and the host environment, in most cases, the extracellular matrix serves as a key modulator. ECM components and their receptors undergo changes during both normal and abnormal growth (17, 18). Localized degradation of matrix components is a prerequisite for tumor cells to migrate through basal laminae. Extracellular proteolytic enzymes (matrix metalloproteinases, urokinase activators, etc.) are responsible for the proteolysis, clearance, turn-over and remodeling of ECM components during tumor invasion. These enzymes have been shown to degrade virtually all the components of the ECM including the basement membrane (19). Proteolytic action releases sequestered growth factors, individual amino acids such as entrapped arginine and glycine, RGD sequences, and other signaling molecules from the ECM.

As noted above, cell-cell and cell-ECM interactions are altered both during malignant transformation and in neoplastic progression (20-22). Further, the ECM components have been found to participate in the regulation of tumor cell invasion (23, 24). Specifically, not only does the ECM serve as a barrier for invading tumor cells but ECM component degradation products participate in signaling selected cellular responses and interact with cell surface receptors. In addition, components of ECM when in contact with host-derived and tumor cell-secreted factors, can generate signals that modulate the motility of cells. The transduction of signals after ligand-receptor coupling is known to be important in the regulation of cell locomotion (25).

4. SIMPLE *IN VITRO* MATRICES TO MEASURE INVASION

4.1 Single ECM Components

Many investigators use only purified ECM components of defined type and concentration to test the invasive ability of tumor cells under experimental conditions. This list is as long as there are ECM components but most often 1 of 4 types (or a combination of 2 or more of these): collagen type I, collagen type IV, fibronectin or laminin is the matrix of choice. However, many of the commercial products are not as pure as anticipated and contain "contaminating" growth factors, metalloproteases or other bioactive molecules (26). Further, animal derived purified ECM and blood product matrices such as rat fibrin and collagen I gels have their own advantages and limitations when dealing with human cells (27-30).

Collagen I, the main substrate for MMP-1, is also the major dry component by weight of the ECM, thus it was reasonable to purify and utilize this matrix in the first invasion assays. Solutions should be prepared at a concentration between 1 and 4 mg/ml, depending on the desired final thickness. The collagen gel should form at 37°C for 30 min, followed by air drying in a tissue culture hood between applications (31). Nejjari et al. dissolved type I collagen in culture mediums and then poured it into culture plates and incubated them overnight at 37°C for gelation. Cells were seeded on the top of the collagen gels. After incubation, the depth of cell migration inside the collagen was measured (32). Japanese investigators prepared Type I collagen diluted to 1.0 mg/ml with ice-cold HCl. Fifty μl alliquots of the matrix solution, after neutralization were applied to an inert filter at 37°C for 15 min. (33). Albo et al. utilized a similar approach whereby 100 μg (1 mg/ml in 60% ethanol) of type I collagen was coated on a filter and dried overnight at 25°C (34). Collagen type I could also be diluted in cold distilled water, added to Transwell chambers (100 μg/well) to examine invasion (35).

Type I collagen could also be directly mixed with cancer cells suspended in culture medium on ice, followed by gelling at 37°C in a CO_2 incubator. Medium supplemented with 5% FCS could be added to the gel and changed on a regular schedule such as every 3 days. After culture for extended periods, changes in cellular morphology and cluster characteristics were observable by either phase contrast or conventional bright field microscope (36, 37). Type I collagen has also been used in three-dimensional (3-D) gel preparations (38).

Type IV Collagen is, as noted above, the major collagenous component of basement membranes. Commercially available collagen IV is derived either from human placenta or EHS mouse tumor. Tumor cells remain viable and can even proliferate when directly seeded onto a substratum consisting of type IV collagen (39). Rooprai et al. applied collagen type IV (10 μg/ml) onto the upper surfaces of polycarbonate filters air-dried overnight, (40) similarly, Kempen et al. coated polycarbonate filters in Transwell inserts with 100 μl of type IV collagen (200 μg/ml) diluted in PBS to examine the invasive capacity of tumor cells. Uniformity of the coating could be assessed by Coomassie blue staining. Ideal rehydration and pH balancing were achieved by washing with water and incubation in serum-free DMEM at 37°C (41).

Commercial entities have developed techniques to directly glue polycarbonate filters to collagen type IV inserts (Becton-Dickinson Labware, Bedford, MA) (42). Collagen IV could also be used in an indirect invasion assay through measuring its degradation by living tumor cells. Tumor cells are inoculated into culture wells coated with biosynthetically (^{14}C)proline-labeled type IV collagen. Soluble degradation products are detected by measuring the radioactivity present in the medium (43).

In a **fibronectin** based invasion assay, cells were added to Fluorescein-fibronectin-coated crosslinked gelatin films on a 15-mm round glass coverslip in a multiwell plate. Since the gelatin film is itself predominantly type I collagen, one might considered this a mixture rather than a single protein system. This technique was shown to reflect the invasiveness of the tumor cell population by measuring invadopodial extensions and surface indentations in the crosslinked gelatin film after culture for 12 h (44).

To detect invasion associated molecules during the process of tumor cells invading a **laminin** matrix, Thirupandiyur et al. precoated Teflon-printed microscope slides with purified human laminin-5 (1 μg/well) at 4°C overnight. Cells were then plated onto the coated laminin-5 and incubated for 48 h. After incubation, cells were removed with EDTA, and the coated laminin was aspirated using a solubilizing buffer for PAGE gel analysis and immunoblotted (45).

4.2 Mixtures of ECM with and without serum components

In an attempt to more closely mimic the native state, investigators have tried to utilize mixtures of ECM components for *in vitro* invasion assays and to select for subpopulations of cells with enhanced invasive ability. One such basement membrane-stromal matrix model was constructed by reconstituting purified laminin and type IV collagen onto type I collagen. The authors were able to show that tumor cells that had the capacity to penetrate such barriers, when isolated and subcultured, were more invasive and able to produce more metastases than the parental population (46). A defined matrix mixture containing laminin (50 μg/ml, type IV collagen (50 μg/ml), and gelatin (2 mg/ml) in a 10 mM glacial acetic acid solution mixture could also be delivered onto a 10-μm pore polycarbonate filter and packaged as a Membrane Invasion Culture System Chamber. Once cells were seeded on to the surface, an invasion assay system was in place (47). Ho et al. used a virtually identical "home-made" system (48).

Polycarbonate filters of slightly smaller pore size (8μm) have also been used as cell culture inserts coated with 5.0 μg laminin or 5.0 $\mu g/cm^2$ fibronectin for similar kinds of experiments (49). Simple ECM component could also be used to adulterate commercial ECM preparations, for example, Farrow et al. mixed laminin with the cells into Matrigel (see below) (50).

To investigate the role that plasminogen activators play in the tumor invasion process, investigators have taken fibrinogen (plasminogen-free 100 μl of a 10 mg/ml stock solution) and 100 μl of 1 unit/ml thrombin, mixed them and immediately delivered them onto an 8 μm pore sized filter. A fibrin gel formed after incubation at 37°C for 30 min. Tumor cells were then trypsinized, centrifuged, resuspended in medium supplemented with 0.1% BSA and 10 μg/ml plasminogen and incubated for 15 min at room temperature. Cells were added to the upper chamber of a Boyden chamber-like invasion assay system and incubated for 24 h before calculation of invasion (51). Similarly, invasion gels composed of fibrin have been created by adding human thrombin (0.2 U/ml) with rabbit fibrinogen (3 mg/ml) dissolved in DME. 0.5 ml of this solution could be added to chambers, swirled to allow uniform distribution and allowed to solidify (30 min, 37°C). Cell invasion was measured by quantifying the number of tubular networks formed by invading microvascular endothelial cells (52).

To evaluate *in vitro* tumor invasiveness, Hori et al. devised a more elaborate mixture, coating microporous membranes (pore size, 12 μm) with laminin (5 μg/ml), type IV collagen (100 μg/ml), hyaluronic acid (50, 100 or 1,000 μg/ml), and fibronectin (1 μg/ml). Cancer cells were then added to the upper transwell chamber and invasion quantified (53). Quax et al. measured the capacity of Dunning rat prostate cells to degrade ECM by culturing them on ^{3}H-labeled ECM produced by bovine smooth muscle cells. Bovine aortic smooth muscle cells were grown to confluency. The cells were then incubated for 4 days with medium containing a ^{3}H-amino-acid mixture (1 mCi/ml). Cells were then lysed using 0.5% (v/v) Triton X-100 in PBS, and the cytoskeleton was removed by 25 mM ammonium hydroxide treatment. Unincorporated ^{3}H-amino acids were washed from the remaining extracellular matrix using H_2O and 75% (v/v) ethanol. The ^{3}H-labeled matrices were soaked with medium for 1 hr before use (54).

5. COMPLEX *IN VITRO* MATRICES TO MEASURE INVASION

5.1 Matrigel™ (BD)

Among the various native, artificial and reconstituted extracellular matrices used in *in vitro* invasion assays, the most widely used is Matrigel, a mixture of basement membrane components isolated from the Engelbreth-Holm-Swarm (EHS) mouse tumor (55-57). Others have used similar natural or induced products derived from animal systems to, for example, characterize the desmoplastic response of human carcinomas (58, 59). Through use of a Matrigel based invasion assay, numerous inhibitors and enhancers of tumor invasion have been identified (60, 61).

Several companies produce ECM extracted from the EHS mouse tumor. In this discussion we refer to the Matrigel produced by BD Biosciences (San Jose, CA) as the model. The major components of BD Matrigel Matrix (Matrigel) including laminin (56%), collagen type IV (31%) and entacin (8%) (62). Matrigel also contains growth factors with EGF (0.7 ng/ml), PDGF (12 ng/ml), IGF-1 (16 ng/ml) and TGF-α (2.3 ng/ml) predominating. At room temperature, Matrigel polymerizes to produce biologically active matrix material resembling basement membrane. There have been several modified preparations of Matrigel, including Growth Factor Reduced (GFR), Phenol-Red Free and HC formated, suitable for special research purposes. Matrigel may be used as thin gel layer (0.5 mm) for plating cells on top of the matrix or cells can be cultured inside the Matrigel using a 1 mm layer. Matrigel provides a suitable ECM environment for studies of cell morphology, biochemical function, migration, invasion, and gene expression as long as the investigator recognizes the presence and potential role of other bioactive molecules within the complex including metalloproteinases. Matrigel support the culture of polarized cells, *in vivo* peripheral nerve regeneration and promotes the differentiation of many cells types. Matrigel has also been used in *in vitro* models for drug toxicity studies, for *in vivo* angiogenesis, for examination of augmentation of growth, maintenance of cell integrity and to localize implanted tumors (60, 63-68).

Matrigel is widely accepted as a biologically active basement membrane mimic, and numerous publications have proved its usefulness for *in vitro* invasion assays (46, 69-73). The advantages of Matrigel invasion assay include its quick, reliable, and easy to quantitate nature and its commercial availability (72, 74). Matrigel has been used to screen cells for malignant potential, to select invasive subclones, and to identify invasive and anti-invasive factors. A positive correlation between the transformed phenotype, tumor cell invasive ability *in vivo*, or metastatic potential and the ability to invade through Matrigel has been demonstrated (72, 74-77). However, others challenge many of these assumptions and thus, interpretation of results may become an issue (78, 79). Matrigel could also be mixed with type I collagen (usually from rat tail) (80) or mixed with fibronectin (250 μg/ml) as a model of a more complex invasion matrix barrier (81). Other combinations have also been offered (82).

As noted above, Matrigel has been found to contain both active and latent MMPs by Western blot analysis. A rough quantitation of MMP-2 and MMP-9 in 100 μg of Matrigel was between 0.3 and 0.6 ng. The presence of these "Contaminants" may blemish the interpretation of information obtained from Matrigel assays (26). Matrigel also contains bio-active plasminogen, which may interfere with the analysis of factors affecting cellular invasion (83). Lastly, Matrigel contains little or no type I collagen, the essential footprint for stromal cell-matrix interactions found under *in vivo* conditions (1-11).

5.2 ECMatrix™ (Chemicon)

Other commercial sources have also developed reconstituted basement membrane/ matrix components derived from the Engelbreth Holm-Swarm (EHS) mouse tumor (72, 84, 85). The Chemicon (Temecula, CA) invasion kit utilizes this matrix as the invasion barrier in their QCM 24-well collagen-based Invasion Assay (86). The utility of this assay kit to measure cell invasiveness have been reported (87). The experiments proceed over long periods of time (48 h) at 37°C and because of the nature of the matrix, is limited by the same bioactive substances as Matrigel. To better assess relative rather than absolute invasive capacity to test cell populations, the highly invasive cell line HT1080 and the noninvasive cell line NIH3T3 are often used as positive and negative controls, respectively (88).

5.3 ECM gel (Sigma)

The ECM gel as developed by Sigma (St. Louis, MO) is also prepared from mouse Engelbreth Holm-Swarm tumor and thus, not surprisingly, its major components include laminin, collagen type IV, heparan sulfate, proteoglycan and entactin. In this system, a filter is coated with 40 μl of the ECM gel and then placed in an invasion chamber (60). The ECM gel undergoes thermally activated polymerization and gels within 5 minutes at 20°C to form a reconstituted basement membrane. Addition of collagen type IV to the ECM gel increases polymerization whereas addition of collagen type I, fibronectin or heparin does not. The ECM gel formed at 37°C maintains its form in culture medium for at least 14 days and viable cells can be recovered from the gel by the use of proteases. ECM gel has a relatively short half-life but may be stored up to 72 hours at 2-8°C.

5.4 PuraMatrix™ (3DM)

PuraMatrix™ (3DM Inc. Cambridge, MA) is a synthetic 16-amino acid polypeptide which assembles into an *in vivo*-like 3-dimensional extracellular matrix hydrogel to support the growth of many different cell types in tissue culture (89-92). PuraMatrix is normally added to the surface of a polylysine-coated dish followed by the addition of media to cause the material to gel. After the PuraMatrix gels, the media is changed twice before plating of the cells at the desired concentration. Gentle handling during media exchange is key to help prevent coating lift off. PuraMatrix peptide gels serve as a synthetic 3-D ECM scaffolding and invasion assays are carried out in combination with cell culture media, serum, and often exogenous growth factors, and cytokines (93, 94). The advantage of PuraMatrix as elucidated by the manufacturer include animal-free production, high reproducibility, approximation of *in vivo* ECM fiber size, high hydration level, sterilization *in situ*, easy cell recovery and straightforward molecular biology analysis.

5.5 Amgel/HuBiogel™ (VBI)

Although animal-derived matrix coatings and scaffolding serve important roles in dissecting the steps of neoplastic progression, these models pose numerous biologic limitations and disadvantages as elucidated above. Major concerns include the fact that although these systems support cell proliferation and monitor invasion they may induce untoward effects including mitogenesis and differentiation and, most critically, they do not recapitulate the defined human biomatrix milieu critical for identifying specific tumorigenic responses.This has therefore retarded our understanding of regulatory steps of tumor invasion (and other critical defined steps including angiogenesis), especially in humans. We, thus recognized in the mid-1980s the need to develop a well-defined human based matrix derived from non-neoplastic tissue for studying steps of neoplastic progression *in vitro*.

Initially we created a human reconstituted basement membrane matrix which we called Amgel, for studying primarily the invasive capabilities of neoplastic cells (95, 96). In more recent years, this product and its process has been reformulated, enhanced and commercialized under the name HuBiogel. This work has extended the usefulness of this material for studies of tumorigenesis (97, 98). A schematic comparison of Amgel/HuBiogel, Matrigel and collagen models as well as their cell-cell and cell-matrix interactions is depicted in Figure 2.

Unlike many of the "ready-to-use" products available, Amgel/HuBiogel contains both stromal and BM matrix proteins: laminin, collagens I and IV, fibronectin, entactin, tenascin and heparan sulfate proteoglycan. Its other advantages include, i) that it is a defined physiological human biomatrix (free of proteases, major growth factors and other immunologic agents), and ii) that it alone is nontumorigenic but supports the promotion of tumor growth in co-culture or co-inoculation models (68). Some unique physical characteristics of Amgel and composition differences with Matrigel are summarized in Table 1.

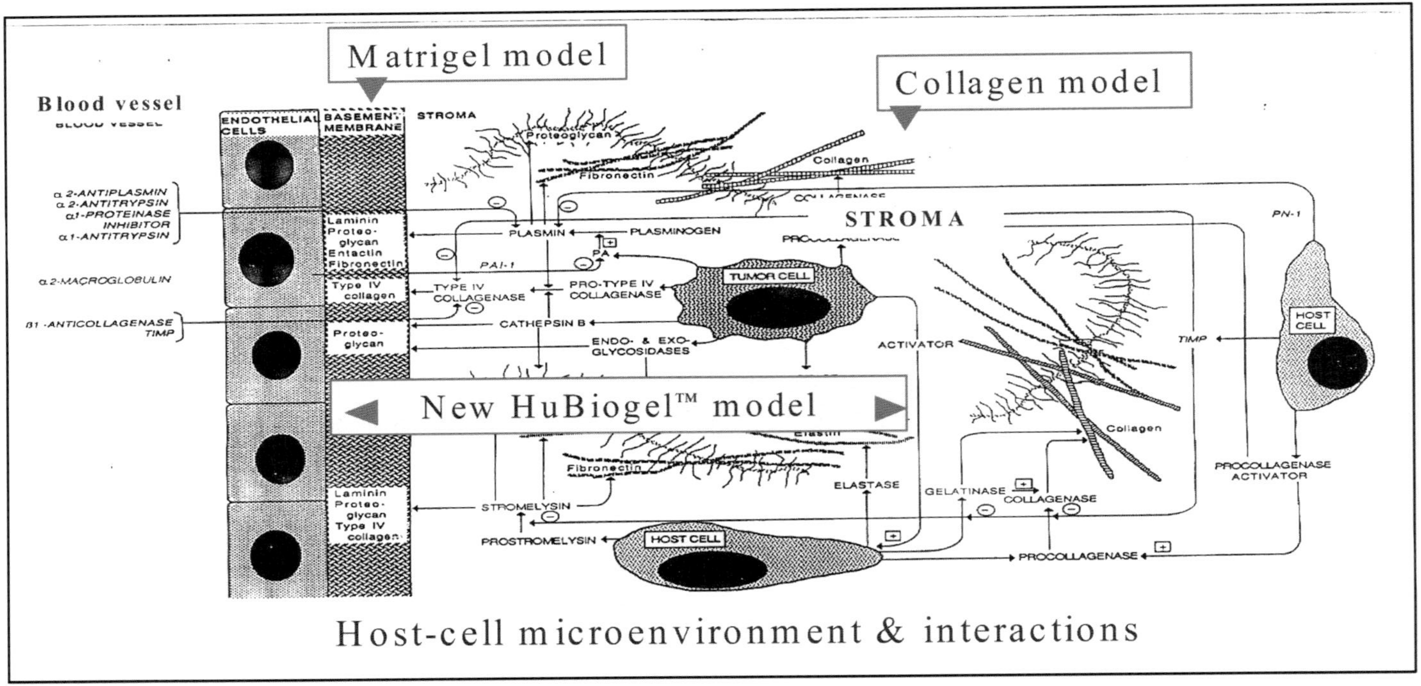

Figure 2. Key sites of action of 3 common *in vitro* matrices to measure cell invasion shown against the cell-cell and cell-matrix interactions involved in this process.

Table 1. Comparison of Amgel(HuBiogel) and Matrigel.

Properties	Amgel	Matrigel
Source/ origin:	Human placentas	Mouse EHS-tumors
Handling:	Semi-gel (4°C to 37°C)	Temperature-sensitive
Matrix environment:	Native biomatrix, Physiological	Reconstituted BM, Tumorigenic, Angiogenic
Major matrix components:	Collagens I & IV, laminin, tenascin, entactin, proteoglycans	Laminin, collagen IV, entactin, proteoglycans
Growth factors:	None, undetectable	TGF, EGF, PDGF, FGF
Proteases:	None, undetectable	Collagenase activity
Biologic advantages:	Normal cell growth, defined or controlled induction	Uncontrolled cell activation, growth, differentiation

Because of its properties, Amgel/HuBiogel mimics more closely an "all-human" physiologic environment for *in vitro* cell growth and maintenance and thus allows one to develop "controllable" bioassays to identify differential cell motility, adhesion and invasive capability. Important too, Amgel/HuBiogel has proved itself useful for identifying single positive and negative modulators (natural or synthetic) of target cell function. Two examples are 1) under defined formulation one can mimic both dose and time dependent stages of endothelial morphogenesis (proliferation, transmigration and differentiation) (98) and 2) Amgel-based tumor invasion assays can be used to determine the phenotypic profiles of transgenic cell constructs. For example, human prostate DU-145 tumor cells (Table 2).

Table 2. Biologic responses elicited by EGFR-cell constructs.

Construct	EGFR-function	Mito	Moto	Detach	Prolif	Invasion
WT (+)	full-length	+	+	+	+	+
c'1000	single P-tyr site	+	+	+	+	+
c'1000F^{992}	mutated P-tyr site	+	-	+	+	+
c'973	truncated, no C-tail	+	-	+	+	+

P- auto phosphorylation; d- dominant-negative; Mito - mitogenesis; Moto - motogenesis; Detach - EGF-induced detachment from ECM; Prolif- proliferation activity; Invasion-Amgel

Transgenic fibroblast constructs expressing EGF-receptors have been analyzed for cell adhesion and transmigration properties (25).

5.6 Multicellular and related models

At present, tumor invasion and angiogenesis are mostly studied using monolayer cell cultures which are, by definition, two dimensional (2D) coculture models. Multicellular models representing a more complex biology have recently been developed using heterogeneous cell lines and biomatrix coculture methodology (99, 100). Using specialized rotary vessel or spinner flask culture technologies, multicellular spheroids and organoids of tumor, liver, bone, prostate and cartilage have been generated with collagen-coated microbeads or Matrigel-embedded cell systems.) Compared to conventional monolayer culture models, these 3D coculture systems more closely resemble the multicellular organization and biologic complexity of target tissues *in situ* (110). Tumor spheroids show promise as experimental models of tumor growth and micrometastasis for studying the effects of anticancer drugs or therapy (111). Unfortunately, heretofore defined micro-tumor models which can replicate the *in vivo* tumorigenesis events are not reported. Matrigel-based cell aggregates in coculture exhibit apoptosis of core cells and cell-matrix signaling is induced by its tumorigenic milieu instead of normal host microenvironment. Moreover, these spheroid models do not support neovascularization or angiogenesis processes. New cell-biomatrix models are now becoming possible utilizing Amgel/HuBiogel to mimic the 3D tumor-host interactions and functional events.

6. INVASION ASSAYS

Various *in vitro* models have been developed to investigate the invasion process, to further understand mechanisms of action and to explore potential anti-invasion drugs. For decades, and still today, investigators obtain most of their information about anti-invasive factors and experimental interventions by use of invasion assay models based on measuring the capability of tumor cells to degrade matrix as elucidated above. Although the *in vitro* invasion assay does not fully mimic the complicated multi-cell-ECM micromilieu *in vivo*, it is still suitable for studying the influence of individual molecules. To better place in context application of the extracellular matrices we introduced above, we overview several common types of *in vitro* tumor invasion assays.

6.1 Chemoinvasion Assays Using Artificial ECMs

The **Boyden-chamber chemo-invasion assay** was first developed by Albini et al. (72) and gradually became the most widely used experimental invasion system *in vitro*. Using different matrices as basement membrane mimics, the invasion chamber system provides a simplified and standardized model to test the cell capacity to invade into ECM towards a chemoattractant (or gradient). Microporous membranes (usually made of polyvinyl-pyrrolidone (PVP)-free polycarbonate, with pore size 8-12 μm) are usually used to separate the chemoattractant in the lower chamber from the cells adherent to the ECM/filter in the upper chamber. Pro- or anti-invasion factors are generally added in the upper chamber containing the tumor cells. The matrices occlude the microporous membrane pores, blocking non-invasive cells from migrating through, while invasive cells actively invade the ECM layer and migrate onto the polycarbonate membrane and eventually into the lower chamber media or floor. Typically after cells are given time to first adhere to and then penetrate the ECM layer over time, non-invaded cells on the upper surface of the filter are removed, and cells successfully invaded onto the lower surface of the filter membrane are fixed, stained and counted (Figure 3). We have utilized this approach to create an Amgel/HuBiogel invasion assay system (Figure 4).

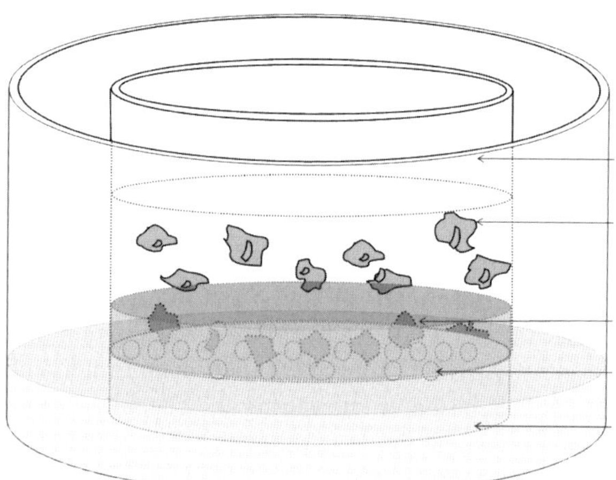

Figure 3. Illustration of a Boyden Chamber Invasion Assay. Tumor cells are suspended in the upper chamber in low-serum medium; various extracellular matrices are coated on a microporous membrane to mimic basement membrane. Most invading tumor cells will be blocked by 8μm pores on the filter and are stained and enumerated; while non-invading cells loosely attached to the upper surface or free-floating are removed with matrix before staining.

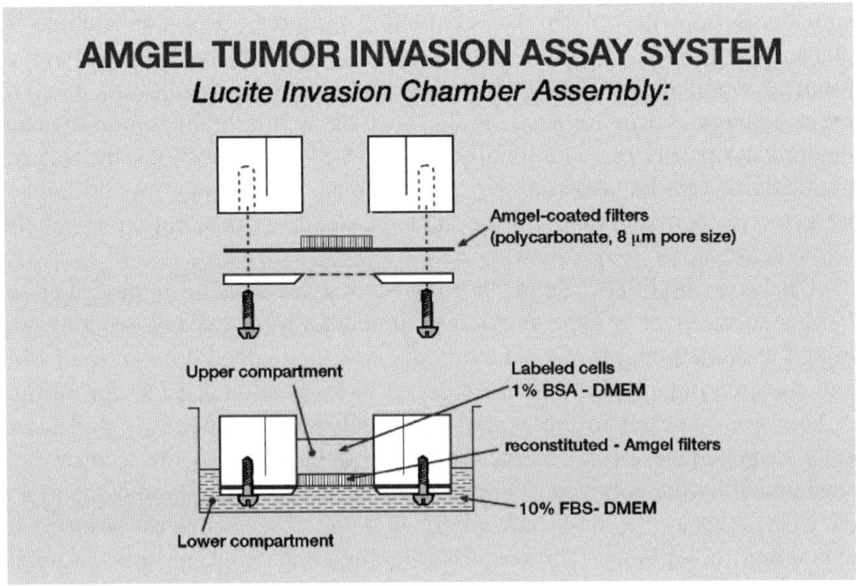

Figure 4. Diagram of disassembled (top) and assembled (bottom) two-chamber HuBiogel Tumor Invasion Assay System.

Most commonly chemotaxis is achieved by providing a serum concentration gradient, with the lower chamber containing the higher concentration (eg. 10% FBS) and the upper chamber containing a lower one (eg. 1%BSA or serum free). In addition, the lower surface of the filter membrane could be coated with different matrix components serving as chemoattractants. These include collagen I (112, 113), fibronectin (114), gelatin (115) and purified type IV collagen (116).

Although the phenomenology of chemoinvasion was perceived as a simple single process, it is now known that the ability of the cells to migrate through the filter is influenced by multiple factors including cell density, locomotive ability, and the pore size. Tumor cells display considerable variation in their ability to adhere, migrate and attach to the lower side of the filter membranes. To adjust for these influences, results should be corrected for these variables. In addition, an internal standard to compensate for variations is often set up by calculating the percentage of invasion relative to a reference cell's invasive ability, which enables comparison of data obtained from independently performed experiments on different cell lines (117).

Three-dimensional invasion assays provide an environment for tumor invasion closer to the natural *in vivo* state as tumor cells are known to grow, form clusters and invade into normal tissue in three dimensions. This can sometimes be achieved by use of rotation cultures. Rotation cultures of chick embryo liver cells and pigmented murine myeloma cells as well as rounded fragments of chick embryo heart have all been used (118). Fetal rat brain aggregates can be used as the host tissue for brain tumor invasion determinations (119). Additionally, invasion of tissue sections by test cell populations can be assessed by conventional microscopy by utilizing a micrometer (ocular reticle) to measure the depth of penetration of labeled tumor cells.

Lastly, a modified "Semi-3D" invasion assay has been described by Perumpanani et al. Type I collagen prepared from rat tail tendons was mixed with medium on ice. The mixture was then added onto an inert filter and collagen polymerization was induced by incubation at 37°C for 60 min. A hole was punched in the center of the gel using a pipette tip and tumor cells suspended in 50 μl of medium were then added to this central hole surrounded by the collagen. One ml of medium was added and incubated at 37°C in a 10% CO_2 environment for 4 days. Invasion was assessed by averaging the distance traveled by the invading front in four mutually perpendicular radial directions (120).

6.2 INVASION SYSTEMS USING TISSUES

Tumor cells that could penetrate natural human or animal membranes, it has been argued, more closely mimic the *in vivo* state, mimicking the complexity of the authentic basement membrane. Several methods have been developed. Chief among these in experimental systems are chick chorioallantoic membrane, mouse urinary bladder and canine blood vessels (121).

Hart et al. measured the rate at which radio-labeled tumor cells migrate through the **chorioallantoic membrane** (CAM) in developing chicken embryos. The chambers were placed in glass vials resting on discs of photographic sponge which were immersed in tissue culture medium. Labeled tumor cells were added to the chamber and quantitation of the radioactivity retained in the sponge provided an indirect measure of cell invasiveness. This invasion assay was shown to correlate with the metastatic potential of the cells *in vivo* (122). Scanning electron microscopy documented that transformed cells passed the chick chorioallantoic membrane ectoderm within 6 hr, while viable NIH/3T3 cells did not penetrate the ectoderm within 24 hr. The transformed cells entered the mesoderm of the CAM and formed tumoUrs aggregationS (123). The invasive potential of human choriocarcinoma cell lines in the CAM was further found to correlate with the tumorigenic potential in hamster cheek pouches. In addition, the invasive capacity of cell lines correlated with the amount of collagenase secreted by them. (1240. Melanoma sublines with highly invasive ability in this assay were also shown to correlate with high rates of spontaneous and experimental metastases *in vivo* (121).

The application of a **sea urchin embryo basement membrane** (**SUECM**) invasion assay was described by Livant et al. and Dyer et al. (125, 126). Tumor cells were layered on top of the SUECM and incubated for 4 hr at 37°C. The percentages of both spreading and adherent cells were evaluated in each assay following fixation in formaldehyde and scoring by microscopy. The numbers of cells adhering to the surfaces of the SUECM and their location with regard to the exterior surfaces were determined. Invasion percentages were determined as a ratio of cells inside (invading the SUECM substrates) as compared to the sum of all cells (both adhering to the exterior and within the interior surfaces) (127).

In contrast with these invasion assays using animal-derived tissue, Liotta and his colleagues used **human amnion** in an invasion assay. When first separated from the chorion and subsequently stripped of the single layer of epithelial cells, basement membrane is exposed to the test cells as the ECM barrier. (128). This succeeds because the amnion is relatively uniform and is composed simply of the aforementioned epithelial layer resting on a continuous basement membrane overlying an avascular collagenous stroma. The protocol for isolation and preparation of human amnion basement membrane have been described in detail by Smith et al. (129). Cells capable of invading the whole thickness of the amnion membrane remain trapped within the microporous membrane filter which is examined after 24 h (130) or will pass into the lower chamber after longer time points (131). Endothelial cells could also be plated onto the amnion basement membrane and thus an artificial vessel wall could be created to study tumor cell intravasation or extravasation *in vitro*.

To investigate the penetration of melanoma cells through the dermal-epidermal junction, an *in vitro* model has been reported by Bechetoille et al. Metastatic melanoma cells were seeded with normal allogenic keratinocytes onto **acellular human de-epidermized dermis** containing a preserved and intact basement membrane and cultured for up to 1 month at an air-liquid interface. The invasion was then identified by histological, immunohistochemical and ultrastructural methods. This model preserved the 3-D architecture of the native dermal-epidermal junction. The authors were able to demonstrate that the active invasion process coincided with the dissolution of ECM components within the basement membrane and that the invasive ability of melanoma cells *in vitro* correlated with metastatic potential *in vivo* (132).

Tumor cells need to penetrate vessel endothelium to gain access to the blood and lymphatic circulation and extravasate across similar endothelial barriers at distant sites to form metastatic foci. To study this critical process, invasion assays have been successfully developed by using a layer of endothelial cells as the barrier for tumor cells to migrate through. These barrier cells including **human bone marrow endothelial cells** (133), **Human umbilical cord endothelium and bovine capillary endothelium** could also be cultured on the basement membrane derived from human amnion. In one series of experiments, squamous carcinoma cells were inoculated onto the basement membrane surface in the presence or absence of such an endothelial monolayer. Tumor cells that invaded the endothelium and underlying basement membrane were captured by a microporous membrane (130).

6.3 INVASION SYSTEMS USING ORGANS

Organ cultures invasion assays measure the ability of tumor cells to invade cultured organ fragments. As the invasion barriers are intact structures containing 1 or more tissues, they are considered by some to more pathophysiologically reflect the conditions *in vivo* during tumor cell invasion. The detailed procedure of performing the **Chicken Heart Invasion Assay** has been described by Engers et al. One major advantage of this approach is that the extent of invasion in three dimensions can quantified by estimating the volume of the host tissue (organ) that had been replaced by the tumor aggregates (134). Using embryonic chick cardiac muscle as the host tissue, Schroyens et al. found that the invasiveness of tumor cell lines correlated with tumorigenicity in nude mice (135). Monitoring the depth of invasion by thin-section electron microscopy, Pauli et al. discovered that salt-extracted **bovine articular cartilage** could be used as an invasion matrix to determine the potential of cells for tumor invasion (136). Fresh **embryonic chick lung fragments** could also be used as the host tissue in this invasion assay (137).

To provide conditions simulating even more closely that of the in-vivo condition, investigators have cocultured myoepithelial cells and fibroblasts in a matrix barrier to which breast or ovarian carcinoma cells were exposed, respectively (138, 139), or cocultured human foreskin fibroblasts with rat-tail collagen and examined the invasive ability of breast cancer cells (140). Nakayama et al. used preformed type I collagen gel overlaid with type IV collagen followed by seeding of calf pulmonary arterial endothelial cells and incubated them until they reached confluence. Highly invasive tumor cell lines displayed a higher penetration rate through the monolayer of endothelial cells and type IV collagen sheet onto the type I collagen gel (141). Wandel et al. prepared 5 mg/ml solutions of type I collagen in complete DMEM which were overlaid with the collagen solution containing fibroblasts and melanoma invasion was examined (142). Similarly, to study the invasion of laryngeal carcinoma cells, an even more complicated reconstituted ECM has been described by Yamada et al. Their invasion matrices incorporated type I collagen, fibroblasts, and fat cells to closely simulated the epithelial-to-submucosal layer of the larynx (143).

7. IMPROVING INVASION ASSAYS

For all intents and purposes an *in vitro* assay of any kind only has value if it is predictive of the *in vivo* state it proposes to represent (Table 3).

This is certainly true for invasion assays. The disadvantages of the invasion assays elucidated above are many. These include, but are not limited to, the fact that they are usually time-consuming and tedious, involving multiple points to insert potential artifacts (for example cotton swabbing), often rely on manual staining and counting, and not the least of all don't, in a uniform and reproducible way, mimic the *in vivo* state where it can be examined. As an example, Le Marer et al. compared the invasiveness of human breast cancer cell lines by Matrigel, Collagen I and Chick Heart Fragment Invasion Assays and revealed contradictory results among the different experiment methods (144). Several improvements have been attempted to address these issues.

Table 3. summarizes the invasiveness of various human cell lines on HuBiogel and Matrigel coatings as compared to *in vivo* findings. Note the Matrigel exhibits higher bioactivity, possibly due to its own mitogenic milieu. A standard 72-hr bioassay employed 75 ug biomatrix coated-Transwells and monolayered tumor cells. Percent invasion represents number of cells present in bottom filters/chambers.

Cell line	Amgel Invasion (%)	Matrigel Invasion (%)	In Vivo Tumor Index
HF	1.0	3.5	-
HT-1080	12.0	17.5	++++
DU-145	5.5	8.5	++
PC-10	1.5	4.0	+
U-87	8.5	14.5	+++
U-251	3.5	5.5	+

Commercially available invasion kits help to improve the reproducibility and decrease the inter-investigator variation. Adaptations and modifications of these *in vitro* invasion systems continue to be used at a high rate today for the study of invasion. With the advent of new techniques, these analyses are becoming more quantitative.

The replacement of manual histological analyses by objective automatic measurement of the labeled cells may represent the future direction for the improvement of invasion assay *in vitro*. Employing the fluorescent dye YO-PRO-1, cells that penetrate Matrigel-coated Transwells are counted on the basis of dye-bound cellular nucleic acid content, measured as fluorescence intensity in 96-well microplates and quantitated against a cell- or DNA-calibration curve (145). The QCM Cell Invasion Assay (Chemicon, Termecula, CA) uses fluorescence methods to measure invasion in 96 well culture plates simultaneously, which provide a more objective and quicker way to measure invasion. This system does not require cell labeling, scraping, washing or counting. The 96-well inserts and homogenous fluorescence detection format allows for large-scale screening and quantitative comparison of multiple samples. Invaded cells on the bottom of the insert membrane are dissociated, lysed and detected by CyQuant GR* dye (Molecular Probes) bound to cellular nucleic acids (146).

FATIMA (Fluorescence-Assisted Transmigration Invasion and Motility Assay) is another invasion assay method using fluorescence labeling. FATIMA offers the hope of providing a rapid, quantitative and large-scale analysis of cellular invasion. Other potential advantages include convenient monitoring of invasion kinetics as well as both more accurate and more reproducible estimates of invasive cells. Fluorescence blocking membrane inserts were also developed to prevent any interference from excessive fluorescence labeling (147).

8. CONCLUSIONS

Beginning with the major component of the ECM, we have explored both simple and complex *in vitro* cell-matrix models of tumor invasion which have been developed. We have briefly reviewed invasion assay systems which are acellular or artificial through those that incorporate cells, tissues and even organs. We have tried to identify both the strengths and weaknesses of these assays while highlighting our own work and those of our colleagues, ending with a brief look to the future. Clearly, cancer remains a major clinical problem in society and any strategy that will allow us to more rapidly identify, prevent, treat or prognosticate has value. These assays are but one step in an ever more complex unraveling of this dreaded constellation of diseases which share the lethal attributes of invasion and distant metastasis.

REFERENCES

1. Beck L, Jr. D'Amore PA. Vascular development: cellular and molecular regulation. Faseb. J. 1997, 11: 365-73.
2. Grant DS, Kleinman HK, Martin GR. The role of basement membranes in vascular development. Ann. N. Y. Acad. Sci. 1990, 588: 61-72.
3. Folkman J, Shing Y. Angiogenesis. J. Biol. Chem. 1992, 267: 10931-4.
4. Furcht LT, Critical factors controlling angiogenesis: cell products, cell matrix, and growth factors. Lab Invest. 1986, 55: 505-9.
5. Ingber DE, Folkman J. How does extracellular matrix control capillary morphogenesis? Cell 1989, 58: 803-5.
6. Paku S, Paweletz N. First steps of tumor-related angiogenesis. Lab Invest. 1991, 65: 334-46.
7. Kumar R, et al. Regulation of distinct steps of angiogenesis by different angiogenic molecules. Int. J. Oncol. 1998, 12: 749-57.
8. Tyagi SC. Vasculogenesis and angiogenesis: extracellular matrix remodeling in coronary collateral arteries and the ischemic heart. J. Cell. Biochem. 1997, 65: 388-94.
9. Liotta LA, Kohn EC. The microenvironment of the tumour-host interface. Nature 2001, 411: 375-9.
10. Liotta LA, Steeg PS, Stetler-Stevenson WG. Cancer metastasis and angiogenesis: an imbalance of positive and negative regulation. Cell 1991, 64: 327-36.
11. Pupa SM, et al. New insights into the role of extracellular matrix during tumor onset and progression. J. Cell Physiol. 2002, 192: 259-67.
12. Siegal GP, et al. Stages of neoplastic transformation of human breast tissue as monitored by dissolution of basement membrane components. An immunoperoxidase study. Invasion Metastasis 1981, 1: 54-70.
13. Liotta LA, et al. Effect of plasminogen activator (urokinase), plasmin, and thrombin on glycoprotein and collagenous components of basement membrane. Cancer Res. 1981, 41: 4629-36.
14. Alberts BB, Dennis; Lewis, Julian; Raff, Martin; Roberts, Keith; Watson, James D., Mol. Biol. Cell. Third ed. 1994, New York and London:: Garland Publishing; c1994. 989-994.
15. Relan NK, Schuger L. Basement membranes in development. Pediatr. Dev. Pathol. 1999, 2: 103-18.
16. Alberts BB, Dennis; Lewis, Julian; Raff, Martin; Roberts, Keith; Watson, James D., Mol. Biol. Cell. Third ed. 1994, New York and London:: Garland Publishing; c1994. 991.
17. Borg TK. It's the matrix! ECM, proteases, and cancer. Am. J. Pathol. 2004, 164: 1141-2.
18. Mareel M, Leroy A. Clinical, cellular, and molecular aspects of cancer invasion. Physiol. Rev. 2003, 83: 337-76.
19. Barsky SH, et al. Loss of basement membrane components by invasive tumors but not by their benign counterparts. Lab Invest. 1983, 49: 140-7.
20. Kiemer AK, Takeuchi K, Quinlan MP. Identification of genes involved in epithelial-mesenchymal transition and tumor progression. Oncogene 2001, 20: 6679-88.

21. Medico E, et al. Osteopontin is an autocrine mediator of hepatocyte growth factor-induced invasive growth. Cancer Res. 2001, 61: 5861-8.
22. van 't Veer LJ, et al. Gene expression profiling predicts clinical outcome of breast cancer. Nature 2002, 415: 530-6.
23. Zhang M, et al. Epidermal growth factor modulates cell attachment to hyaluronic acid by the cell surface glycoprotein CD44. Clin. Exp. Metastasis 1996, 14: 268-76.
24. Zhang M, et al. Epidermal growth factor induces CD44 gene expression through a novel regulatory element in mouse fibroblasts. J. Biol. Chem. 1997, 272: 14139-46.
25. Xie H, et al. In vitro invasiveness of DU-145 human prostate carcinoma cells is modulated by EGF receptor-mediated signals. Clin. Exp. Metastasis 1995, 13: 407-19.
26. Mackay AR, et al. Identification of the 72-kDa (MMP-2) and 92-kDa (MMP-9) gelatinase/type IV collagenase in preparations of laminin and Matrigel. Biotechniques 1993, 15: 1048-51.
27. Satake S, et al. Angiogenic stimuli are essential for survival of vascular endothelial cells in three-dimensional collagen lattice. Biochem. Biophys. Res. Commun. 1998, 244: 642-6.
28. Brown KJ, et al. A novel in vitro assay for human angiogenesis. Lab Invest. 1996, 75: 539-55.
29. Montesano R, Orci L, Vassalli P. In vitro rapid organization of endothelial cells into capillary-like networks is promoted by collagen matrices. J. Cell Biol. 1983, 97: 1648-52.
30. Nicosia RF, Ottinetti A. Growth of microvessels in serum-free matrix culture of rat aorta. A quantitative assay of angiogenesis in vitro. Lab Invest. 1990, 63: 115-22.
31. Fujimoto K, et al. Transforming growth factor-beta1 promotes invasiveness after cellular transformation with activated Ras in intestinal epithelial cells. Exp. Cell Res. 2001, 266: 239-49.
32. Nejjari M, et al. Inhibition of proprotein convertases enhances cell migration and metastases development of human colon carcinoma cells in a rat model. Am. J. Pathol. 2004, 164: 1925-33.
33. Takeuchi S, et al. Isolation of differentiated squamous and undifferentiated spindle carcinoma cell lines with differing metastatic potential from a 4-nitroquinoline N-Oxide-induced tongue carcinoma in a F344 rat. Jpn. J. Cancer Res. 2000, 91: 1211-21.
34. Albo D, et al. Tumour cell thrombospondin-1 regulates tumour cell adhesion and invasion through the urokinase plasminogen activator receptor. Br. J. Cancer 2000, 83: 298-306.
35. Koike N, et al. Invasive potentials of gastric carcinoma cell lines: role of alpha 2 and alpha 6 integrins in invasion. J. Cancer Res. Clin. Oncol. 1997, 123: 310-6.
36. Benbow U, et al. Confocal assay for invasion: use of propidium iodide fluorescence and laser reflectance to quantify the rate of migration of cells through a matrix. Cytometry 2000, 40: 253-9.
37. Nishimura K, et al. Prostate stromal cell-derived hepatocyte growth factor induces invasion of prostate cancer cell line DU145 through tumor-stromal interaction. Prostate 1999, 41: 145-53.
38. Schor SL. Cell proliferation and migration on collagen substrata in vitro. J. Cell Sci. 1980, 41: 159-75.

39. McCarthy JB, et al. The role of cell adhesion proteins–laminin and fibronectin–in the movement of malignant and metastatic cells. Cancer Metastasis Rev. 1985, 4: 125-52.
40. Rooprai HK, et al. Evaluation of the effects of swainsonine, captopril, tangeretin and nobiletin on the biological behaviour of brain tumour cells in vitro. Neuropathol. Appl. Neurobiol. 2001, 27: 29-39.
41. Kempen I, et al. 3-Bromophenyl 6-acetoxymethyl-2-oxo-2H-1-benzopyran-3-carboxylate inhibits cancer cell invasion in vitro and tumour growth in vivo. Br. J. Cancer 2003, 88: 1111-8.
42. Kuniyasu H, Chihara Y, Kondo H. Differential effects between amphoterin and advanced glycation end products on colon cancer cells. Int. J. Cancer 2003, 104: 722-7.
43. Garbisa S, et al. Quantitation of basement membrane collagen degradation by living tumor cells in vitro. Cancer Lett. 1980, 9: 359-66.
44. Nakahara H, et al. Transmembrane/cytoplasmic domain-mediated membrane type 1-matrix metalloprotease docking to invadopodia is required for cell invasion. Proc. Natl. Acad. Sci. USA 1997, 94: 7959-64.
45. Udayakumar TS, et al. Membrane type-1-matrix metalloproteinase expressed by prostate carcinoma cells cleaves human laminin-5 beta3 chain and induces cell migration. Cancer Res. 2003, 63: 2292-9.
46. Terranova VP, et al. Use of a reconstituted basement membrane to measure cell invasiveness and select for highly invasive tumor cells. Proc. Natl. Acad. Sci. USA 1986, 83: 465-9.
47. Chen JJ, et al. Global analysis of gene expression in invasion by a lung cancer model. Cancer Res. 2001, 61: 5223-30.
48. Ho CC, et al. Up-regulated caveolin-1 accentuates the metastasis capability of lung adenocarcinoma by inducing filopodia formation. Am. J. Pathol. 2002, 161: 1647-56.
49. Sawai H, et al. Enhancement of integrins by interleukin-1alpha, and their relationship with metastatic and invasive behavior of human pancreatic ductal adenocarcinoma cells. J. Surg. Oncol. 2003, 82: 51-6.
50. Farrow B, et al. Butyrate inhibits pancreatic cancer invasion. J. Gastrointest. Surg. 2003, 7: 864-70.
51. Trikha M, et al. Multiple roles for platelet GPIIb/IIIa and alphavbeta3 integrins in tumor growth, angiogenesis, and metastasis. Cancer Res. 2002, 62: 2824-33.
52. Henke CA, et al. CD44-related chondroitin sulfate proteoglycan, a cell surface receptor implicated with tumor cell invasion, mediates endothelial cell migration on fibrinogen and invasion into a fibrin matrix. J. Clin. Invest. 1996, 97: 2541-52.
53. Hori T, et al. A novel orthotopic implantation model of human esophageal carcinoma in nude rats: CD44H mediates cancer cell invasion in vitro and in vivo. Int. J. Cancer 2001, 92: 489-96.
54. Quax PH, et al. Plasminogen activator and matrix metalloproteinase production and extracellular matrix degradation by rat prostate cancer cells in vitro: correlation with metastatic behavior in vivo. Prostate 1997, 32: 196-204.
55. Inoue S, Leblond CP. The basement-membrane-like matrix of the mouse EHS tumor: I. Ultrastructure. Am. J. Anat. 1985, 174: 373-86.
56. Grant DS, et al. The basement-membrane-like matrix of the mouse EHS tumor: II. Immunohistochemical quantitation of six of its components. Am. J. Anat. 1985, 174: 387-98.

57. Inoue S, Leblond CP. The basement-membrane-like matrix of the mouse EHS tumor: III. Immunodetection of the amyloid P component in basotubules. Am. J. Anat. 1985, 174: 399-407.
58. Barsky SH, Gopalakrishna R. An experimental model for studying the desmoplastic response to tumor invasion. Cancer Lett. 1987, 35: 271-9.
59. Barsky SH, et al. Two human tumors with high basement-membrane-producing potential. Cancer 1988, 61: 1798-806.
60. Kleinman HK, et al. Basement membrane complexes with biological activity. Biochemistry 1986, 25: 312-8.
61. Baatout S, Endothelial differentiation using Matrigel (review). Anticancer Res. 1997, 17: 451-5.
62. Kleinman HK, et al. Isolation and characterization of type IV procollagen, laminin, and heparan sulfate proteoglycan from the EHS sarcoma. Biochemistry 1982, 21: 6188-93.
63. Hadley MA, et al. Extracellular matrix regulates Sertoli cell differentiation, testicular cord formation, and germ cell development in vitro. J. Cell Biol. 1985, 101: 1511-22.
64. Kubota Y, et al. Role of laminin and basement membrane in the morphological differentiation of human endothelial cells into capillary-like structures. J. Cell Biol. 1988, 107: 1589-98.
65. Li ML, et al. Influence of a reconstituted basement membrane and its components on casein gene expression and secretion in mouse mammary epithelial cells. Proc. Natl. Acad. Sci. USA 1987, 84: 136-40.
66. Barcellos-Hoff MH, et al. Functional differentiation and alveolar morphogenesis of primary mammary cultures on reconstituted basement membrane. Development 1989, 105: 223-35.
67. Madison R, et al. Increased rate of peripheral nerve regeneration using bioresorbable nerve guides and a laminin-containing gel. Exp. Neurol. 1985, 88: 767-72.
68. Goodly LJ, et al. In vivo modulation of human tumor cell growth by normal human extracellular matrix. Tumour Biol. 1994, 15: 326-36.
69. Liotta LA, Thorgeirsson UP, Garbisa S. Role of collagenases in tumor cell invasion. Cancer Metastasis Rev. 1982, 1: 277-88.
70. Melchiori A, et al. Inhibition of tumor cell invasion by a highly conserved peptide sequence from the matrix metalloproteinase enzyme prosegment. Cancer Res. 1992, 52: 2353-6.
71. Taniguchi S, et al. High invasiveness associated with augmentation of motility in a fos-transferred highly metastatic rat 3Y1 cell line. Cancer Res. 1989, 49: 6738-44.
72. Albini A, et al. A rapid in vitro assay for quantitating the invasive potential of tumor cells. Cancer Res. 1987, 47: 3239-45.
73. Hendrix MJ, et al. A simple quantitative assay for studying the invasive potential of high and low human metastatic variants. Cancer Lett. 1987, 38: 137-47.
74. Bae SN, et al. Molecular and cellular analysis of basement membrane invasion by human breast cancer cells in Matrigel-based in vitro assays. Breast Cancer Res. Treat. 1993, 24: 241-55.
75. Hendrix MJ, et al. Retinoic acid inhibition of human melanoma cell invasion through a reconstituted basement membrane and its relation to decreases in the expression of proteolytic enzymes and motility factor receptor. Cancer Res. 1990, 50: 4121-30.

76. Parish CR, Jakobsen KB, Coombe DR. A basement-membrane permeability assay which correlates with the metastatic potential of tumour cells. Int. J. Cancer 1992, 52: 378-83.
77. Janiak M, Hashmi HR, Janowska-Wieczorek A. Use of the Matrigel-based assay to measure the invasiveness of leukemic cells. Exp. Hematol. 1994, 22: 559-65.
78. Noel AC, et al. Invasion of reconstituted basement membrane matrix is not correlated to the malignant metastatic cell phenotype. Cancer Res. 1991, 51: 405-14.
79. Vukicevic S, et al. Identification of multiple active growth factors in basement membrane Matrigel suggests caution in interpretation of cellular activity related to extracellular matrix components. Exp. Cell Res. 1992, 202: 1-8.
80. Li Z, et al. Macrophage migration inhibitory factor enhances neoplastic cell invasion by inducing the expression of matrix metalloproteinase 9 and interleukin-8 in nasopharyngeal carcinoma cell lines. Chin. Med. J. (Engl). 2004, 117: 107-14.
81. Zand L, et al. Differential Effects of Cellular Fibronectin and Plasma Fibronectin on Ovarian Cancer Cell Adhesion, Migration, and Invasion. In Vitro Cell Dev. Biol. Anim. 2003, 39: 178-182.
82. Passaniti A, et al. A simple, quantitative method for assessing angiogenesis and antiangiogenic agents using reconstituted basement membrane, heparin, and fibroblast growth factor. Lab Invest. 1992, 67: 519-28.
83. Farina AR, et al. Identification of plasminogen in Matrigel and its activation by reconstitution of this basement membrane extract. Biotechniques 1996, 21: 904-9.
84. Repesh LA. A new in vitro assay for quantitating tumor cell invasion. Invasion Metastasis 1989, 9: 192-208.
85. Buchholz M, et al. SERPINE2 (protease nexin I) promotes extracellular matrix production and local invasion of pancreatic tumors in vivo. Cancer Res. 2003, 63: 4945-51.
86. Suyama E, et al. Identification of genes responsible for cell migration by a library of randomized ribozymes. Cancer Res. 2003, 63: 119-24.
87. Srikantan V, et al. HEPSIN inhibits cell growth/invasion in prostate cancer cells. Cancer Res. 2002, 62: 6812-6.
88. Zhang B, et al. IL-18 increases invasiveness of HL-60 myeloid leukemia cells: up-regulation of matrix metalloproteinases-9 (MMP-9) expression. Leuk Res. 2004, 28: 91-5.
89. Zhang S, et al. Spontaneous assembly of a self-complementary oligopeptide to form a stable macroscopic membrane. Proc. Natl. Acad. Sci. USA 1993, 90: 3334-8.
90. Zhang S, et al. Self-complementary oligopeptide matrices support mammalian cell attachment. Biomaterials. 1995, 16: 1385-93.
91. Zhang S. Emerging biological materials through molecular self-assembly. Biotechnol. Adv. 2002, 20: 321-39.
92. Zhang S. Fabrication of novel biomaterials through molecular self-assembly. Nat. Biotechnol. 2003, 21: 1171-8.
93. Altman M, et al. Conformational behavior of ionic self-complementary peptides. Protein Sci. 2000, 9: 1095-105.
94. Zhang S, et al. Biological surface engineering: a simple system for cell pattern formation. Biomaterials 1999, 20: 1213-20.
95. Siegal GP, et al. Development of a novel human extracellular matrix for quantitation of the invasiveness of human cells. Cancer Lett. 1993, 69: 123-32.

96. Zhang M, GC, Gillespie GY, Wang MH, Singh RK, Siegal GP. A model system for in vitro angiogenesis using novel human ECM. Proc. Am. Assoc. Cancer Res. 1994, 35: 185.
97. Singh RK, CP, Alapati V, Chen X, Siegal GP. Multiple defined human cell-biomatrix models for tissue growth & differentiation. In Vitro Cell Develop. Biol. 2002, 38:11.
98. Siegal GP, Singh RK. A new defined human angiogenesis bioassay system. Pharma-Transfer, and DSI-UABRF (FY00-0034, 2000); HuBiogel bioassays: patent/license pending. 2000, 2003.
99. Hamilton G. Multicellular spheroids as an in vitro tumor model. Cancer Lett. 1998, 131: 29-34.
100. Desoize B. Contribution of three-dimensional culture to cancer research. Crit. Rev. Oncol. Hematol. 2000, 36: 59-60.
101. Ingram et al. NASA Proceedings-Bioreactor Workshop on Regulation of Cell & Tissue differentiation. In vitro Cell Dev. Med. 1997, 33: 325-405.
102. Schwarz RP, Goodwin TJ, Wolf DA. Cell culture for three-dimensional modeling in rotating-wall vessels: an application of simulated microgravity. J. Tissue Cult. Methods 1992, 14: 51-7.
103. Jessup JM, Goodwin TJ, Spaulding G. Prospects for use of microgravity-based bioreactors to study three-dimensional host-tumor interactions in human neoplasia. J. Cell Biochem. 1993, 51: 290-300.
104. Granet C, et al. Rotating-wall vessels, promising bioreactors for osteoblastic cell culture: comparison with other 3D conditions. Med. Biol. Eng. Comput. 1998, 36: 513-9.
105. Sanford GL, et al. Three-dimensional growth of endothelial cells in the microgravity-based rotating wall vessel bioreactor. In Vitro Cell Dev. Biol. Anim. 2002, 38: 493-504.
106. Okubo H, et al. A novel method for faster formation of rat liver cell spheroids. Artif. Organs. 2002, 26: 497-505.
107. Horino K, et al. Tumor cell invasion of model 3-dimensional matrices: demonstration of migratory pathways, collagen disruption, and intercellular cooperation. Faseb. J. 2001, 15: 932-9.
108. Lang SH, et al. Experimental prostate epithelial morphogenesis in response to stroma and three-dimensional matrigel culture. Cell Growth Differ. 2001, 12: 631-40.
109. Torok E, et al. Optimization of hepatocyte spheroid formation for hepatic tissue engineering on three-dimensional biodegradable polymer within a flow bioreactor prior to implantation. Cells Tissues Organs 2001, 169: 34-41.
110. Hamilton GA, Westmorel C, George AE. Effects of medium composition on the morphology and function of rat hepatocytes cultured as spheroids and monolayers. In Vitro Cell Dev. Biol. Anim. 2001, 37: 656-67.
111. Lam JT, et al. Replication of an integrin targeted conditionally replicating adenovirus on primary ovarian cancer spheroids. Cancer Gene. Ther. 2003, 10: 377-87.
112. Nakajima Y, et al. Inhibition of metastasis of tumor cells overexpressing thymidine phosphorylase by 2-deoxy-L-ribose. Cancer Res. 2004, 64: 1794-801.
113. Nishikawa H, et al. The role of cathepsin B and cystatin C in the mechanisms of invasion by ovarian cancer. Gynecol. Oncol. 2004, 92: 881-6.
114. Ueda M, et al. Correlation between vascular endothelial growth factor-C expression and invasion phenotype in cervical carcinomas. Int. J. Cancer. 2002, 98: 335-43.

115. Jung HJ, et al. Cochlioquinone A1, a new anti-angiogenic agent from Bipolaris zeicola. Bioorg. Med. Chem. 2003, 11: 4743-7.
116. Kang JA, et al. Anti-angiogenic and anti-tumor invasive activities of tissue inhibitor of metalloproteinase-3 from shark, Scyliorhinus torazame. Biochim. Biophys. Acta. 2003, 1620: 59-64.
117. Sieuwerts AM, Klijn JG, Foekens JA. Assessment of the invasive potential of human gynecological tumor cell lines with the in vitro Boyden chamber assay: influences of the ability of cells to migrate through the filter membrane. Clin. Exp. Metastasis 1997, 15: 53-62.
118. Bindels EM, et al. Influence of the microenvironment on invasiveness of human bladder carcinoma cell lines. Virchows Arch. 2001, 439: 552-9.
119. Yanamandra N, et al. Activation of caspase-9 with irradiation inhibits invasion and angiogenesis in SNB19 human glioma cells. Oncogene 2004, 23: 2339-46.
120. Perumpanani AJ, Byrne HM. Extracellular matrix concentration exerts selection pressure on invasive cells. Eur. J. Cancer 1999, 35: 1274-80.
121. Poste G, et al. In vitro selection of murine B16 melanoma variants with enhanced tissue-invasive properties. Cancer Res. 1980, 40: 1636-44.
122. Hart IR, Fidler IF. An in vitro quantitative assay for tumor cell invasion. Cancer Res. 1978, 38: 3218-24.
123. Scher C, Haudenschild C, Klagsbrun M. The chick chorioallantoic membrane as a model system for the study of tissue invasion by viral transformed cells. Cell 1976, 8: 373-82.
124. Sekiya S, et al. Invasion potential of human choriocarcinoma cell lines and the role of lytic enzymes. Gynecol. Oncol. 1985, 22: 324-33.
125. Livant DL, et al. Invasion of selectively permeable sea urchin embryo basement membranes by metastatic tumor cells, but not by their normal counterparts. Cancer Res. 1995, 55: 5085-93.
126. Dyer ES, et al. Phenylbutyrate inhibits the invasive properties of prostate and breast cancer cell lines in the sea urchin embryo basement membrane invasion assay. Int. J. Cancer 2002, 101: 496-9.
127. Livant DL, et al. Anti-invasive, antitumorigenic, and antimetastatic activities of the PHSCN sequence in prostate carcinoma. Cancer Res. 2000, 60: 309-20.
128. Thorgeirsson UP, et al. Protein synthesis but not DNA synthesis is required for tumor cell invasion in vitro. Invasion Metastasis 1984, 4: 73-83.
129. Smith HS, et al. Invasiveness and ploidy of human mammary carcinomas in short-term culture. Proc. Natl. Acad. Sci. USA 1985, 82: 1805-9.
130. Russo RG, Foltz CM, Liotta LA. New invasion assay using endothelial cells grown on native human basement membrane. Clin. Exp. Metastasis 1983, 1: 115-27.
131. Siegal GP, et al. Interferon enhancement of the invasive capacity of Ewing sarcoma cells in vitro. Proc. Natl. Acad. Sci. USA 1982, 79: 4064-8.
132. Bechetoille N, et al. Penetration of human metastatic melanoma cells through an authentic dermal-epidermal junction is associated with dissolution of native collagen types IV and VII. Melanoma Res. 2000, 10: 427-34.
133. Sikes RA, et al. Cellular interactions in the tropism of prostate cancer to bone. Int. J. Cancer 2004, 110: 497-503.
134. Engers R, et al. Protein kinase C in human renal cell carcinomas: role in invasion and differential isoenzyme expression. Br. J. Cancer 2000, 82: 1063-9.

135. Schroyens W, et al. Comparison of invasiveness and non-invasiveness of two epithelial cell lines in vitro. Invasion Metastasis 1984, 4: 160-70.
136. Pauli BU, Memoli VA, Kuettner KE. In vitro determination of tumor invasiveness using extracted hyaline cartilage. Cancer Res. 1981, 41: 2084-91.
137. Storme GA, et al. Antiinvasive effect of racemic 1-O-octadecyl-2-O-methylglycero-3-phosphocholine on MO4 mouse fibrosarcoma cells in vitro. Cancer Res. 1985, 45: 351-7.
138. Jones JL, et al. Primary breast myoepithelial cells exert an invasion-suppressor effect on breast cancer cells via paracrine down-regulation of MMP expression in fibroblasts and tumour cells. J. Pathol. 2003, 201: 562-72.
139. Sedlak J, et al. Cell surface phenotype and increased penetration of human multidrug-resistant ovarian carcinoma cells into in vitro collagen-fibroblasts matrix. Neoplasma 1996, 43: 389-95.
140. Tong D, et al. Association of in vitro invasiveness and gene expression of estrogen receptor, progesterone receptor, pS2 and plasminogen activator inhibitor-1 in human breast cancer cell lines. Breast Cancer Res. Treat. 1999, 56: 91-7.
141. Nakayama Y, et al. An in vitro invasion model for human renal cell carcinoma cell lines mimicking their metastatic abilities. Clin. Exp. Metastasis 1996, 14: 466-74.
142. Wandel E, et al. Fibroblasts enhance the invasive capacity of melanoma cells in vitro. Arch. Dermatol. Res. 2002, 293: 601-8.
143. Yamada S, et al. Effects of stromal fibroblasts and fat cells and an environmental factor air exposure on invasion of laryngeal carcinoma (HEp-2) cells in a collagen gel invasion assay system. Arch. Otolaryngol Head Neck Surg. 1999, 125: 424-31.
144. Le Marer N, Bruyneel E. Comparison of in vitro invasiveness of human breast carcinoma in early or late stage with their malignancy in vivo. Anticancer Res. 1996, 16: 2767-72.
145. Gohla A, Eckert K, Maurer HR. A rapid and sensitive fluorometric screening assay using YO-PRO-1 to quantify tumour cell invasion through Matrigel. Clin. Exp. Metastasis 1996, 14: 451-8.
146. Gildea JJ, et al. Transmembrane motility assay of transiently transfected cells by fluorescent cell counting and luciferase measurement. Biotechniques 2000, 29: 81-6.
147. Yin Z, et al. Cooperative activity of alpha4beta1 and alpha4beta7 integrins in mediating human B-cell lymphoma adhesion and chemotaxis on fibronectin through recognition of multiple synergizing binding sites within the central cell-binding domain. Blood 1999, 93: 1221-30.

Chapter 3

IMAGING INVASION AND METASTASIS *IN VIVO*

Wendy Ingman[1], Jeffrey Wyckoff[2], Chengsen Xue[2], Elaine Y. Lin[1], Weigang Wang[2], Sumanta Goswami[2], Jeffrey W. Pollard[1], John Condeelis[2], and Jeffrey E. Segall[2]

[1]Departments of Developmental and Molecular Biology and Obstetrics and Gynecology and Women's Health, Center for the Study of Reproductive Biology and Women's Health, and [2]Department of Anatomy and Structural Biology, Albert Einstein College of Medicine, Bronx, NY, USA

Abstract: Current clinical imaging methods for breast cancer do not visualize individual cells moving in the primary tumor microenvironment. In animal models of cancer, confocal imaging using multiphoton excitation provides reduced photobleaching, deeper imaging into thick specimens, and observation of some extracellular matrix fibers. Individual cells expressing fluorescent proteins such as GFP can be observed in order to evaluate the mechanisms of tumor cell invasion and metastasis. Tumor cells *in vivo* can show orientation towards blood vessels, and rapid movement along extracellular matrix fibers. To directly measure tumor cell invasion in the primary tumor in response to imposed gradient of growth factors, needles containing EGF or CSF-1 have been found to stimulate invasion into the needle. Transgenic models of breast cancer which have lost expression of CSF-1 show reduced invasiveness and metastasis. These results indicate that tumor cells can collaborate with macrophages during invasion. Such paracrine interactions between epithelial cells and macrophages likely reflect interactions that occur during ductal outgrowth in normal mammary gland development. The combination of *in vivo* imaging with genetic manipulation of xenograft and transgenic models of breast cancer will enable direct examination of mechanisms of invasion and metastasis.

Key words: *in vivo* imaging, animal models, live imaging, microscopy

A. Wells (ed.), Cell Motility in Cancer Invasion and Metastasis, 55-72.

1. RATIONALE FOR *IN VIVO* IMAGING STUDIES

Improved understanding of the events that occur during tumor invasion and metastasis will greatly assist the development of successful strategies of cancer therapy and diagnosis. Tumor cell metastasis can be dissected into stages (1-3); broadly speaking, tumor cells must leave the primary tumor, enter the circulation, survive in the circulation, arrest at a target site, and then grow to form a new metastasis at that site. Molecules that play an important role in invasion or metastasis, whether they are proteins, glycosaminoglycans, or small compounds, will be important targets for novel cancer therapies. These targets may initially be identified by molecular biology and *in vitro* culture approaches, however due to the complexity of tumor invasion and metastasis, testing needs to be conducted in live animal models. *In vivo* imaging allows researchers to view cells moving about inside a live tissue in real time, providing a means to test the efficacy of new therapies at the level of cell migration. Additionally, and perhaps most importantly, the simple process of observation can stimulate significant advances in our understanding of tumor biology, and generate new hypotheses which can be tested.

2. TECHNIQUES FOR *IN VIVO* IMAGING

Although the ideal approach is to image invasive tumors *in situ* in patients, at our current level of expertise and ability only certain, limited technologies can be performed on patients safely. Technologies such as PET, MRI and CT are the most common methods that are used to image tumors in patients (4).

Positron emission tomography (PET) and the related single photon emission tomography (SPECT) are highly sensitive methods that make use of radioactive compounds that accumulate in tumors to identify both the primary tumor and metastases (5). As the radioactive nuclei decay, the emitted high energy photons can be detected by imaging systems designed to determine the location of the source. ^{18}F fluorodeoxyglucose (FDG), the most often used compound, accumulates in many tumors due to high rates of glycolysis. The metabolic dependence of the signal makes FDG also useful for evaluating the effects of treatments designed to kill tumor cells - effective treatments are reflected in reduced FDG uptake rates. However, the resolution of PET is on the order of 4 – 5 mm in patients, or 2 mm in special animal systems (6, 7). This low resolution limits the imaging capability to mainly one of determining the location of a tumor, and does not provide much information on morphology or invasiveness.

Magnetic resonance imaging (MRI) uses magnetic fields to distinguish variations in relaxivity among internal structures (8-10). It can provide remarkable images of anatomical structures, but for tumor identification, injected iron or gadolinium compounds are typically used to identify the increased blood volume or permeability associated with the tumor periphery. Metabolic studies using magnetic resonance spectroscopy may provide additional capabilities similar to PET imaging, but a fundamental limitation of MRI currently is the low sensitivity of the technique. This low signal to noise feature also limits the resolution that is easily achieved without extremely long imaging times: on the order of millimeters in humans to 0.2 mm in animals. This provides improved resolution over PET with some limited information about the tumor borders, but millimeter scale determination of the irregularity of tumor borders together with high vessel permeability provides only a limited capability of identifying an underlying mechanism.

Computed tomography (CT) converts one of the staples of clinical imaging, x-ray radiography, into a 3 dimensional data set (11, 12). As with MRI, CT can help identify anatomic structures directly, or use injectable compounds to discriminate between soft tissues (such as tumors) more accurately. CT can rapidly provide anatomic location of PET signals for more accurate diagnosis. Radiation concerns have thus far limited higher resolution CT stand alone analysis, although for specific targets that minimize radiation dose the resolution of CT is similar to or higher than MRI. Nevertheless, the resolution currently provided is inadequate for directly evaluating the invasive properties of tumors in humans.

As the current imaging modalities available for studying human tumors in the clinic are unable to provide the resolution necessary for studying the cellular mechanisms of invasion directly, most research in this area is focused on animal models (13, 14). The most fully developed high resolution approach for the study of cell invasion in animal models makes use of confocal light microscopy (15). Light microscopy provides the resolution necessary for following the behavior of single cells, making it possible to directly visualize the invasion process both in *in vitro* models as well as in *in vivo* animal models. However, the penetration level of light-based imaging methods is limited, imposing requirements such as partial dissection to expose the tumor surface or the use of a viewing chamber, and imaging is then constrained to near the surface of the tumor. Certain imaging modalities such as multiphoton confocal imaging extend the capabilities significantly, as described in the next section.

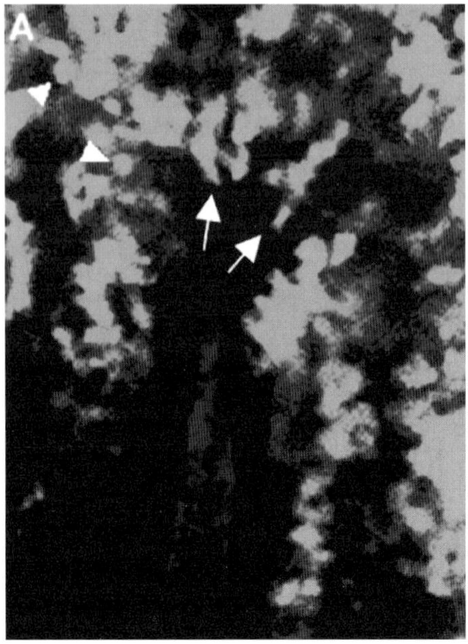

Figure 1. Single photon imaging of a tumor formed by MTLn3 cells expressing GFP and a blood vessel visualized using rhodamine dextran. From (1). See also Color Plate Section page 340.

3. SINGLE VS MULTIPHOTON CONFOCAL IMAGING

Confocal light microscopy is important for the analysis of tumor cell motility *in vivo*, providing the capability of submicron resolution and imaging of tumor cells in the local microenvironment. In our experience, single photon confocal imaging enables imaging of tumor cells within 50 um of the surface of the tumor, with optimal imaging occurring within 30 um of the surface. This provides an opportunity to evaluate tumor cell behavior several cell layers into the tumor, and begins to afford a view of tumor cell interactions with other tumor cells as well as with blood vessels (16) (Figure 1). However, bleaching occurs relatively rapidly, and out of focus planes are simultaneously bleached, effectively limiting data collection to a single plane of view.

With multiphoton confocal imaging, two photons of ½ the excitation energy (or twice the wavelength) are used to excite the fluorophore (17). Typical multiphoton systems use pulsed infrared lasers to achieve the high photon densities required to produce near simultaneous absorption of two photons. The plane upon which the excitation light is focused has the highest photon density, and the density then falls off as the square of the distance from the plane of focus. The net effect is that mainly fluorophores in the plane of focus are excited, the thickness of the plane of focus varying with the lens magnification. In essence, this is confocal excitation of the fluorophores. The result is reduced stimulation of regions not in the plane of focus, and increased sensitivity due to the use of all the emitted light that is collected. A potential limitation is that the depth of the confocal plane is fixed by the lens and is not able to be varied, but we have not found this to be a problem. The greatest problems with multiphoton imaging are the higher cost of the excitation laser and the limitation to a single excitation wavelength. Because the two photon absorption spectra tend to be broader, a wide range of fluorophores can be stimulated. Conversely, selective fluorophore stimulation could be achieved by rapidly varying the excitation wavelength, but for most live imaging applications we have not found this to be practical. However, only the wavelength of the emitted light can be used to identify the fluorophore being detected. One of the consequences of using a single wavelength for excitation is that one cannot cleanly separate CFP from GFP into separate emission channels: CFP emission is present in both blue and green channels while GFP is present in only in green.

Nevertheless, for intravital imaging of tumors, the advantages of multiphoton microscopy significantly outweigh the limitations (18). The two major enhancements provided by multiphoton microscopy are increased imaging depth and reduced bleaching (19) (Figure 2). These two features enable time lapse z series collection as a routine method. In our xenograft and transgenic models of breast cancer, we find that practical imaging depths of roughly 100 um are accessible with 5 – 10 um z steps over a 512x512 um field of view at roughly 1 um/pixel. Higher magnifications are possible, with the corresponding reduction of size of the field of view. Typically a single series takes around 1 minute, thus we routinely use a 1 minute time lapse interval, although by reducing the number of z steps taken or the image of view, one can shorten the interval if needed.

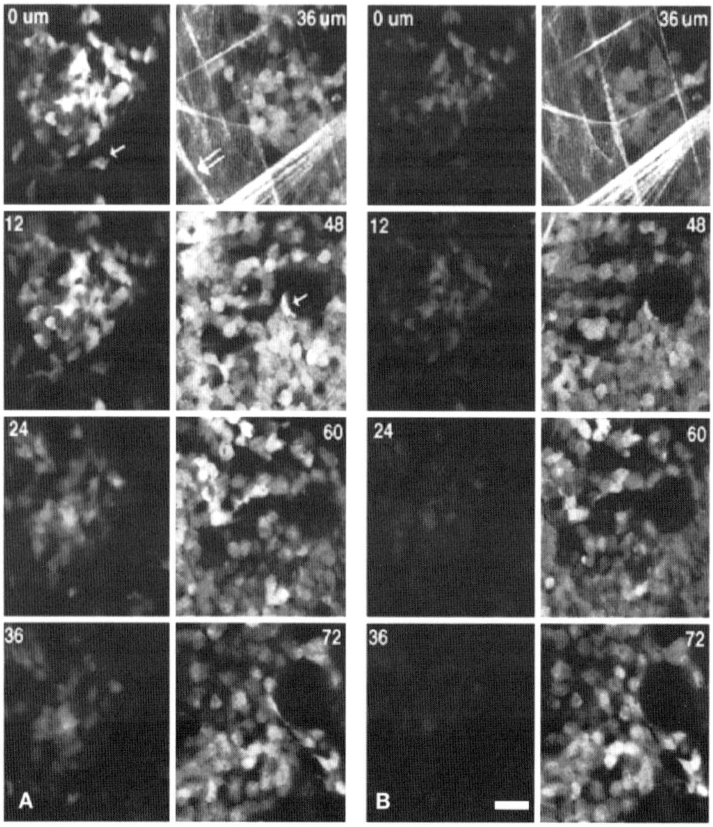

Figure 2. Multiphoton microscopy is superior in imaging primary tumors more deeply and with less photodamage. A, initial z-series of images taken with the confocal (left panels) and multiphoton (right panels) microscope showing the greater depth of imaging possible with multiphoton excitation. The confocal microscope can only image to approximately 36 um. The first panel of the multiphoton images (top right panel) begins at 36 um and shows an image of better quality than that of the confocal image at 0 um (top left panel). Arrows point to a single cell. Double arrow points to ECM seen only in multiphoton microscopy due to second harmonic generation. B, second z-series of the same focal planes illustrates dramatic improvement in GFP stability (right panels) in the multiphoton microscope compared with (left panels) the confocal microscope. From (19).

An additional advantage of multiphoton microscopy is the ability to image some of the extracellular matrix fibers that are present (19, 20) (Figure 3). Elastin fibers show fluorescence when illuminated with <800 nm wavelength light. The highly repetitive organization of collagen fibers (types I, II or III), enables second harmonic scattering, and allows discrimination of these fibers from elastic fibers in the blue channel by illumination with >800 nm light.

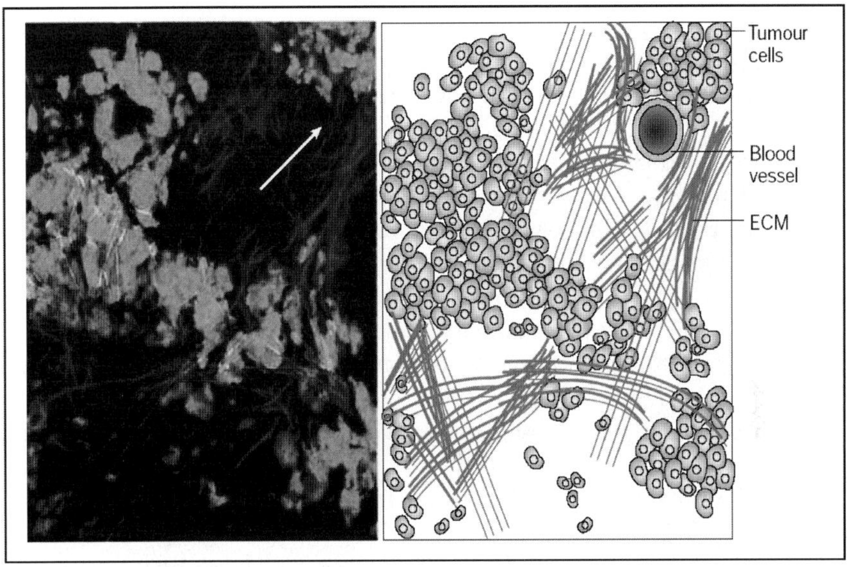

Figure 3. ECM fibres converge on blood vessels in mammary tumours. Multiphoton microscopy shows that carcinoma cells (green) move on extracellular-matrix fibres (purple), some of which converge on blood vessels (arrow). From (18). See also Color Plate Section page 340.

4. IMAGING OF INVASION IN ANIMAL MODELS

Imaging of tumor cell invasion *in vivo* provides the opportunity to examine living tumor cells in the tumor microenvironment, mimicking the clinical situation in humans as closely as is currently possible. Initial studies using fluorescent microscopy of tumor cell invasion utilized chemical labeling of cells followed by intravenous injection and imaging in easily accessible target organs such as the liver (21). With the advent of GFP as a stable fluorescent cell label, it became possible to image both primary tumors as well as metastases long after injection of tumor cells. Observation of both local invasion around the primary tumor as well as arrest and extravasation at sites of metastasis may yield valuable findings. For reviews of important work being performed by other groups please see (22-27).

Our studies of tumor cell invasion and metastasis have focused on imaging mammary tumor cell invasion around the primary tumor and imaging developing mammary glands. Particularly valuable is the ability to simultaneously observe multiple features of interest. Cell motility and morphology are typically observed by using stable expression of fluorescent proteins such as GFP or CFP. In parallel with tumor or mammary epithelial cells, features that can be observed include host cells, collagen fibers, and blood vessels. Collagen fibers are imaged using second harmonic generation, and are detectable in the blue channel. Injection of Texas-Red labeled compounds provides labeling of blood vessels and macrophages in the red channel. Transgenic animals provide the opportunity to label host cells as well as tumor cells. We will focus on three types of animal models that we have been using to study mammary tumor cell invasion and metastasis *in vivo*: xenograft studies, transgenic tumor studies, and elements of normal physiology that are relevant to invasion.

Our initial intravital imaging studies utilized injection of GFP-expressing rat mammary adenocarcinoma cells into syngeneic female Fisher rats and observation using one photon confocal microscopy (16). GFP expression did not alter tumor growth or metastasis rates. The single photon imaging technology limited our observations to single plane time lapse data collection near the surfaces of tumors. Nevertheless, these studies disproved our initial hypothesis regarding the role of cell motility in metastasis (1). We had hypothesized that increased metastasis would reflect increased cell motility. However, we found little difference in overall motility between metastatic MTLn3 and nonmetastatic MTC derivatives of the 13762NF tumor. Rather, the more metastatic MTLn3 line showed increased polarization of cells towards nearby blood vessels identified by intravenous injection of rhodamine dextran (Figure 1). In addition, the overall frequency of cases of tumor cell movement was low, with few cells in an entire field of view showing net translocation.

Multiphoton imaging has extended our ability to learn about tumor cell morphology *in situ* by allowing time lapse z series to be acquired. Thus we are now examining tumor cell polarization and motility as a function of depth into the tumor. Blood vessel structure and properties as a function of depth into the tumor is now easier to evaluate (Figure 4). Questions that are currently under study include evaluating whether specific regions of the tumor show increased invasive properties.

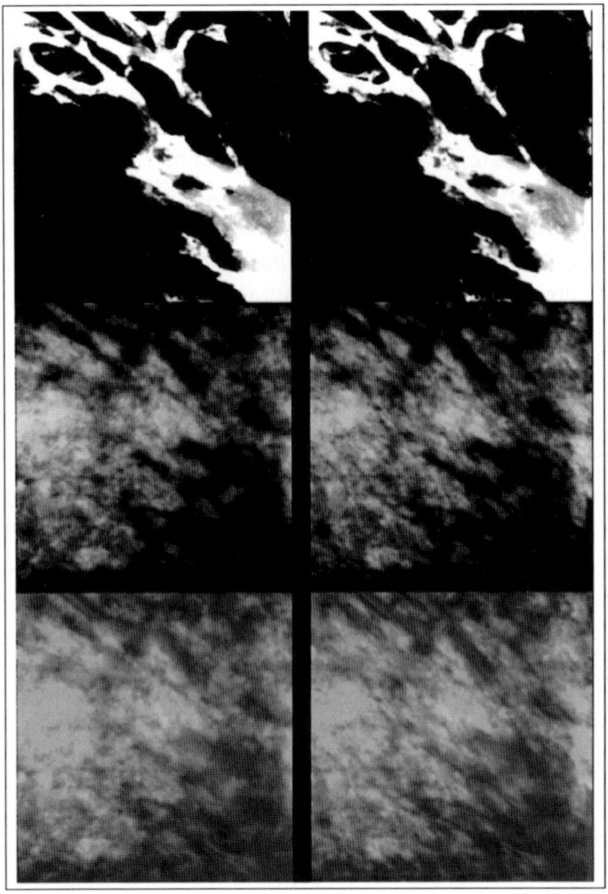

Figure 4. Blood vessel structure determined by multiphoton microscopy. Animals were injected with Texas Red dextran (70,000 MW) and then the tumors were imaged taking a slice every 5 microns. The z series were then projected to generate a stereo image showing blood vessels (top and red) and tumor cells (middle and green). See also Color Plate Section page 341.

Once such regions have been identified, simultaneous imaging of other tumor microenvironment features will be used to identify potential mechanisms for enhancing tumor invasiveness. One limitation of the current cytoplasmic GFP labeling methods is that borders of tumor cells can only be easily identified if they are not in contact with other tumor cells. Simultaneous membrane labeling or mixtures of cells with different labels may aid in determining the morphology of individual tumor cells in tumor masses. Simultaneous labeling of the nucleus with a different color from the cytoplasm will also be useful for positively identifying individual cells (28).

Second harmonic generation has enabled the identification of interactions between tumor cells and extracellular matrix fibers during tumor cell invasion of neighboring loose connective tissues. Rapidly moving tumor cells can follow collagen I fibers away from the primary tumor (18). These contacts may guide tumor cell invasion towards the periphery and blood vessels. However, not all tumor cell movement is in contact with matrix fibers detectable via second harmonic imaging. This does not rule out that cells are moving along matrix fibers that are too small to detect or along matrix structures that are not sufficiently periodic to generate a second harmonic signal. Alternatively, in the absence of matrix fibers or appropriate matrix metalloproteinase activity, amoeboid motion may be utilized to squeeze through extracellular matrix structures (29). Higher magnification imaging together with imaging of matrix metalloproteinase activity will aid in confirming the presence of amoeboid motility. We hypothesize that chemoattractant gradients may stimulate polarized tumor cell movement, and the nature of movement, mesenchymal or amoeboid, may be determined by the availability of fibers.

The sources of such chemoattractant gradients might be the tumor cells themselves, neighboring nontumor cells, blood vessels, or a combination of the above. In our initial single photon imaging studies, some host cells were seen as shadows rapidly moving on the border of the fluorescent tumor (1). The size and speed of movement were consistent with host immune cells such as neutrophils or macrophages. The number of such host cells associated with the tumor was significantly increased in the more metastatic MTLn3 tumors compared to the poorly metastatic MTC tumors. Multiphoton imaging inside these xenograft tumors has enabled us to image host cells inside the tumor as well, often moving along matrix fibers running through the tumor. A subset of phagocytic host cells, presumably macrophages, can be imaged by preinjecting fluorescently labeled dextrans intravenously, or by labeling using cell-specific promoters in transgenic animals. Colocalization of these cells with tumor cells can be used to determine whether their presence is correlated with tumor cell motility.

An important approach for studying the roles of host cells in tumor invasion and metastasis makes use of transgenic animals. Tumors can be generated in transgenic animals through the targeted expression of oncogenes in specific tissues. We have been particularly interested in the MMTV-PyMT model, in which the MMTV promoter targets expression of the Polyoma middle T protein to the mammary epithelium (30). Such animals develop tumors with a well defined time course and sequence of stages, beginning with hyperplasias which then progress to carcinomas.

Analysis of the progression of MMTV-PyMT tumors in animals lacking production of the colony stimulating factor 1 (CSF1) protein (*Csf1^op^/Csf1^op^*) has demonstrated an important role of macrophages in tumor invasion and metastasis (31). CSF1 is an important growth stimulator of macrophages, and in $Csf1^{op}/Csf1^{op}$ animals, there are reduced numbers of macrophages present in and around the PyMT tumors. Such tumors show reduced invasion and metastasis, and expression of CSF1 specifically in the tumor cells can enhance progression to malignancy and metastasis (32). Through the use of cell-type specific promoters, tumor cells, macrophages, fibroblasts, or endothelial cells can be visualized in tumors. Imaging of macrophages (using GFP driven by the macrophage specific lysozyme M or CSF1 receptor promoter) enables us to directly track their movement around tumors, and we find that they show motility characteristics consistent with the host cells that we imaged in our xenograft models (Figure 5) (33, 34).

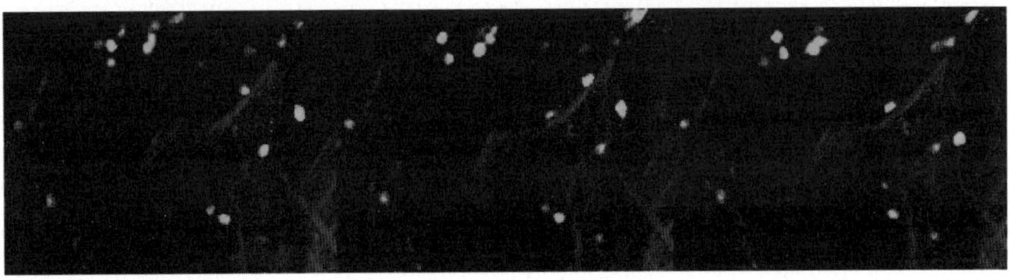

Figure 5. Movement of macrophages in tumors. A knockin of GFP at the lys promoter (courtesy of Dr. T. Graf) (34) was crossed with the MMTV-PyMT strain to generate tumors containing GFP labeled macrophages. The tumors were imaged for GFP (green) and matrix fibers (2nd harmonic, blue). Three frames from a time-lapse sequence are shown, demonstrating the high motility of the GFP labeled cells. See also Color Plate Section page 342.

To directly evaluate invasive potential *in vivo*, we have developed an *in vivo* invasion assay in which needles filled with Matrigel and chemoattractants are inserted into tumors (Figure 6) (35). The diffusion of the chemoattractants out of the needle stimulates invasion into the Matrigel, and by removing the needle, extracting the Matrigel and counting the number of cells that have entered, we can quantitate cell invasive properties in the tumor microenvironment. Remarkably, we find that macrophages and tumor cells both enter needles that contain chemoattractants that stimulate either tumor cells or macrophages, suggesting a paracrine loop between these two cell types is important for tumor cell invasion (36). Imaging of either tumor cells or macrophages during the *in vivo* invasion assay confirms that cells are actively moving towards the needle.

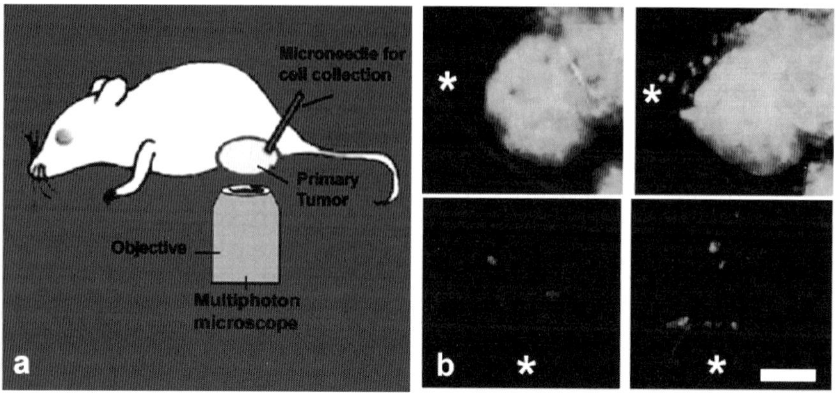

Figure 6. Movement of macrophages and carcinoma cells into collection needles in response to growth factors. A, Model shows how the experiment is performed. The primary tumor in the right number 5 mammary gland is selected to minimize breathing motion, and the position of the collection needle is controlled with a micromanipulator (not shown). B, movement of fluorescent carcinoma cells (top, WAP-Cre/CAG-CAT-EGFP/MMTV-PyMT tumor) and macrophages (bottom, MMTV-PyMT/lys-GFPKi tumor) toward EGF-containing collecting needles. The approximate opening of the collection needle is shown in each field. Each image is a 50-um z-projection and is from a time-lapse series. Images on the right were recorded 90 minutes after images on the left; bar, 25 um. From (36). See also Color Plate Section page 342.

The synergy of epithelial tumor cells with macrophages during invasion and metastasis may reflect normal developmental events that occur during ductal invasion of the mammary fat pad. Our previous studies have shown that macrophages have a key role in assisting ductal invasion (37-39). In the developing mouse mammary gland at around three weeks of age, the rudimentary ducts form club-shaped multi-laminate epithelial structures at their ends known as terminal end buds (TEBs), surrounded by a unique stroma. These TEBs are sites of high proliferative activity. They invade and bifurcate through the fat pad to give the branched ductal structure observed in the non pregnant adult. During development of the TEB, macrophages accumulate around the neck of the club shaped TEB, and have been found over and ahead of the TEB as well (Figure 7). This places them in a location to influence epithelial outgrowth and ECM organization. In the absence of macrophages, the rate of developmental outgrowth is slowed and the branching complexity is reduced (39). This gives an atrophic poorly branched mammary gland. Furthermore, the dramatic side branching and outgrowth found in pregnancy is significantly retarded in the absence of macrophages (40).

Figure 7. Example of terminal end bud imaging from Mac green female showing green fluorescent macrophages and collagen structures surrounding a terminal end bud (stained with propidium iodide, red). Macrophages (green) and collagen (blue) appear to be co-localized. See also Color Plate Section page 343.

It is now possible with multiphoton technology to study this process in much greater detail and even to image outgrowth in real time. As the multiphoton microscope allows greater depth of imaging through this high fat content tissue, compared to single photon conventional confocal imaging, we are able to reconstruct 3 dimensional images of the terminal end bud to analyze relationships between structure and function. Once a z-series is taken of the structure of interest, the image can be resliced in any plane, and measurements pertaining to diameter, volume and epithelial thickness easily obtained (Figure 8). This method is further assisted by second harmonic generation, as we have discovered that the stroma immediately surrounding the terminal end bud has a distinctive collagen morphology. Not only does this pose interesting questions as to the role of extracellular matrix in terminal end bud growth and invasion, this finding has proven to be hugely beneficial when imaging live mammary gland development, as methods to stain the fat pad in living mice have proven elusive. Mammary tumor cells may be using cell-cell interactions similar to those used to facilitate epithelial cell invasion into the mammary fat pad during development for invasion during metastasis. One of the goals of our studies of TEB growth and invasion of the mammary fat pad is to identify additional mechanisms that contribute to tumor cell invasion.

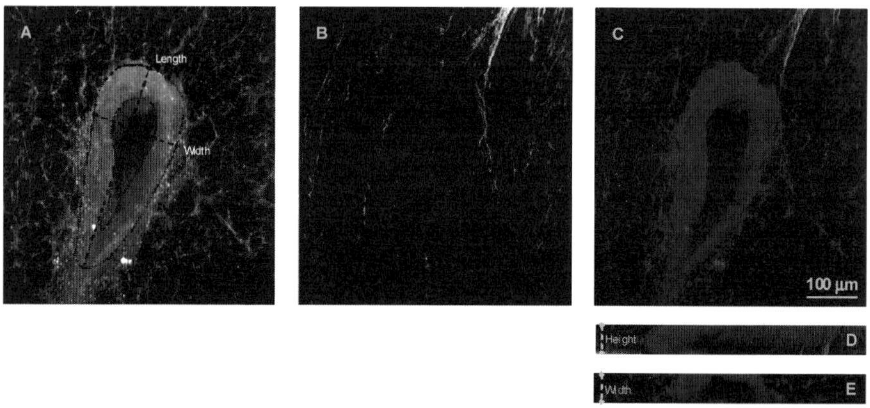

Figure 8. Example of 3-dimensional imaging of terminal end bud in op/+ mice. Fixed whole mounts were imaged with a red filter (A) for propidium iodide staining, and a blue filter (B) for collagen matrix by second harmonic generation, and merged using Image J software (C). These images are 22 um stacks through the largest part of the terminal end bud. The images were resliced in 2 um steps from left to right (D) and top to bottom (E) to produce images showing the length and height, and width and height respectively. See also Color Plate Section page 343.

5. FUTURE STUDIES AND MODELS

Combining multiphoton microscopy with both tumor cell and developmental biology models provides the capability to dissect the invasion process *in vivo*. Our initial characterization of the process has revealed that mammary epithelial cells show motility that is limited to specific sites of invasion. Host cells such as macrophages can be highly motile, and contribute to the ability of the epithelial cells to invade the surrounding stromal components. Gradients of chemotactic factors such as epidermal growth factor can enhance invasion *in vivo*. Matrix fibers may guide the invasion process.

The field is now poised to make use of genetic manipulations and transplantation studies to evaluate the contributions of specific cell types and proteins on ductal invasion and breast cancer progression. For xenograft models involving injection of cell lines, regulatable siRNA as well as traditional gene expression constructs will provide the ability to determine the roles of specific proteins in invasion. For transgenic models of cancer, recent advances in gene modification approaches allow the targeting of gene overexpression or null mutation to a cell or tissue of interest in a highly regulated manner. As discussed earlier, protein overexpression can be targeted to the mammary gland epithelium using the MMTV promoter, and regulated using gene manipulation approaches, including tetracycline operated promoters. Cell and tissue specific knockouts are also now available, employing floxed genes and cre driven by a cell specific promoter. Cells are then tracked *in vivo* using eGFP tags and the multiphoton microscope. To further concentrate studies on the tissue and cell type of interest, genetically modified mammary epithelium, bone marrow or even whole mammary gland can be transplanted into normal host mice, and the effects observed in real time.

An important advance to *in vivo* imaging technology will be the simultaneous imaging of multiple cell types during the invasion process. Imaging of both stromal and tumor cells during invasion will help to more precisely define roles for specific cell types in enhancing metastasis. At present eGFP is commonly expressed in transgenic animals, allowing only one cell type to be fluorescently labeled, and analyzing interactions between two cell types requires a second, different fluorescent tag. Currently, macrophages can be marked using labeled dextrans. It is anticipated that transgenic animals carrying eCFP, eYFP and RFP will be available in the near future, and will greatly improve the opportunities for viewing how specific cell types interact *in vivo* with tumor cells during invasion.

ACKNOWLEDGEMENTS

The authors were supported by CA100324 (JC, JWP, and JES), CA94173 (JWP) and NHMRC Australia 250473 (WI). We thank the Analytical Imaging Facility for aid in image acquisition and analysis.

REFERENCES

1. Wyckoff JB, Jones JG, Condeelis JS, et al. A critical step in metastasis: in vivo analy of intravasation at the primary tumor. Cancer Res. 2000, 60: 2504-2511.
2. Chambers AF, Groom AC, MacDonald IC. Metastasis: Dissemination and growth of cancer cells in metastatic sites. Nat. Rev. Cancer 2002, 2: 563-572.
3. Fidler IJ. The organ microenvironment and cancer metastasis. Differentiation 2002, 70: 498-505.
4. McDonald DM, Choyke PL. Imaging of angiogenesis: from microscope to clinic. Nat. Med. 2003, 9: 713-725.
5. Rohren EM, Turkington TG, Coleman RE. Clinical applications of PET in oncolog Radiology 2004, 231: 305-332.
6. Lewis JS, Achilefu S, Garbow JR, et al. Small animal imaging. current technology and perspectives for oncological imaging. Eur. J. Cancer 2002, 38: 2173-2188.
7. Rudin M, Weissleder R. Molecular imaging in drug discovery and development. Nat. Rev. Drug Discov. 2003, 2: 123-131.
8. Degani H, Chetrit-Dadiani M, Bogin L, et al. Magnetic resonance imaging of tumor vasculature. Thromb. Haemost. 2003, 89: 25-33.
9. Knopp MV, Tengg-Kobligk H, Choyke PL. Functional magnetic resonance imaging oncology for diagnosis and therapy monitoring. Mol. Cancer Ther. 2003, 2: 419-426.
10. Sinha S, Sinha U. Functional magnetic resonance of human breast tumors: diffusion an perfusion imaging. Ann. N. Y. Acad. Sci. 2002, 980: 95-115.
11. Uematsu T, Sano M, Homma K, et al. Three-dimensional helical CT of the breas accuracy for measuring extent of breast cancer candidates for breast conserving surger Breast Cancer Res. Treat. 2001, 65: 249-257.
12. Boone JM, Nelson TR, Lindfors KK, et al. Dedicated breast CT: radiation dose an image quality evaluation. Radiology 2001, 221: 657-667.
13. Weissleder R. Scaling down imaging: molecular mapping of cancer in mice. Nat. Re Cancer 2002, 2: 11-18.
14. Blasberg RG. In vivo molecular-genetic imaging: multi-modality nuclear and optic combinations. Nucl. Med. Biol. 2003, 30: 879-888.
15. Brelje TC, Wessendorf MW, Sorenson RL. Multicolor laser scanning confocal immunofluorescence microscopy: practical application and limitations. Methods Cell Biol. 2002, 70: 165-244.
16. Farina KL, Wyckoff J, Rivera J, et al. Cell motility of tumor cells visualized in living intact primary tumors using green fluorescent protein. Canc. Res. 1998, 58: 2528-2532.
17. Zipfel WR, Williams RM, Webb WW. Nonlinear magic: multiphoton microscopy in the biosciences. Nat. Biotechnol. 2003, 21: 1369-1377.
18. Condeelis J, Segall JE. Intravital imaging of cell movement in tumours. Nat. Rev. Cancer 2003, 3: 921-930.
19. Wang W, Wyckoff JB, Frohlich VC, et al. Single cell behavior in metastatic primar mammary tumors correlated with gene expression patterns revealed by molecular profiling. Cancer Res. 2002, 62: 6278-6288.
20. Zipfel WR, Williams RM, Christie R, et al. Live tissue intrinsic emission microscop using multiphoton-excited native fluorescence and second harmonic generation. Proc. Natl. Acad. Sci. USA 2003, 100: 7075-7080.

21. Chambers AF, MacDonald IC, Schmidt EE, et al. Steps in tumor metastasis: new concepts from intravital videomicroscopy. Canc. Met. Rev. 1995, 14: 279-301.
22. Hoffman R. Green fluorescent protein imaging of tumour growth, metastasis, and angiogenesis in mouse models. Lancet Oncol. 2002, 3: 546-556.
23. Jain RK, Munn LL, Fukumura D. Dissecting tumour pathophysiology using intravital microscopy. Nat. Rev. Cancer 2002, 2: 266-276.
24. MacDonald, IC, Groom AC, Chambers AF. Cancer spread and micrometastasis development: quantitative approaches for in vivo models. Bioessays. 2002, 24: 885-893.
25. Qiu H, Orr FW, Jensen D, et al. Arrest of B16 Melanoma Cells in the Mouse Pulmonary Microcirculation Induces Endothelial Nitric Oxide Synthase-Dependent Nitric Oxide Release that Is Cytotoxic to the Tumor Cells. Am. J. Pathol. 2003, 162: 403-412.
26. Shan S, Sorg B, Dewhirst MW. A novel rodent mammary window of orthotopic breast cancer for intravital microscopy. Microvasc. Res. 2003, 65: 109-117.
27. Friedl P, Wolf K. Tumour-cell invasion and migration: diversity and escape mechanisms. Nat. Rev. Cancer 2003, 3: 362-374.
28. Yamamoto N, Jiang P, Yang M, et al. Cellular dynamics visualized in live cells in vitro and in vivo by differential dual-color nuclear-cytoplasmic fluorescent-protein expression. Cancer Res. 2004, 64: 4251-4256.
29. Wolf K, Mazo I, Leung H, et al. Compensation mechanism in tumor cell migration: mesenchymal-amoeboid transition after blocking of pericellular proteolysis. J. Cell Biol. 2003, 160: 267-277.
30. Lin EY, Jones JG, Li P, et al. Progression to malignancy in the polyoma middle T oncoprotein mouse breast cancer model provides a reliable model for human diseases. Am. J. Pathol. 2003, 163: 2113-2126.
31. Pollard JW. Tumour-educated macrophages promote tumour progression and metastasis. Nat. Rev. Cancer 2004, 4: 71-78.
32. Lin EY, Nguyen AV, Russell RG, et al. Colony-stimulating factor 1 promotes progression of mammary tumors to malignancy. J. Exp. Med. 2001, 193: 727-740.
33. Sasmono RT, Oceandy D, Pollard JW, et al. A macrophage colony-stimulating factor receptor-green fluorescent protein transgene is expressed throughout the mononuclear phagocyte system of the mouse. Blood 2003, 101: 1155-1163.
34. Faust N, Varas F, Kelly LM, et al. Insertion of enhanced green fluorescent protein into the lysozyme gene creates mice with green fluorescent granulocytes and macrophages. Blood 2000, 96: 719-726.
35. Wyckoff JB, Segall JE, Condeelis JS. The collection of the motile population of cells from a living tumor. Cancer Res. 2000, 60: 5401-5404.
36. Wyckoff J, Wang W, Lin EY, et al. A paracrine loop between tumor cells and macrophages is required for tumor cell migration in mammary tumors. Cancer Res. 2004, 64: 7022-7029.
37. Lin EY, Gouon-Evans V, Nguyen AV, et al. The macrophage growth factor CSF-1 in mammary gland development and tumor progression. J. Mammary. Gland. Biol. Neoplasia. 2002, 7: 147-162.
38. Gouon-Evans V, Lin EY, Pollard JW. Requirement of macrophages and eosinophils and their cytokines/chemokines for mammary gland development. Breast Cancer Res. 2002, 4: 155-164.

39. Gouon-Evans V, Rothenberg ME, Pollard JW. Postnatal mammary gland development requires macrophages and eosinophils. Development. 2000, 127: 2269-2282.
40. Pollard JW, Hennighausen L. Colony stimulating factor 1 is required for mammary gland development during pregnancy. Proc. Natl. Acad. Sci. USA 1994, 91: 9312-9316.

Chapter 4

IMAGING INVASION AND METASTASIS *EX VIVO*

Clayton Yates[1], Donna Beer Stolz[2], Linda Griffith[3]
[1] Department of Pathology, University of Pittsburgh School of Medicine, Pittsburgh, PA, USA, [2]Center for Biologic Imaging, Cell Biology and Physiology, University of Pittsburgh School of Medicine, Pittsburgh, PA, USA, [3]Division of Biological Engineering and Center for Biomedical Engineering, and Biotechnology Process Engineering Engineering, Center, Massachusetts Institute of Technology, Cambridge, MA, USA

Abstract: Recent advances in imaging and three dimensional cancer models have sought to expose the metastatic behavior of cancer by recreating appropriate tumor-host environment, *ex vivo*. In this review we will highlight some of the most recent advances in imaging using multiphoton microscopy, with a discussion on contributions of 2D and 3D systems most commonly as tumor assays for dissecting tumor biology, and pharmalogical efficacy. Lastly, we will discuss our previously described organotypic liver perfusion system and it's potential impact to cancer research.

Key words: bioreactors, 3-D models, microscopy, liver metastasis, tumor-host

1. INTRODUCTION

According to the 2004 American Society for Cancer, cancer death rates have not changed from 1950 to 2001 (1). This sobering fact motivates continued research in all steps of cancer initiation and progression. In most advance cancers, metastases are a primary contributor to overall mortality and thus therapies directed at eliminating establishment and growth of metastases are expected to have impact on overall mortality. There remains however a critical deficit of accessible *in vitro* assays and model systems that can capture the full spectrum of environmental cues that dictate the behavior of cancer cells at the metastatic site.

A. Wells (ed.), Cell Motility in Cancer Invasion and Metastasis, 73-88.

A variety of *in vivo* animal models are used to study metastasis. The most common models involve tumors in animal hosts; these are often xenografts of human tumor cells or fragments, animal allografts, or *in situ* tumors. *In vivo* assays provide for the integrated process or subparts thereof (such as for tail-vein seeding of lung fields) and have revealed many general phenomena involved in metastasis, such as the roles of endothelial cells and resident macrophages in host tissue invasion and for evaluation of potential inhibitors (2-5). The advent of new imaging methods to follow behavior of individual cells and metastases *in vivo* in real time offers even greater possibilities. Still, *in vivo* models do not completely fulfill all the desired features for replicating human metastasis. First, in xenografts, many key cytokine and matrix signals do not cross species barriers, and most rodent tumors are of limited generalizability to human tumors (6). Thus, some phenomena representative of human metastasis may be easily found and studied in animal models, while others may be missed. *In vivo* assays, especially those that involve imaging at the individual cell level, are cumbersome, often inefficient for metastases, and time-consuming limiting the number of parameters that can be studied. Still, we have learned much from these models as a result of great effort on the part of the individual investigators.

Intravital and whole body imaging are increasingly being used to study establishment and early growth of metastases *in situ* in animals. Intravital imaging relies on confocal or multi-photon imaging to follow the behavior of individual fluorescently-labeled cells within a particular target organ, allowing dissection of cell-cell interactions involved in penetration of the target tissue and providing a dynamic picture of tumor cell morphology as it moves between the vascular space and the tissue. The depth of focus is less than 0.5 mm and thus to gain images a portion of a target organ is exposed and placed on the microscopy viewing platform. This approach has been especially useful for characterizing primary-tumor, properties, growth rates and mechanisms of metastasis to target organs (7). With fluorescence markers, direct imaging of intravasation at the single-cell level within the primary tumor has been observed and related mechanistically to metastatic potential. For example, this approach has revealed that tumor cell arrest in the vasculature is distinct from the hematopoietic rolling mechanism (8). Segall and Condeelis and colleagues (7, 9) have imaged cells with confocal microscopy after establishment of a mass in the tissue and demonstrated that tumor cell lamellipodia are oriented towards vasculature and have increased activity and motility. They have also further used intravital imaging to reveal cross talk between macrophages and tumor cells in tissue invasion (4). Multiphoton laser scanning microscope allows for deeper penetration of tissue than does confocal imaging, and has been used to follow vascular and lymphatic vessel size and tortuosity, quantitative during growth of a tumors arising in a dorsal skin flap from single cells injected into the animal (10).

Whole body imaging is also emerging as a powerful tool in determining metastatic behavior of cancer cells and is becoming more widely used as the instrumentation becomes more available and the probes and host systems become more varied and well-characterized (11-14). Typically the cancer cells are manipulated to express a reporter gene that allows tracking position, and in some cases functional behavior, following metastasis from the primary tumor. Reporter genes include GFP, RFP and other variants detected by fluorescence imaging; luciferase genes from firefly and other organisms detected by luminescence of the products of the oxidative cleavage of their substrates; endogenous transferrin receptor, detected by magnetic resonance imaging of iron accumulation; and a variety of metabolic genes that allow intracellular trapping of radiolabelled substrates that can be detected with high resolution by positron emission spectroscopy (14, 15). Reporter genes can also provide a real-time window into the effectiveness of treatment modalities, including gene therapy approaches (16). Whole body imaging can also provide insights into the role of host stromal and immune tissues, either by introduction of labeled cells that home to tumor (17), tissue level imaging of nonluminous angiogenic blood vessels that appear as sharply defined dark networks (18-20) or using animals engineered to express a contrasting fluorescent label in stromal cells, (21).

Our focus here is the advent of *ex vivo* metastasis assays that seek to recapitulate features of the *in vivo* environment. Such assays allow often provided enhanced access to molecular-level information, are more accessible (particularly in a context of drug discovery and development), and offer the potential to provide a window into how human tumor cells behave in the context of underlying human tissue. We highlight recent advances in complex 3D *in vitro* models involving heterotypic cell cultures with an emphasis on tissue engineering approaches.

1.1 Static Culture Invasion and Growth Assays

The most accessible approaches to modeling metastasis employ static culture assays as a means to dissect molecular events in a reasonably controlled way. An example of an invasion assay is the invasion chamber, which in its simplest form is a thin (~0.1 mm) matrix barrier (typically collagen or Matrigel) on a large pore-containing support in a modified Boyden chamber, with the tumor cells placed on top of this barrier. After a period of time, and in response to factors in the originating top or targeting bottom compartment, the number of tumor cells that transmigrate this barrier is determined. Such assays often allow enhanced assess response to soluble factors or to parse the roles of various cell-matrix interactions; for example, Mercurio and co-workers used fluorescent beads embedded in Matrigel to assess the tractile forces exerted by cancer cells on the extracellular matrix, thus implicating basement remodeling during tumor invasion (22). Although, a predominance of *in vitro* metastasis assays have concentrated on metastatic cancer cells invading the basement membrane *in situ*, a similar design has been successful in identifying molecular cues during extravasation. Extravasation is evaluated by forming an endothelial cell monolayer, typically with intact endothelial cells; tight junctions are verified by electrical resistance or dye exclusion. Both assays have been useful in defining cells that have invasive potential and parsing key regulatory switches and cell behaviors. Retraction of endothelia away from the contact point when contacted by cancer cells has been observed for endothelia cultured on collagen gels and overseeded with mammary tumor spheroids (23) and this behavior has been observed *in vivo* (2). For example, the HNF-4 – mediated expression of Fas ligand by endothelia has been shown to kill tumor cells that express high levels of Fas, and this observation has been related back to expression *in vivo* to describe the possible differences in sites of metastasis of certain tumors (24).

Obviously, each *in vitro* assay is limited by simplification. The matrices presented are not truly representative of target organs having different components and growth factors. The endothelial cell barriers are not necessarily organ-specific. Other cellular elements, chiefly stromal cells consisting of underlying epithelial cells, are lacking. Lastly, only invasiveness is evaluated since there is no underlying parenchyma for assessment of metastatic growth.

Beyond initial tumor cell invasion events, subsequent events in the early stages of metastatic growth are also being examined in 3D monocultures and cocultures of normal cells with tumor cells. One of the most universally accepted forms of 3D model systems for cancer research are the monotypic 3D cell culture assays, which are possible to reproduce in almost any laboratory. Spheroidal aggregates of cells have been used for decades to create 3-D spheroids of tumor cells, using the spheroids as models of the primary tumors and also to model the process of metastasis by shedding of tumor cells. (25-29). Spheroidal aggregates of tumor cells in suspension are an enduring model, and whereas early methods typically resulted in a broad range of spheroid sizes, new methods of creating near monodisperse-sized spheroids in a reasonably high throughput fashion have recently been described (30), and these and other methods are being adapted to early stage screening of anticancer drug efficacy (31).

While monocultures of spheroids in suspension provide a more realistic phenotype than monolayer culture, the interplay between tumor cells and matrix is also a strong determinant of tumor phenotype and thus culture systems that provide an *in vivo*-like matrix milieu are becoming favored for dissection of basic disease processes (32, 33) A central theme of these assays is the use of ECM to create a tissue relevant environment. Matrigel is typically used for assays of epithelial behavior as includes many components of basement membrane. A number of cell models have been coupled with appropriate 3D matrices and show promising results in recapitulating tissue functions in 3D (34). Studies have been reported for liver, salivary gland, vasculature, bone, lung, skin, intestine, kidney and mammary and thyroid glands, but arguably the most well characterized models have been with the mammary gland. Both mouse and human mammary cells embedded in or cultured on Matrigel adopt a spherical, polarized structure that resembles the normal mammary alveolus (or acinus) that is capable of mammary-gland-specific function (e.g., producing milk in response to lactogenic hormones) (35-39). Pathophysiological behaviors that are similar to those observed *in vivo* are also captured in this *in vitro* system, for example loss of polarization and aberrant proliferation in the center of acini when signaling by epidermal growth factor receptor (EGFR) family members is perturbed through overexpression or mutations (33, 40). The mechanical environment – rigidity of the matrix – can also be systematically controlled using synthetic gels to which matrix proteins are cross-linked, and changes in cell signaling and down-stream behaviors resulting from matrix compliance changes are now being correlated with *in vivo* changes in tumor mechanical properties (Val Weaver, personal communication). A powerful aspect of this assay is the potential to adapt it to moderate or high-throughput screens for metastasis (41).

Tumor-stromal interactions are emerging as a critical factor in growth of metastatic tumors (42, 43). At a basic histological level, stroma appears as a matrix-rich tissue populated by fibroblasts and permeated by a blood vessel network. Many approaches have been described in the literature for creating 3D cultures of either fibroblasts or endothelial cells and using these cultures to examine the interactions with cancer cells. To model tumors, the stromal cells are typically mixed in a gel or cultured in a 3D scaffold, and the tumor cell might be added directly on top of the gel containing fibroblasts or endothelial cells, mixed in another gel layer on top, or various other configurations (44-48). 3D *in vitro* co-cultures have been particularly useful in revealing the profound effects of tumor-derived versus normal stroma in contributing to malignant behavior of epithelial tumors (31, 43, 49), and in identifying profiles of molecules secreted by tissue-specific stroma that may foster tissue-specific growth of certain kinds of carcinomas (50, 51).

1.2 3D Organoids cultured under flow conditions

Most tissues comprise a hierarchical arrangement of cells permeated by capillary blood vessels. Tissue homeostasis is maintained in part by a symphony of communication between the different cell types in tissue; each cell receives signals from neighbors via direct cell-cell interactions, cell-matrix interactions, and via soluble signaling molecules (cytokines and growth factors). In addition, mechanical forces – such as shear stress on endothelium from flowing blood – are converted to chemical signals that are necessary for normal tissue function. As discussed above, both 3D culture and heterotypic cell cultures are useful tools in dissecting dynamic processes in tumor progression. Such models fulfill an important connection between the well-defined cultures of single cell types and the complexity of the whole animal. They also provide experimental models of human tissue responses, where *in vivo* models are usually unavailable. A distinguishing feature of the *in vivo* cell environment is that cells are typically within a few tens or hundreds of microns from a nutrient capillary perfused with blood.

Thus various bioreactor configurations have been developed to provide enhancement of mass transfer, shear stress, or both, by providing fluid flow on or through cells. Bioreactors also offer the possibility of monitored and tightly controlled environmental and operating conditions (e.g. pH, temperature, pressure, nutrient supply and waste removal) (52). They also have been proposed as the excellent models to more fully recreate the complex *in vivo* environment *ex vivo.*

Tissue engineered model systems provide and promote a fundamental understanding of structure–function relationships in normal and pathological conditions, with possible commercial applications in molecular therapeutics (e.g. drug screening) (53). Recent advances have been made in the development of engineered tissue for repair of cartilage, bone, liver, kidney, skeletal muscle, blood vessels, the nervous system, and urological disorders (54-58). These system, repair being their primary function, also offer relevant target organs study cancer metastasis and invasion.

Among the simplest bioreactors are fluid-filled spinner flasks, which have been used for decades to create 3-D spheroids of tumor cells and create environments which provide controlled mass transfer to the outside of the spheroid (25-29, 34, 59). Spheroid formation of cancer cells are cell type specific and time-dependent, since smaller spheroids may be relatively homogenous when small or may have a necrotic center due to nutrient and oxygen deprivation when larger (29, 60) (Figure1).

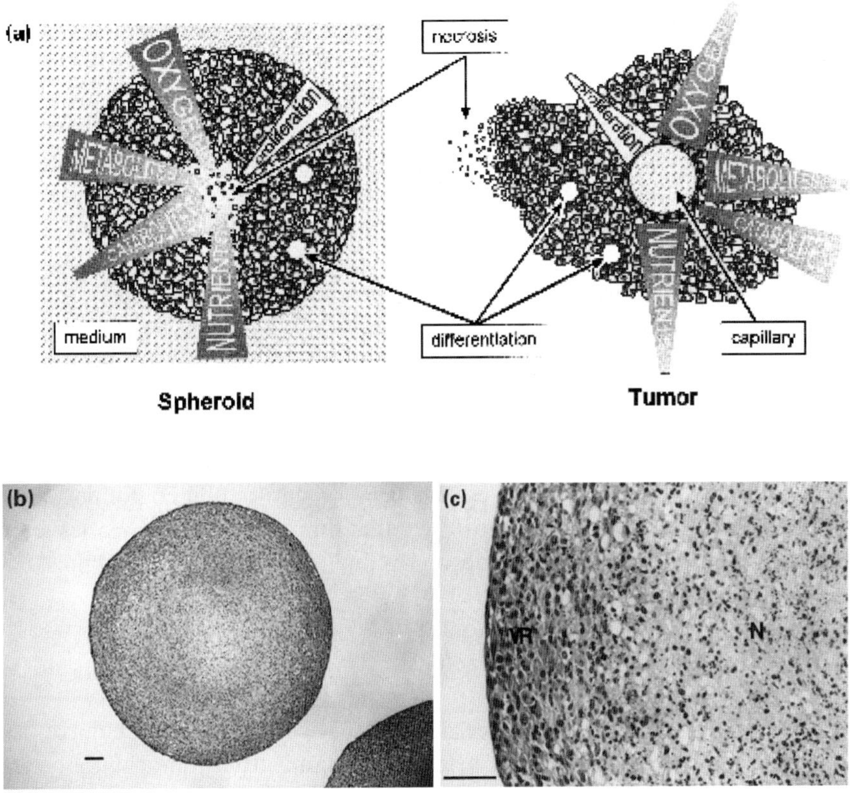

Figure 1. (a) Schematic illustration of the analogy between tumor micromilieu and multicellular tumor spheroids (modified from Nederman et al., 1983. In Vitro 19: 479–488). (b) representative H&E stained paraffin section through the center of a spheroid consisting of large-T-antigen transfected, transformed human fibroblasts that express an extensive ECM. N, necrosis; VR, viable cell rim; bar=150 _m. (c) sector magnification of (b).

The rotating wall vessel bioreactor (RWV) is an alternative spinner flask design for three-dimensional culture of cells, and was developed to simulate microgravity conditions. In this system, a cylindrical vessel filled with cells and culture medium is rotated about an axis parallel to the ground, thus balancing gravity with fluid drag forces and creating a low-shear stress, high mass transfer environment (61). The environment in the RWV has been shown to foster tissue-like structures of mullerian tumor cells of the ovary and prostate (59) and melanoma cancer cells (62) as assessed by histology. Inclusion of human umbilical vein endothelial cells with cervical tumor cell cultures resulted in tubular structures penetrating the tumor cell masses, forming aggregates larger in size than the monocultures and typically with greater cell mass and number (63) cervical tumor cells and human umbilical vein endothelial cells (HUVEC). Of particular interest was that these 3D cell cultures to show biochemical markers known to be involved with cellular function and cancer progression. Cell aggregates formed in RWV bioreactors have also been used as model systems to test therapeutic options which include radiation resistance, phenotypic differentiation, and response to anti-cancer drugs. Among the potential limitations of the RWV are the polydispersity of spheroid sizes and the barriers to *in situ* imaging of cell behavior.

Microfluidic bioreactors offer the potential for more homogeneous, controlled formation of 3D structures and thus potentially better-controlled exposure of cells to agents that affect cell behavior in the context of a 3D environment. A step in this direction is the recent report by Torisawa and co-workers, where 3D culture of MCF-7 breast cancer cells in a collagen gel was controlled in tiny chambers in a microfluidic device, and could be used in a multiplexed format to assess efficacy of anti-tumor compounds (64). Although it remains to be demonstrated that this approach will work well on a large scale, the success of "lab on a chip" chemical assays suggests the barriers are not insurmountable.

1.2.1 Flow Perfusion Bioreactors

Although a rich array of three dimensional co-culture models have been developed for investigations of behaviors ranging from tumor angiogenesis and embryonic differentiation to skin permeation of drugs, the available models lack one crucial feature of most tissues – a perfused microvasculature. Inclusion of endothelial cells in the RWV or in static cultures provides some features of tissue structure, but the lack of flow through the vessels limits the ability of these systems to mimic tissue physiology.

Thus an organotypic system would ideally include several features 1) an integrated epithelial/stromal/endothelial cell architecture representing the key target organ; 2) long-term (weeks) stability to allow tumor cell survival and growth; 3) controlled local perfusion of the organoid structures over length scales comparable to a capillary bed; 4) direct visualization throughout the process to discern subprocesses such as extravasation versus growth; 5) easy manipulation and intervention; and 6) assay robustness in reproducibility and moderate or high throughput.

As a step toward creating true physiological mimics of human and animal tissues that recapitulate the features of a capillary bed, we have developed a microfabricated bioreactor system that facilitates perfusion of 3D heterotypic co-cultures at the length scale of the capillary bed in an arrangement that also allows *in situ* analysis of cell behavior via microscopy (58, 64, 65).

This system circumvents rapid loss of liver specific functions that normally occurs when hepatocytes are maintained under standard culture conditions thus providing a reasonable model system for the testing of tumor-host interactions in *ex vivo* environment. The liver is a major site of metastasis for many carcinomas, and its anatomically relevant for metastasis models due to a simplified architecture of hepatocytes covered with an endothelial lining, and can be recreated *in vivo* via selective cell adhesion and cell self-assembly (58, 65, 66).

Our cross-flow perfusion reactor is designed to address several needs for 3D liver tissue culture (58, 65). The classical challenges in reactor design for three dimensional perfusion culture – ensuring a relatively homogeneous distribution of flow and mass transfer throughout the system to meet the metabolic demands of the cells – are augmented in the case of three dimensional cultures of primary cells by the need to provide a scaffold appropriate for tissue morphogenesis. Varying degrees of histotypic reorganization have been observed in several types of three dimensional liver cell cultures, particularly those incorporating perfusion through the tissue mass (67-71). Distinguishing features of our design include: an appropriate scaffold for tissue morphogenesis; uniform distribution of fluid flow and nutrients throughout the 3D culture; and an optical window to allow repeated *in situ* observation of cells via light or 2-photon microscopy during perfusion culture.

A photo and schematic cross-section depicting the major design features of our current reactor (as used for liver culture) is shown in Figure 2. The heart is the cell scaffold, comprising a thin (~230 μm) silicon sheet permeated from top to bottom by a regular array of ~300 μm channels (created by deep reactive ion etching) and seated atop a microporous filter, which is in turn mechanically supported by a second scaffold. The morphogenesis of cells into tissue-like structures following seeding into the channels is guided in part by scaffold surface chemistry, which controls the relative values of cell-cell and cell-substrate adhesion strength, and by the channel geometry and dimensions. The scaffold is maintained between the upper and lower chambers of a flow-through housing. Each chamber has a pair of ports to allow flow of culture medium across the surfaces of the chip. The arrangement of the ports allows for several modes of operation. Under our current mode of operation for liver tissue, the fluid in the upper chamber is initially maintained at a higher pressure than that in the lower chamber, thus creating a driving force for perfusion of culture medium through the tissue in the channels immediately after cells are introduced. Cells seeded into the channels are initially held in place by the filter, and after initial attachment and reorganization (~ 1 day), by adhesion to the channel walls.

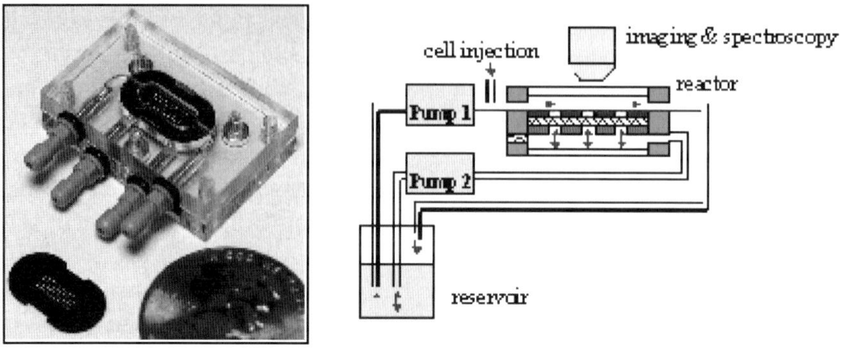

Figure 2. Photo of the silicon-chip scaffold (lower left) and polycarbonate microreactor housing that holds the scaffold and provides four ports for continuous flow of culture medium across the top of the scaffold and through the tissue mass held within the scaffold. An optical window at the top allows in-situ observation of the tissues by two-photon microscopy. Right: Schematic of scaffold and reactor housing shown in cross-section, indicating the connections to recirculation loop and observation window. The main flow of culture medium is across the top of the chip in the upper chamber (pump 1). Flow through the tissue mass in the chip is maintained at a constant rate by either pulling medium through or pumping medium in reverse flow through with the second pump. See alos Color Plate Section page 344.

After 1-2 days in culture, fluid flow through the channel is reversed (i.e., to flow upward through the filter and then through the tissue mass), dislodging any dead or unattached cells. This reverse flow is controlled by a peristaltic pump throughout the remainder of the culture period, and in the case of liver, stable tissue structures are maintained for weeks, thus creating appropriate environment for cancer cells.

Metastasis is a highly orchestrated series of events comprised of many "sub-processes" whereby cancer cells must attach, proliferate, and invade the parenchyma of specific target tissue (Figure 3a). Using our cross-flow perfusion bioreactor we were able to recreate the progression of invasion and metastasis with the added advantage of real-time visualization by fluorescence microscopy over several weeks. GFP expressing cancer cells were introduced into previously seeded, and established hepatic parenchymal spheroids. Initially, individual cancer cells occupy the hepatic wells. However cancer cells overgrow the hepatic wells and the rest of the chamber by day 14 (Figure 3b). During this rapid and visually apparent growth of cancer cells, invasion into hepatic parenchyma also occurs, with very distinct heterotypic cell-cell interactions observed upon histology and electron microscopy (72, 73).

Unlike enclosed bioreactors, this bioreactor allows for repeated visualization and affords the opportunity to examine critical processes underlying the metastatic program over time, such as cell specific proliferation, death and tissue organization. Interaction of metastases with the parenchymal cells of the target organ, as well as resident ECM and endothelial cells are required to fully recapture the microenvironment presented to and modified by the metastatic cancer. As more advanced tissue engineered models are generated to determine cellular responses of many organs, research should include alternative applications, as these systems could possibly represent the leading models for metastasis and invasion.

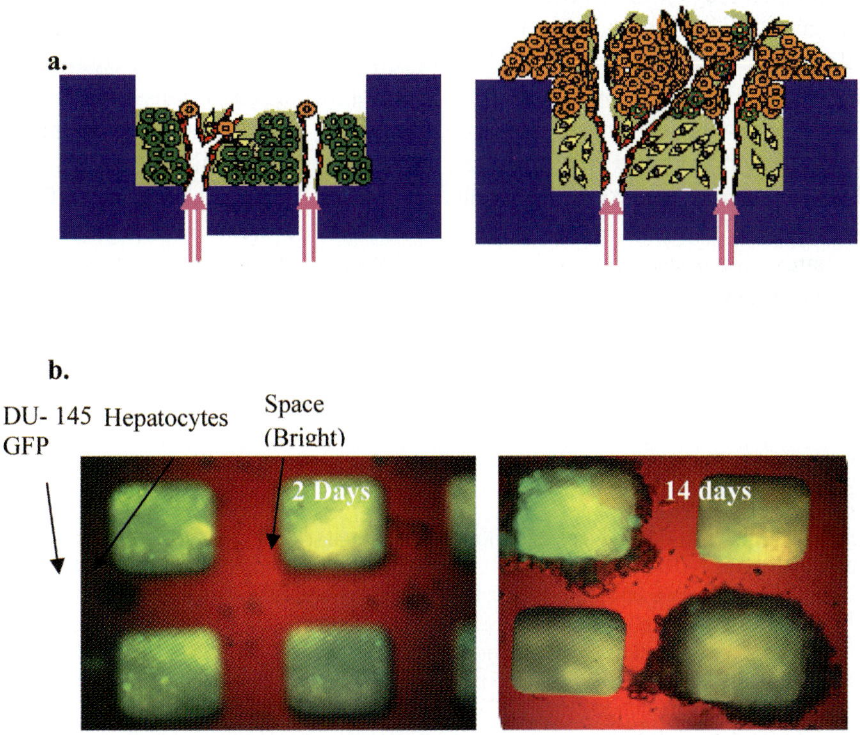

Figure 3. (a) Schematic illustration of the tumor related events that occur in bioreactor. (b) Light micrograph of 4 channels of coculture from 2 to 14 days. Dark grass green area is hepatic tissue, bright yellow is autofluorescence from underling filter, and bright green are GFP DU-145 human prostate cancer cells.

2. SUMMARY

The use of tissue engineered cancer cells may have a profound impact on advancements in cancer research. *In vitro* generation of 3D tissue from isolated cells has attempted to overcome the limitations of individual and co culture systems. The level of control over specific environmental factors in 3D cultures can provide the systematic means to perform controlled studies aimed at understanding which specific biological, chemical or physical parameter plays which function in a defined environment. As we continue cancer research, researchers must develop more specific models to derive clinically relevant conclusions. Bioreactors offer this advantage, creating an interdisciplinary approach to identify environmental conditions required for the progression of cancer in specific tissues. These systems will prove invaluable not only to understanding tumor biology, but will provide insights into the tumor micro milieu, thus increasing our understanding of the dynamic cancer process. Imaging within these systems also significantly enhances the opportunity to investigate biological changes within the context of the metastasis environment. Therefore the development of new tissue engineered systems must be multifunctional with variable applications possible.

REFERENCES

1. Arias E, et al. Death: Final report for 2001. Nat vital statistics report 2003, 52: 1-116.
2. Mook OR, et al. Visualization of early events in tumor formation of eGFP-transfected rat colon cancer cells in liver. Hepatology 2003, 38: 295-304.
3. Timmers M, et al. Interactions between rat colon carcinoma cells and Kupffer cells during the onset of hepatic metastasis. Int. J. Cancer 2004, 112: 793-802.
4. Wyckoff J, et al. A paracrine loop between tumor cells and macrophages is required for tumor cell migration in mammary tumors. Cancer Res. 2004, 64: 7022-9.
5. Liang Z, et al. Inhibition of breast cancer metastasis by selective synthetic polypeptide against CXCR4. Cancer Res. 2004, 64: 4302-8.
6. Rangarajan A, et al. Species- and cell type-specific requirements for cellular transformation. Cancer Cell 2004, 6: 171-83.
7. Condeelis J, Segall JE. Intravital imaging of cell movement in tumours. Nat. Rev. Cancer 2003, 3: 921-30.
8. Chambers AF, et al. Steps in tumor metastasis: new concepts from intravital videomicroscopy. Cancer Metastasis Rev. 1995, 14: 279-301.
9. Condeelis JS, et al. Lamellipodia in invasion. Semin. Cancer Biol. 2001, 11: 119-28.
10. Jain RK, Munn LL, Fukumura D. Dissecting tumour pathophysiology using intravital microscopy. Nat. Rev. Cancer 2002, 2: 266-76.

11. Iyer M, et al. Noninvasive imaging of enhanced prostate-specific gene expression using a two-step transcriptional amplification-based lentivirus vector. Mol. Ther. 2004, 10: 545-52.
12. Hoffman RM. Imaging tumor angiogenesis with fluorescent proteins. Apmis. 2004, 112: 441-9.
13. Michalet X, et al. Quantum dots for live cells, in vivo imaging, and diagnostics. Science 2005, 307: 538-44.
14. Gross S, Piwnica-Worms D. Spying on cancer; Molecular imaging in vivo with genetically encoded reporters. Cancer Cell 2005, 7: 5-15.
15. Piwnica-Worms D, Luker KE. Imaging protein-protein interactions in whole cells and living animals. Ernst. Schering Res. Found Workshop 2005, (49): 35-41.
16. Yaghoubi SS, et al. Imaging progress of herpes simplex virus type 1 thymidine kinase suicide gene therapy in living subjects with positron emission tomography. Cancer Gene. Ther. 2004.
17. Kim YJ, et al. Multimodality imaging of lymphocytic migration using lentiviral-based transduction of a tri-fusion reporter gene. Mol. Imaging Biol. 2004, 6: 331-40.
18. Hoffman R. Green fluorescent protein imaging of tumour growth, metastasis, and angiogenesis in mouse models. Lancet Oncol. 2002, 3: 546-56.
19. Hoffman RM. Orthotopic metastatic mouse models for anticancer drug discovery and evaluation: a bridge to the clinic. Invest. New Drugs 1999, 17: 343-59.
20. Hoffman RM. Whole-body fluorescence imaging with green fluorescence protein. Methods Mol. Biol. 2002, 183: 135-48.
21. Yamamoto N, et al. Cellular dynamics visualized in live cells in vitro and in vivo by differential dual-color nuclear-cytoplasmic fluorescent-protein expression. Cancer Res. 2004, 64: 4251-6.
22. Rabinovitz I, Gipson IK, Mercurio AM. Traction forces mediated by alpha6beta4 integrin: implications for basement membrane organization and tumor invasion. Mol. Biol. Cell 2001, 12: 4030-43.
23. Zamora PO, Danielson KG, Hosick HL. Invasion of endothelial cell monolayers on collagen gels by cells from mammary tumor spheroids. Cancer Res. 1980, 40: 4631-9.
24. Osanai M, et al. Hepatocyte nuclear factor (HNF)-4alpha induces expression of endothelial Fas ligand (FasL) to prevent cancer cell transmigration: a novel defense mechanism of endothelium against cancer metastasis. Jpn. J. Cancer Res. 2002, 93: 532-41.
25. Durand, RE, Olive PL. Irradiation of multi-cell spheroids with fast neutrons versus X-rays: a qualitative difference in sub-lethal damage repair capacity or kinetics. Int. J. Radiat. Biol. Relat. Stud. Phys. Chem. Med. 1976, 30: 589-92.
26. Deakin AS. Model for the growth of a solid in vitro tumor. Growth 1975, 39: 159-65.
27. Mueller-Klieser WF, Sutherland RM. Influence of convection in the growth medium on oxygen tensions in multicellular tumor spheroids. Cancer Res. 1982, 42: 237-42.
28. Franko AJ, Koch CJ. The radiation response of hypoxic cells in EMT6 spheroids in suspension culture does model data from EMT6 tumors. Radiat. Res. 1983, 96: 497-504.
29. Santini MT, Rainaldi G. Three-dimensional spheroid model in tumor biology. Pathobiology 1999, 67: 148-57.

30. Kelm JM, et al. Method for generation of homogeneous multicellular tumor spheroids applicable to a wide variety of cell types. Biotechnol. Bioeng. 2003, 83: 173-80.
31. Kunz-Schughart LA, et al. The use of 3-D cultures for high-throughput screening: the multicellular spheroid model. J. Biomol. Screen. 2004, 9: 273-85.
32. Jacks T, Weinberg RA. Taking the study of cancer cell survival to a new dimension. Cell 2002, 111: 923-5.
33. Muthuswamy SK, et al. ErbB2, but not ErbB1, reinitiates proliferation and induces luminal repopulation in epithelial acini. Nat. Cell Biol. 2001, 3: 785-92.
34. Schmeichel KL, Bissell MJ. Modeling tissue-specific signaling and organ function in three dimensions. J. Cell Sci. 2003, 11: 2377-88.
35. Petersen OW, et al. Interaction with basement membrane serves to rapidly distinguish growth and differentiation pattern of normal and malignant human breast epithelial cells. Proc. Natl. Acad. Sci. USA 1992, 89: 9064-8.
36. Weaver VM, et al. Reversion of the malignant phenotype of human breast cells in three-dimensional culture and in vivo by integrin blocking antibodies. J. Cell Biol. 1997, 137: 231-45.
37. Howlett AR, et al. Cellular growth and survival are mediated by beta 1 integrins in normal human breast epithelium but not in breast carcinoma. J. Cell Sci. 1995, 108: 1945-57.
38. Fata JE, Werb Z, Bissell MJ. Regulation of mammary gland branching morphogenesis by the extracellular matrix and its remodeling enzymes. Breast Cancer Res. 2004, 6: 1-11.
39. Stoker AW, et al. Designer microenvironments for the analysis of cell and tissue function. Curr. Opin. Cell Biol. 1990, 2: 864-74.
40. Muthuswamy SK, Gilman M, Brugge JS. Controlled dimerization of ErbB receptors provides evidence for differential signaling by homo- and heterodimers. Mol. Cell Biol. 1999, 19: 6845-57.
41. Martin SS, et al. A cytoskeleton-based functional genetic screen identifies Bcl-xL as an enhancer of metastasis, but not primary tumor growth. Oncogene 2004, 23: 4641-5.
42. Bhowmick NA, et al. TGF-beta signaling in fibroblasts modulates the oncogenic potential of adjacent epithelia. Science 2004, 303: 848-51.
43. Bhowmick NA, Moses HL. Tumor-stroma interactions. Curr. Opin. Genet. Dev. 2005, 15: 97-101.
44. Donovan D, et al. Comparison of three in vitro huiman 'angiogenesis' assays with capillaries formed in vivo. Angiogenesis 2001, 4: 113-121.
45. Feraud O, Cao Y, Vittet D. Embryonic stem cell-derived embryoid bodies development in collagen gels recapitulates sprouting angiogenesis. Lab Invest. 2001, 81: 1669-1681.
46. Masso-Welch PA, et al. Inhibition of angiogenesis by the cancer chemopreventive agent conjugated linoleic acid. Cancer Res. 2002, 62: 4383-4389.
47. Parmar H, et al. A novel method for growing human breast epithelium in vivo using mouse and human mammary fibroblasts. Endocrinology 2002, 143: 4886-4896.
48. Velazquez OC, et al. Fibroblast-dependent differentiation of human microvascular endothelial cells into capillary-like 3-dimensional networks. FASEB J. 2002, 16: 1316-1318.
49. Atula S, Grenman R, Syrjanen S. Fibroblasts can modulate the phenotype of malignant epithelial cells in vitro. Exp. Cell Res. 1997, 235: 180-7.

50. Oh HS, et al. Bone marrow stroma influences transforming growth factor-beta production in breast cancer cells to regulate c-myc activation of the preprotachykinin-I gene in breast cancer cells. Cancer Res. 2004, 64: 6327-36.
51. Nakashiro K, et al. Hepatocyte growth factor secreted by prostate-derived stromal cells stimulates growth of androgen-independent human prostatic carcinoma cells. Am. J. Pathol. 2000, 157: 795-803.
52. Martin I, Wendt D, Heberer M. The role of bioreactors in tissue engineering. Trends Biotechnol. 2004, 22: 80-6.
53. Griffith LG, Naughton G. Tissue engineering--current challenges and expanding opportunities. Science 2002, 295: 1009-14.
54. Atala A. Tissue engineering in urology. Curr. Urol. Rep. 2001, 2: 83-92.
55. MacNeill BD, et al. Toward a new blood vessel. Vasc. Med. 2002, 7: 241-6.
56. Vunjak-Novakovic G. The fundamentals of tissue engineering: scaffolds and bioreactors. Novartis Found Symp 2003, 249: 34-46; discussion 46-51, 170-4, 239-41.
57. Sharma V, et al. Sensitization of tumor cells toward chemotherapy: enhancing the efficacy of camptothecin with imidazolines. Chem. Biol. 2004, 11: 1689-99.
58. Powers MJ, et al. Functional behavior of primary rat liver cells in a three-dimensional perfused microarray bioreactor. Tissue Eng. 2002, 8: 499-513.
59. Margolis L, et al. Long term organ culture of human prostate tissue in a NASA-designed rotating wall bioreactor. J. Urol. 1999, 161: 290-7.
60. Kunz-Schughart LA. Multicellular tumor spheroids: intermediates between monolayer culture and in vivo tumor. Cell Biol. Int. 1999, 23: 157-61.
61. Schwarz RP, Goodwin TJ, Wolf DA. Cell culture for three-dimensional modeling in rotating-wall vessels: an application of simulated microgravity. J. Tissue Cult. Methods 1992, 14: 51-7.
62. Goodwin TJ, et al. Three-dimensional culture of a mixed mullerian tumor of the ovary: expression of in vivo characteristics. In Vitro Cell Dev. Biol. Anim. 1997, 33: 366-74.
63. Chopra V, Dinh TV, Hannigan EV. Three-dimensional endothelial-tumor epithelial cell interactions in human cervical cancers. In Vitro Cell Dev. Biol. Anim. 1997, 33: 432-42.
64. Torisawa YS, et al. Multi-channel 3-D cell culture device integrated on a silicon chip for anticancer drug sensitivity test. Biomaterials 2005, 26: 2165-72.
65. Powers MJ, et al. A microfabricated array bioreactor for perfused 3D liver culture. Biotechnol. Bioeng. 2002, 78: 257-69.
66. Powers MJ, Griffith LG. Adhesion-guided in vitro morphogenesis in pure and mixed cell cultures. Microsc. Res. Tech. 1998, 43: 379-84.
67. Griffith LG, et al. In vitro organogenesis of liver tissue. Ann. N. Y. Acad. Sci. 1997, 831: 382-97.
68. Michalopoulos GK, et al. Morphogenetic events in mixed cultures of rat hepatocytes and nonparenchymal cells maintained in biological matrices in the presence of hepatocyte growth factor and epidermal growth factor. Hepatology 1999, 29: 90-100.
69. Kaihara S, et al. Silicon micromachining to tissue engineer branched vascular channels for liver fabrication. Tissue Eng. 2000, 6: 105-17.
70. Gerlach JC, et al. Improved hepatocyte in vitro maintenance in a culture model with woven multicompartment capillary systems: electron microscopy studies. Hepatology 1995, 22: 546-52.

71. Bader A, et al. 3-D coculture of hepatic sinusoidal cells with primary hepatocytes-design of an organotypical model. Exp. Cell Res. 1996, 226: 223-33.
72. Yates C, et al. Direct Visualization of Prostate Cancer Progression utilizing a Novel Organotypic Liver Bioreactor as Metastastic Target Organ. Submitted.
73. Yates C, et al. Direct Visualization of Prostate Cancer Progression utilizing a Novel Organotypic Liver Bioreactor as Metastastic Target Organ. Submitted, 2005.

Chapter 5

AUTOCRINE GROWTH FACTOR SIGNALING IN MOTILITY

Elizabeth J. Joslin, Douglas A. Lauffenburger
Massachusetts Institute of Technology, Cambridge, MA, USA

Abstract: The objective of this chapter is to summarize our current understanding of the role of growth factor autocrine signaling in cell motility. The Epidermal Growth Factor Receptor (EGFR) was chosen as a central example system for motivating autocrine operation as important in both normal physiological processes as well as pathological conditions such as cancer. We provide specific evidence from research literature for autocrine stimulation of cell motility measured *in vitro*, from several cell lines, and *in vivo* from mouse models. Relevant growth factors include the EGFR ligands TGF-alpha, HB-EGF, and Amphiregulin, as well as members of other growth factor families including FGF, G-CSF, GM-CSF, HGF and VEGF. Results from autocrine systems that involve EGFR crosstalk are also discussed, including GPCR and IL-6 mediated activation. We outline the cellular parameters, determined by experimental and computational work, that govern autocrine operation and the spatial range of autocrine ligands. We also present intercellular signaling pathways relevant to cell motility that may be involved in propagating localized signals from the cell surface. Finally, we review experimental findings that demonstrate how the mode of growth factor presentation can affect cell migration behavior.

Key words: autocrine signaling, EGF receptor, erbB receptors, mathematical modeling

1. AUTOCRINE GROWTH FACTOR SIGNALING

The term autocrine signaling was first proposed by Sporn and Todaro to describe the type of self stimulation that could occur if a cell produced both a hormone-like factor and the cognate receptor (1). Normal cell growth in

A. Wells (ed.), Cell Motility in Cancer Invasion and Metastasis, 89-109.

culture required the addition of specific growth factors, whereas malignant cells could grow autonomously without the same exogenous supplements. While growth factor autocrine amplification or dysregulation has become one of the hallmarks of cancer, autocrine signaling is also now known to be a mode of signaling involved in normal physiological processes (2-4). For example, growth factor autocrine signaling has been implicated in wound healing (5), angiogenesis (6), and tissue organization during development (7).

A prominent example of autocrine growth factor signaling that is involved in numerous cell types and cellular responses is the Epidermal Growth Factor Receptor (EGFR) system. EGFR belongs to the ErbB family of receptor tyrosine kinases, which has several known EGF-like ligands (8). ErbB signaling plays an important role in normal epithelial cell proliferation, development, survival, adhesion and migration (8-10). Beyond its role in normal physiology, dysregulated ErbB signaling is also extremely important in cancer progression (11, 12). The EGFR and the EGFR ligands are expressed or overexpressed in several human cancers, for example lung, breast, gastric, pancreatic, colon, head and neck, prostate, ovarian, and brain cancers (13-21). Altered ErbB signaling can lead to aberrant, growth promoting signals that may have a major impact on tumor initiation, progression, and metastasis.

EGFR ligands include EGF, TGF-alpha, HB-EGF, betacellulin, amphiregulin, epiregulin, and epigen (22). These ligands are produced as membrane-bound peptides and proteolytic cleavage between the ligand and the membrane-spanning region releases the ligand from the membrane. The region between the receptor binding domain and the transmembrane domain, the ligand releasing cleavage site, is highly variable among the EGFR ligands, although they are key determinates of ectodomain shedding (22, 23). The ligands have the potential to signal in an autocrine, paracrine, and/or a juxtracrine mode depending on extracellular proteolytic processing. While an autocrine-signaling mode is characterized by ligands that are released from the cell surface in an active form and bind to receptors on the same cell, paracrine signaling involves the capture of ligand released from a neighboring cell, see Figure 1. Juxtacrine signaling involves ligand-receptor binding between a cell expressing the membrane-bound ligand and an adjacent cell that expresses the correct receptor, an event that does not involve ligand cleavage and requires direct contact between the stimulatory and target cell.

Members of the ADAM family of metalloproteases proteolytically release the EGFR ligands from the cell surface. Although TACE/ADAM 17 is known to be involved in TGF-alpha, HB-EGF, and AR release, the full identities and regulation of the metalloproteases involved in EGFR ligand shedding are currently under investigation (22, 24-26). There is evidence that TGF-alpha and HB-EGF shedding is stimulated by signals downstream of the EGF receptor, such as PKC and MAPK activation, creating a positive feedback mechanism that may involve metalloprotease phosphorylation (27-30). Activation of the ERK pathway has also been implicated in increasing the transcription of EGFR ligands, creating an additional positive feedback (27, 31).

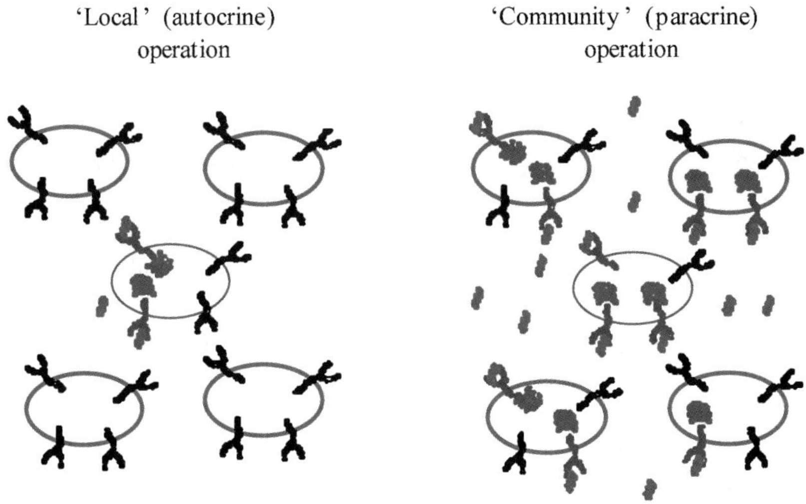

Figure 1. Illustration of autocrine vs. paracrine signaling. The green circle represents a cell that produces a ligand and expresses the receptor for the ligand. The blue circles are neighboring cells that also express the receptor. Red receptors are occupied by ligand, which leads to intracellular signling. (Left) Case of autocrine operation, essentially all of the ligand produced is captured locally. (Right) Case of paracrine operation, with very little ligand captured by the producing cell. (Adapted from DeWitt et al., 2002). See also Color Plate Section page 344.

Due to the intrinsic and closed-loop nature of autocrine signaling systems, identification and experimental analysis of true autocrine effects on cellular behavior is a challenge (32). While there seems to be lots of evidence for the role of autocrine signaling in cell proliferation, less is known about whether autocrine mechanisms may be involved in the transition to, or control of a motile cellular phenotype. To this end, several cell types and growth factor families have been studied *in vitro* using various migration assays, while fewer studies have addressed this question *in vivo*. Experimental results that provide evidence for the role of growth factor autocrine signaling specifically involved in cell motility are presented below with emphasis on the EGF receptor system as an example.

2. EVIDENCE FOR ROLE OF GROWTH FACTOR AUTOCRINE SIGNALING IN CELL MIGRATION

2.1 EGFR System

2.1.1 *In Vitro*

Gavrilovic et al. transfected a rat bladder carcinoma cell line (NBTII) with a gene encoding TGF-alpha that resulted in an epithelial to fibroblast-like morphological transition (33). Clones were plated on glass coverslips and cell migration was measured using videomicroscopy and cell tracking. TGF-alpha synthesizing cells and parental cells stimulated with conditioned media achieved higher migration speeds compared to the parental cells stimulated with exogenous TGF-alpha. The transfected cells also acquired an increase in gelatinase synthesis.

Elevated levels of the EGF receptor and expression of its ligands have been found in human prostate carcinomas (34). The DU145 prostate carcinoma cell line expresses TGF-alpha and EGFR, which are involved in growth stimulation (35). Xie et al. investigated the involvement of this autocrine stimulation in transmigration of an *in vitro* matrix (36). The parental DU145 cells were transfected to overexpress EGFR. The EGFR overexpressing variant showed an increased migration through the matrix compared to the parentals. The increased matrix invasion of the EGFR overexpressing cells was inhibited upon addition of an anti-EGFR antibody. The EGFR overexpressors and parentals secreted similar levels of proteolytic activity and thus in this invasion assay the increase in migration was determined to be autocrine stimulated and potentially independent of matrix degradation.

TGF-alpha expression has been found in human gliomas and has been correlated with tumor grade (37, 38). El-Obeid et al. demonstrated that the expression of TGF-alpha in an EGFR expressing human malignant glioma cell line (U-1242 MG) induced cell locomotion (39). ^{125}I-labeled EGF binding studies showed that 60% of EGF specifically bound to cells induced to express TGF-alpha compared to the same clone that was repressed using a tetracycline regulated gene expression system. Receptor phosphorylation of derived clones was similar to repressed cells also stimulated with exogenous EGF. EGFR phosphorylation levels were also found to be independent of cell seeding density. Phagokinetic track area of individual cells in an *in vitro* migration assay was higher for autocrine-induced cells. An EGFR blocking antibody only slightly decreased cell track area of the induced TGF-alpha expressing clones, whereas exogenously added TGF-alpha was completely neutralized by the antibody. These results showed that co-expression of EGFR and TGF-alpha enabled individual glioma cells to operate in an independent, migratory manner, where EGFR activation may have occurred in cell regions inaccessible to blocking antibodies.

Dong et al. investigated the effect of a broad-spectrum matrix metalloprotease inhibitor on EGFR ligand release and migration of a human mammary epithelial cell line (hMEC) (40). The TGF-alpha concentration in hMEC-conditioned media and the speed of individually tracked cells seeded at low density decreased in presence of the inhibitor. Exogenous EGF reversed the inhibitory effect on migration speed, suggesting that the release of the endogenous ligands, such as TGF-alpha or amphiregulin, was necessary for EGFR stimulated motility in this cell system.

The migration of sheets of corneal epithelial cells in a wounding assay, which involved lifting an agarose strip to induce a gap without cell damage, was shown to be dependent on HB-EGF signaling (41). Addition of an EGFR kinase inhibitor abolished the wound closure in this *in vitro* system. Incubation with an EGFR or a HB-EGF neutralizing antibody also inhibited wound closure. Inhibition of the biological activity of HB-EGF with a non-toxic diphtheria toxin analog also reduced wound closure of the corneal epithelial cell layers.

EGFR autocrine signaling has also been found to stimulate cell motility as a result of cell signaling crosstalk. Interleukin (IL)-6 is a member of the IL-6-type cytokine family and is involved in the immune response, inflammation, and hematopoiesis (42). Badache et al. investigated the mechanism by which IL-6 increased breast carcinoma cell migration in an *in vitro* transwell assay (43). Activation of the MAPK and PI3K signaling pathways were required for IL-6 induced cell migration. Both the addition of an anti-EGFR antibody and an EGFR kinase inhibitor also decreased the IL-6 induced migration. Although the identity of the EGFR ligand(s) involved were not determined, these results showed that IL-6 stimulated breast carcinoma cell migration through transactivation of an EGFR autocrine pathway and downstream activation of MAPK and PI3K.

EGFR activation has also been implicated in crosstalk with G Protein-Coupled Receptors (GPCR) (44). Gschwind et al. examined GPCR stimulated migration of squamous cell carcinoma cells using an *in vitro* transwell assay (45). Lysophosphatidic acid induced activation of GPCR resulted in proteolytic cleavage of membrane bound pro-amphiregulin (proAR). RNA silencing of proAR decreased LPA stimulated migration. Addition of an AR neutralizing antibody inhibited LPA aggravated EGFR tyrosine phosphorylation. LPA induced AR release and cell migration was also inhibited by RNA silencing of the membrane bound metalloprotease TACE. Although the mechanism of TACE activation was unknown, the shedding of AR was shown to be involved in GPCR-EGFR crosstalk and cell migration in this experimental system.

EGFR and GPCR crosstalk has also been implicated in bombesin stimulated migration of prostate cancer cells *in vitro*. Bombesin is a neuropeptide that was first isolated from the skin of the frog *Bombin bombina* and later found to have a mammalian equivalent, gastrin-releasing peptide (GRP), that binds a family of GPCRs (46). GRP receptor overexpression has been identified in prostate cancer and bombesin has been shown to influence prostate cancer cell migration *in vitro* (47, 48). Madarame et al. investigated in involvement of HB-EGF shedding in bombesin stimulated prostate cell migration (49). The addition of a metalloprotease inhibitor decreased bombesin induced HB-EGF shedding and EGFR tyrosine phosphorylation. Addition of either an anti-EGFR antibody or the metalloprotease inhibitor partially reduced migration of bombesin stimulated cells in an *in vitro* wound assay.

EGFR autocrine signaling has recently been implicated in the role of TNF-alpha induced cell motility (50). TNF-alpha is a cytokine that has been shown to induce cytotoxicity and apoptosis in transformed cells while it is also thought to illicit pro-survival signals in normal cells (51, 52). Chen et al. investigated the role of EGFR crosstalk in mediating TNF induced signals in a normal hMEC cell line (50). TNF stimulated hMEC proliferation and migration in an *in vitro* transwell migration assay. The addition of TNF also stimulated shedding of TGF-alpha in a dose-dependent manner. TNF stimulated cell migration was decreased in the presence of a metalloprotease inhibitor or an EGFR kinase inhibitor. A late phase of ERK activation was also inhibited in the presence of the metalloprotease inhibitor suggesting that autocrine activation of the EGFR was involved in this secondary ERK signaling peak. These results implicated a novel role for EGFR transactivation in mediating TNF induced cell responses in this hMEC system.

2.2 *In Vivo*

As discussed earlier, Xie et al. demonstrated that endogenous EGFR autocrine signaling stimulated engineered DU-145 prostate carcinoma cell migration *in vitro* (36). Subsequently, Turner et al. inoculated athymic mice with the parental cells and the EGFR overexpressing cells to determine if the results of increased transwell migration *in vitro* corresponded to *in vivo* tumor progression (53). Both the parental and EGFR overexpressing DU-145 cells formed tumors that metastasized to the lung when inoculated into the prostate and peritoneal cavity, although the EGFR overexpressing tumors were more invasive. Injections of a PLC-gamma inhibitor reduced tumor invasiveness, as measured by the extent of tumor cell penetration of the diaphragm, suggesting a role for EGFR-mediated cell migration in this *in vivo* tumor invasion model.

Pilcher et al. demonstrated that keratinocyte migration depended on EGFR signaling using a phagokinetic assay (54). Inhibition of the EGFR using either a kinase inhibitor or an anti-EGFR antibody decreased the relative migration area and the production of collagenase-1 in a primary keratinocyte cell system shown to express EGF, TGF-alpha, amphiregulin, and HB-EGF when plated on type I collagen. These results were tested in an *in vivo* porcine burn wound-healing assay. Wound incubation with an EGFR inhibitor resulted in significant decrease in burn re-epithelialization compared to control treatments, implicating a role for EGFR autocrine signaling in burn wound closure. In another study, Tokumaru et al. used a murine punch biopsy wound model to investigate the importance of EGFR ligand shedding in *in vivo* cutaneous wound healing (5). Incubation with a metalloprotease inhibitor resulted in a lack of keratin staining in excised skin wound histology in contrast to the control treatment. These inhibitory effects were reversed upon addition of recombinant HB-EGF, suggesting that the shedding of EGFR ligands may play an important role in keratinocyte migration and would healing.

The AP-1 stress response transcription factor is composed of a heterodimer of Fos and Jun proteins and has been implicated in the normal development of the epidermis (55, 56). The EGFR and its ligand HB-EGF are both known AP-1 target genes (57, 58). Li et al. created a transgenic mouse lacking *c-jun* in the epidermis (59). These transgenic mice were born with open eyelids, a phenotypic result also seen in mice born with gene disruption of the EGFR or TGF-alpha (60, 61). The conditionally *c-jun* null mice also had decreased EGFR expression and phosphorylation at the epidermal eyelid tip, and had slower wound closure rates of punch biopsies. Addition of exogenous HB-EGF rescued the impaired motility of *c-jun* null keratinocytes in an *in vitro* scratch assay. Together these results suggest that the activation of positive EGFR feedback is involved in leading edge epidermal sheet migration (62).

2.3 Examples of Other Growth Factor Families

2.3.1 FGF

Basic Fibroblast Growth Factor (bFGF, also known as FGF-2) is a member of a large family of heparin-binding growth factors and is known to have various influences on several cell types (63). In particular, FGF signaling is known to be involved in the migration of endothelial cells and is a therapeutic target for inhibition of angiogenesis in tumor progression (64). Endogenous bFGF was shown to regulate endothelial cell movement in an *in vitro* scratch assay (65). Migration of confluent bovine aortic endothelial cells past a razor-induced wound edge was inhibited upon addition of an anti-bFGF antibody. The addition of soluble recombinant bFGF reversed the inhibitory effects of the ligand-neutralizing antibody. Although this work was the first documented evidence of autocrine activity of the bFGF ligand, it was not known how the ligand reached the extracellular space and whether the method for cell removal may have caused cytoplasmic ligand leakage from dead cells at the wound edge.

Mignatti et al. later validated that bFGF stimulated cell motility of cultured fibroblasts in an autocrine manner (66). NIH 3T3 cells were transfected with bFGF cDNA and single cells were plated on colloidal gold-coated coverslips. Increasing phagokinetic cell track areas were correlated with the level of endogenous bFGF. The cell clone that produced the highest amount of bFGF did not show an increase in migration upon addition of exogenous, recombinant bFGF, which suggests that the cells had reached a state of receptor saturation or a maximum migratory stimulation at this level of endogenous ligand production. Addition of anti-bFGF reduced the motility of the isolated, transfected cells. These results indicated that extracellular bFGF could stimulate cell motility of the cell producing the ligand, experimentally indicating that the bFGF ligand can act in a 'true' autocrine manner.

2.3.2 G-CSF/GM-CSF

Granulocyte colony-stimulating factor (G-CSF) and granulocyte macrophage colony-stimulating factor (GM-CSF) stimulate growth and differentiation in the hematopoietic system, are produced by many cell types, and have also been found to be expressed by non-hematopoietic tumor cells suggesting their involvement in tumor growth and invasion (67-70). Mueller et al. analyzed RNA and protein expression of G-CSF, GM-CSF, and their receptors in 22 human gliomas and derived cell cultures (71). The co-expression of these ligands and their cognate receptors correlated with advanced tumor stage. Receptor and ligand expression also influenced the migration of the glioblastoma cell lines as shown in an *in vitro* wound assay. A derived cell line expressing both ligands and receptors had an increase in both the distance traveled and number of cells moving into the wounded area compared to a similar cell line expressing only G-CSF and both receptors. The addition of either an anti G-CSF or an anti GM-CSF monoclonal antibody reduced migration of the wound edge. These results suggest that G-CSF and GM-CSF autocrine signaling may contribute to increased cell migration in late stage glioblastomas.

2.3.3 HGF

Hepatocyte growth factor (HGF) is secreted by cells of mesodermal and mesenchymal origin, stimulates growth and migration of epithelial and endothelial cells that express the HGF receptor (HGFR), and HGFR signaling has been shown to be involved in processes such as placental development and liver regeneration (72, 73). Overexpression of HGFR, also known as the *met* proto-oncogene, lead to aberrant growth formation in a mouse model and HGFR has been found to be overexpressed in several types of cancer (74-76). Vadnais et al. used Moloney sarcoma virus (MSV) transformed polarized epithelial MDCK cells to determine that the autocrine activation of HGFR is associated with the pseudopodial protrusion and acquisition of a motile, invasive phenotype *in vitro* (77). The invasive variant derived from this cell line showed constitutive HGFR phosphorylation and increased migration distances as measured using videomicroscopy and nuclear tracking. The addition of an anti-HGF-alpha antibody decreased cell motility, increased pseudopodial retraction and the number of cell-cell contacts of the MSV-MDCK invasive variant.

2.3.4 VEGF

The vascular endothelial growth factor (VEGF) is known to play an important role in angiogenesis and increased expression has been associated with cancer progression (78-81). A recent study looked at the role of VEGF and other neurophilin-1 (NP1) ligands in the regulation of breast carcinoma cell migration towards conditioned media *in vitro* (82). An antagonistic relationship exists between VEGF and the migratory inhibitor SEMA3A. The autocrine production ratio of these competing ligands correlated with chemotactic potential of three breast carcinoma cell lines. RNAi reduction of VEGF and expression of a mutant, constitutively active NP1 coreceptor plexin-A1 decreased migration while RNAi reduction of NP1 and SEMA3A expression increased transwell migration. This work demonstrated the involvement of VEGF autocrine signaling in breast carcinoma chemotaxis through the inhibition of an endogenous migration suppressor (81).

3. SPATIAL LOCALIZATION ISSUES

Up to this point, we have focused on evidence in the literature for involvement of autocrine growth factor signaling in cell motility phenomena. We wish to now consider key aspects of mechanisms underlying these phenomena, focusing our attention on features of autocrine signaling systems that might be of particular importance for governing cell migration behavior. These features primarily revolve around the spatial distribution of autocrine ligands, their receptors, and generated downstream signals, because the central characteristic of autocrine systems distinguishing them from paracrine and/or endocrine systems is their potential action on the producing cell or proximal neighbors.

3.1 Cellular Control of Growth Factor Spatial Distribution

Autocrine signaling has typically been difficult to investigate using standard techniques since the production, release, and binding of a ligand within a cell operate in a 'closed loop' fashion (32). Therefore, several modeling approaches have been used to gain insight about the parameters that govern the dynamics of autocrine operation complemented by a growing body of experimental work. Two modeling approaches, one that used Brownian-motion theory and computer simulations to calculate the trajectories of the released ligands, and another that used continuum reaction-diffusion equations to model the fluxes of the ligands, both suggest that EGFR autocrine loops could be highly localized on the order of less than a cell diameter (83, 84). Using a microphysiometer assay to measure receptor ligand binding, Lauffenburger et al. showed that anti-receptor antibodies were far more effective than anti-ligand antibodies in inhibiting autocrine signaling, suggesting experimentally that these signals may operate in a spatially-restricted, local manner (85). The modeling work by Shvartsman et al. also implicated that altering the ligand diffusion coefficient, the density of cell surface receptors, the ligand secretion rate, and the rate constants for ligand binding and endocytic internalization could modulate the spatial range of the growth factors (84). Oehrtman et al. used a tetracycline-controlled TGF-alpha expression system in mouse B82L fibroblasts transfected with an EGFR gene to validate a model for the escape of autocrine ligands into the extracellular bulk media (86). Their experimental results suggested that the ligand secretion rate, receptor availability, and cell density in culture controlled the rate of ligand accumulation in the media.

DeWitt et al. performed a rigorous set of experiments to show that the fraction of ligand that is captured by the producing cells is a function of ligand secretion rate and receptor synthesis rate, using a human EGF/EGFR autocrine loop engineered into mouse fibroblasts (87). In this study, the ligand production rate was varied using the tet-off expression system and the ligand release rate was modulated by addition of metalloprotease inhibitors. The number of surface accessible receptors was varied using an EGFR blocking antibody. The ligand secretion rate, VL, was measured using an EGF ELISA, and the appearance rate of the receptors, VR, was calculated under specific antibody concentrations by solving a kinetic model describing receptor-ligand, receptor-antibody binding, and trafficking at steady state. The fraction of total and surface receptors occupied was measured by analyzing microphysiometer-based assay results using a quantitative model of ligand release and receptor dynamics (85). At VL/VR ratios of less than 0.3, almost no ligand was found in the conditioned medium, however, 30-40% of the receptors were occupied. At levels of ligand secretion sufficient to occupy >90% of the receptors, the fraction of ligand captured dropped to <10% and the VL/VR ratio increased to >1. These results suggested that a significant amount of autocrine signaling could occur even when the amount of ligand in the conditioned medium is close to the limits of experimental detection. Later work by DeWitt et al. showed that ligand-receptor affinity also regulates the spatial range of the fraction of ligand captured using an EGF mutant with a lower affinity for the EGFR (88). The mutant ligand was captured less efficiently, shifting the relationship between VL/VR and the fraction of ligand captured.

Recent computational work by Maly et al. also predicted that single cells are capable of achieving autocrine signaling on a dimension smaller than the cell diameter (89). Depending on the state of several cellular parameters, from an initial state of uniform ligand release and intracellular signaling, this dynamic model of extracellular and intracellular EGFR signaling could achieve a state of no signaling, uniform signaling, or a state of steady polarized ligand release and downstream signaling on a subcellular scale. In this model ligand shedding rate, receptor density, the strength of intracellular negative feedback, and the concentration of adaptor molecules in the EGFR signaling cascade were all sensitive parameters that determined the final signaling state.

Together these modeling and experimental results begin to describe how the spatial range of growth factor signaling is dependent on several cellular parameters. *Drosophila* oogenesis is an example of a complex process that is thought to require the production of multiple growth factors and both paracrine and autocrine signaling events for proper development of two dorsal eggshell appendages (7, 90). While the spatial range and fraction of extracellular ligand that is captured by the producing cell may vary, it is important to also consider how the spatial distribution of available ligand translates to the distribution of receptor activation and intracellular signaling events and subsequent cell behavior.

3.2 Spatial Range Of Ligand Signaling Events

Does the spatial distribution of ligand signaling affect downstream signaling and potentially cell behavior? Since previous work had implicated that EGF-induced de-adhesion during cell motility required the activation of calpain, Glading et al. investigated the importance of intermediate ERK localization on this regulation (91, 92). An internalization-deficient EGFR construct was able to activate calpain, while both membrane and cytosolic localized EGFR was shown to activate ERK. Membrane-targeted ERK was sufficient for calpain activity and cell de-adhesion, however when membrane-associated ERK was sequestered, EGFR-mediated calpain activation and de-adhesion was reduced. Results from Kempiak et al. showed that cells stimulated locally with EGF coated beads had less diffuse activated ERK when compared to cells stimulated with exogenous EGF, which suggests that active ERK may also remain somewhat localized in response to a polarized stimulus (93). Work by Haugh et al. revealed that PLC-gamma, another important signaling pathway downstream of the EGFR that is involved in cell motility, could be activated both at the cell membrane and by internalized receptors but its action of PIP2 hydrolysis appeared to be localized to the cell membrane (94). Together these results suggest that ERK, calpain and PLC-gamma, all key signaling components downstream of EGFR that are involved in cell motility, may play mechanistic roles in effectively transferring localized growth factor signaling inputs into cellular outcomes (95, 96).

Short range activation and long range inhibition of specific signaling molecules is thought to be another mechanism by which cells are able to receive and interpret an asymmetric stimulus such as a chemotactic gradient (97, 98). Phosphoinositide-3 Kinase (PI3K) is activated by the EGFR and is known to be a major player in cell polarization as well as membrane protrusion and migration (98, 99). While PI3K may act locally downstream of active EGF receptors, PTEN, which is the primary phosphatase for the PI3K product phosphatidylionsitol (3,4,5)-triphosphate, is thought to work as the global inhibitor (100). For example, PI3K activation is found localized at the cell membrane closest to the chemoattractant, while PTEN is localized to the cell rear in *Dictyostelium discoideum* amoebae chemotaxis (98). PI3K has also been implicated in cell polarity and chemotaxis of mammalian cells (99, 101, 102).

It is also interesting to point out that while growth factors may obtain various spatial distributions and may be polarized at the cell membrane, and downstream signals may potentially propagate this input, the EGFR itself may also play a role in signaling distribution. Verveer et al. first proposed that ligand-independent EGFR activation could spread laterally on the cell surface when cells were stimulated locally with EGF-coated beads (103). Work by Sawano et al. later showed that local activation of the EGFR, achieved by stimulating the edge of single cells in laminar flows containing EGF, could lead to lateral spread of downstream signaling over the entire cell, but that this process only occurred at high receptor densities or when receptor endocytosis was inhibited (104). More recent studies have shown that local stimulation with EGF-coated beads led to actin polymerization and membrane protrusion at the point of bead contact, suggesting that under these circumstances local signaling lead to a local response that is relevant to cell migration (93, 105). It is still under investigation whether stimulation passes through the membrane via EGFR activation locally or ligand-independent lateral propagation of EGFR activation occurs, and this mechanism could be dependent on cellular parameters such as receptor density (106, 107).

3.3 The Mode of Growth Factor Presentation Affects Cell Migration Behavior

Maheshwari et al. used a human mammary epithelial cell line transfected with one of two different EGF chimeras or the addition of exogenous EGF to investigate the effect ligand presentation mode (i.e. autocrine vs. intracrine vs. paracrine) on cell motility (108). The first chimera, EGF-Ct, encoded for the receptor binding EGF ligand domain as well as the cytoplasmic and transmembrane domains of pro-EGF, which required proteolytic cleavage at the plasma membrane prior to receptor binding and was capable of stimulating cells in an autocrine manner. The EGF-Ct cells did not stimulate the migration of neighboring parental cells, while the addition of exogenous EGF resulted in stimulated parental cell migration, see Figure 2. The second chimera, sEGF, encoded for the mature EGF ligand but lacked a membrane-anchoring domain and was thought to stimulate the producing cells in an intracrine fashion. Maheshwari et al. measured the migration behavior of the parental, EGF-Ct, and sEGF hMEC cell lines *in vitro*, analyzed individual cell tracks, and found that the mode of ligand presentation affected the cell speed and migration persistence time.

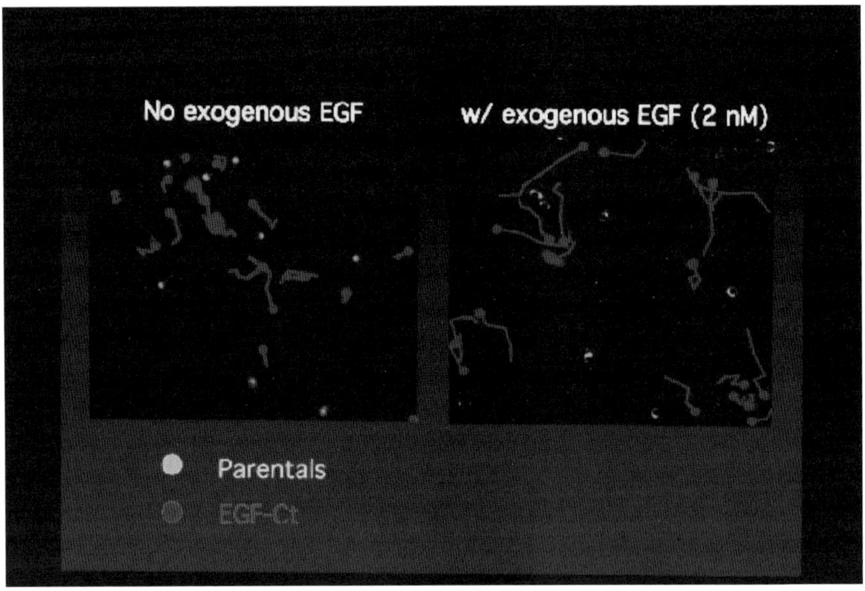

Figure 2. Autocrine producing EGF-Ct cells do not stimulate migration of neighboring parental hMEC cells. The initial location of EGF-Ct (orange) and parental hMEC (green) cells and their subsequent cell paths are shown in the absence (left, Control) and presence (right, +EGF) of 2 nM exogenous EGF. (Adapted from Maheshwari et al., 2001). See also Color Plate Section page 345.

Expression of membrane bound EGF (EGF-Ct) and the addition of exogenous EGF both increased hMEC cell speed in 2D, which was inhibited with an EGFR blocking antibody. The sEGF-expressing cells had a basal migration speed that was decreased upon anti-EGFR antibody addition, suggesting that the surface receptor-ligand complexes governed cell migration speed. The most intriguing result from this study was that the migration persistence, or the approximate time a cell traveled before changing direction significantly, of the autocrine EGF-Ct expressing cells was dramatically increased compared to the parental cells in the presence of exogenous EGF. This increased persistence time was abolished upon addition of an EGFR blocking antibody or exogenous EGF.

The results from Maheshwari et al. suggest that the expression of membrane-bound EGF in the EGF-Ct hMEC cell line may lead to spatially restricted EGFR signaling that drives persistent migration behavior. The persistent migration may be the result of an asymmetrical distribution of EGFR signaling that stimulates further directed migration. The asymmetry could result from polarized distributions of the ligand, the receptor, or an asymmetric release of the ligand. EGFR ligand autocrine signaling spatially restricted to the front of the cell could stimulate membrane protrusion, thus producing what might be thought of as a local "pseudo-chemotaxis' response" (99, 105). Although it was shown that the EGF-Ct chimera-expressing, autocrine cells had persistent migration behavior that was abolished under exogenous EGF/paracrine-like conditions, it is not yet known if the signaling is indeed spatially restricted or how intermediate signaling modes would alter the migration behavior of this experimental system. While previous results have shown that ligand presentation mode can affect cell behaviors such as cell organization (109), these results conclude that the mode of ligand presentation can alter cell migration behavior.

4. FUTURE PERSPECTIVE

Although the number of growth factor receptor systems in which autocrine signaling has been implicated to date is relatively small, we anticipate that in the coming decade many more will become identified. This expectation is based on the unusual difficulty inherent in ascertaining the presence of autocrine ligands, which we believe will be overcome by the development and application of new experimental methodologies. These should include transcriptional profiling techniques which can discover the simultaneous expression of a ligand and its cognate receptor, mass spectrometry proteomic techniques which can analogously elucidate simultaneous ligand and receptor expression at the protein level, and new micron-scale biochemical assays. Moreover, we speculate that a significant proportion of these autocrine circuits will be found to play a role in driving and/or orienting cell migration, because of the attendant spatial localization capabilities. These effects will likely be important in development, tissue homeostasis, and pathological dysregulation, similarly to the currently identified autocrine circuits, and thus will be worthy of substantial further research.

ACKNOWLEDGEMENTS

This contribution was partially supported by an NIGMS grant to D.A. Lauffenburger along with H.S. Wiley, an NCI grant to D.A. Lauffenburger, and a Whitaker Foundation Graduate Fellowship to E.J. Joslin. The authors thank Steve Wiley, Alan Wells, Greg Oehrtman, Ann Dewitt, Gargi Maheshwari, Stas Shvartsman, and Ivan Maly for helpful discussions on this topic over the past number of years.

REFERENCES

1. Sporn MB, Todaro GJ. Autocrine secretion and malignant transformation of cells. N. Engl. J. Med. 1980, 303: 878-80.
2. Hanahan D, Weinberg RA. The hallmarks of cancer. Cell 2000, 100: 57-70.
3. Sporn MB, Roberts AB. Autocrine growth factors and cancer. Nature 1985, 313: 745-7.
4. Sporn MB, Roberts AB. Autocrine secretion–10 years later. Ann. Intern. Med. 1992, 117: 408-14.
5. Tokumaru S, et al. Ectodomain shedding of epidermal growth factor receptor ligands is required for keratinocyte migration in cutaneous wound healing. J. Cell Biol. 2000, 151: 209-20.
6. Seghezzi G, et al. Fibroblast growth factor-2 (FGF-2) induces vascular endothelial growth factor (VEGF) expression in the endothelial cells of forming capillaries: an autocrine mechanism contributing to angiogenesis. J. Cell Biol. 1998, 141: 1659-73.
7. Wasserman JD, Freeman M. An autoregulatory cascade of EGF receptor signaling patterns the Drosophila egg. Cell 1998, 95: 355-64.
8. Yarden Y, Sliwkowski MX. Untangling the ErbB signalling network. Nat. Rev. Mol. Cell Biol. 2001, 2: 127-37.
9. Caric D, et al. EGFRs mediate chemotactic migration in the developing telencephalon. Development 2001, 128: 4203-16.
10. Kumar R, Wang RA. Protein kinases in mammary gland development and cancer. Microsc. Res. Tech. 2002, 59: 49-57.
11. Holbro T, Civenni G, Hynes NE. The ErbB receptors and their role in cancer progression. Exp. Cell Res. 2003, 284: 99-110.
12. Arteaga CL. Epidermal growth factor receptor dependence in human tumors: more than just expression? Oncologist 7 Suppl. 2002, 4: 31-9.
13. Hsieh ET, Shepherd FA, Tsao MS. Co-expression of epidermal growth factor receptor and transforming growth factor-alpha is independent of ras mutations in lung adenocarcinoma. Lung Cancer 2000, 29: 151-7.
14. Suo Z, et al. The expression of EGFR family ligands in breast carcinomas. Int. J. Surg. Pathol. 2002, 10: 91-9.
15. Cai YC, et al. Expression of transforming growth factor-alpha and epidermal growth factor receptor in gastrointestinal stromal tumours. Virchows Arch. 1999, 435: 112-5.
16. Kopp R, et al. Clinical implications of the EGF receptor/ligand system for tumor progression and survival in gastrointestinal carcinomas: evidence for new therapeutic options. Recent Results Cancer Res. 2003, 162: 115-32.

17. Yamanaka Y, et al. Coexpression of epidermal growth factor receptor and ligands in human pancreatic cancer is associated with enhanced tumor aggressiveness. Anticancer Res. 1993, 13: 565-9.
18. Zhu Z, et al. Epiregulin is Up-regulated in pancreatic cancer and stimulates pancreatic cancer cell growth. Biochem. Biophys. Res. Commun. 2000, 273: 1019-24.
19. Damstrup L, et al. Amphiregulin acts as an autocrine growth factor in two human polarizing colon cancer lines that exhibit domain selective EGF receptor mitogenesis. Br. J. Cancer 1999, 80: 1012-9.
20. Rubin Grandis J, et al. Levels of TGF-alpha and EGFR protein in head and neck squamous cell carcinoma and patient survival. J. Natl. Cancer Inst. 1998, 90: 824-32.
21. Salomon DS, et al. Epidermal growth factor-related peptides and their receptors in human malignancies. Crit. Rev. Oncol. Hematol. 1995, 19: 183-232.
22. Harris RC, Chung E, Coffey RJ. EGF receptor ligands. Exp. Cell Res. 2003, 284: 2-13.
23. Arribas J, Lopez-Casillas F, Massague J. Role of the juxtamembrane domains of the transforming growth factor-alpha precursor and the beta-amyloid precursor protein in regulated ectodomain shedding. J. Biol. Chem. 1997, 272: 17160-5.
24. Merlos-Suarez A, et al. Metalloprotease-dependent protransforming growth factor-alpha ectodomain shedding in the absence of tumor necrosis factor-alpha-converting enzyme. J. Biol. Chem. 2001, 276: 48510-7.
25. Borrell-Pages M, et al. TACE is required for the activation of the EGFR by TGF-alpha in tumors. Embo. J. 2003, 22: 1114-24.
26. Hinkle CL, et al. Multiple metalloproteinases process protransforming growth factor-alpha (proTGF-alpha). Biochemistry 2003, 42: 2127-36.
27. Baselga J, et al. Autocrine regulation of membrane transforming growth factor-alpha cleavage. J. Biol. Chem. 1996, 271: 3279-84.
28. Fan H, Derynck R. Ectodomain shedding of TGF-alpha and other transmembrane proteins is induced by receptor tyrosine kinase activation and MAP kinase signaling cascades. Embo. J. 1999, 18: 6962-72.
29. Gechtman Z, et al. The shedding of membrane-anchored heparin-binding epidermal-like growth factor is regulated by the Raf/mitogen-activated protein kinase cascade and by cell adhesion and spreading. J. Biol. Chem. 1999, 274: 28828-35.
30. Fan H, Turck CW, Derynck R. Characterization of growth factor-induced serine phosphorylation of tumor necrosis factor-alpha converting enzyme and of an alternatively translated polypeptide. J. Biol. Chem. 2003, 278: 18617-27.
31. Schulze A, et al. Analysis of the transcriptional program induced by Raf in epithelial cells. Genes. Dev. 2001, 15: 981-94.
32. Wiley HS, Shvartsman SY, Lauffenburger DA. Computational modeling of the EGF-receptor system: a paradigm for systems biology. Trends Cell Biol. 2003, 13: 43-50.
33. Gavrilovic J, et al. Expression of transfected transforming growth factor alpha induces a motile fibroblast-like phenotype with extracellular matrix-degrading potential in a rat bladder carcinoma cell line. Cell Regul. 1990, 1: 1003-14.
34. Kim HG, et al. EGF receptor signaling in prostate morphogenesis and tumorigenesis. Histol. Histopathol. 1999, 14: 1175-82.
35. Tillotson JK, Rose DP. Endogenous secretion of epidermal growth factor peptides stimulates growth of DU145 prostate cancer cells. Cancer Lett. 1991, 60: 109-12.
36. Xie H, et al. In vitro invasiveness of DU-145 human prostate carcinoma cells is modulated by EGF receptor-mediated signals. Clin. Exp. Metastasis 1995, 13: 407-19.

37. Ekstrand AJ, et al. Genes for epidermal growth factor receptor, transforming growth factor alpha, and epidermal growth factor and their expression in human gliomas in vivo. Cancer Res. 1991, 51: 2164-72.
38. Schlegel U, et al. Expression of transforming growth factor alpha in human gliomas. Oncogene 1990, 5: 1839-42.
39. El-Obeid A, et al. Cell scattering and migration induced by autocrine transforming growth factor alpha in human glioma cells in vitro. Cancer Res. 1997, 57: 5598-604.
40. Dong J, et al. Metalloprotease-mediated ligand release regulates autocrine signaling through the epidermal growth factor receptor. Proc. Natl. Acad. Sci. U S A 1999, 96: 6235-40.
41. Block ER, et al. Wounding induces motility in sheets of corneal epithelial cells through loss of spatial constraints: role of heparin-binding epidermal growth factor-like growth factor signaling. J. Biol. Chem. 2004, 279: 24307-12.
42. Hirano T. Interleukin 6 and its receptor: ten years later. Int. Rev. Immunol. 1998, 16: 249-84.
43. Badache A, Hynes NE. Interleukin 6 inhibits proliferation and, in cooperation with an epidermal growth factor receptor autocrine loop, increases migration of T47D breast cancer cells. Cancer Res. 2001, 61: 383-91.
44. Gschwind A, et al. Cell communication networks: epidermal growth factor receptor transactivation as the paradigm for interreceptor signal transmission. Oncogene 2001, 20: 1594-600.
45. Gschwind A, et al. TACE cleavage of proamphiregulin regulates GPCR-induced proliferation and motility of cancer cells. Embo. J. 2003, 22: 2411-21.
46. Anastasi A, Erspamer V, Bucci M. Isolation and structure of bombesin and alytesin, 2 analogous active peptides from the skin of the European amphibians Bombina and Alytes. Experientia 1971, 27: 166-7.
47. Markwalder R, Reubi JC. Gastrin-releasing peptide receptors in the human prostate: relation to neoplastic transformation. Cancer Res. 1999, 59: 1152-9.
48. Aprikian AG, et al. Bombesin stimulates the motility of human prostate-carcinoma cells through tyrosine phosphorylation of focal adhesion kinase and of integrin-associated proteins. Int. J. Cancer. 1997, 72: 498-504.
49. Madarame J, et al. Transactivation of epidermal growth factor receptor after heparin-binding epidermal growth factor-like growth factor shedding in the migration of prostate cancer cells promoted by bombesin. Prostate 2003, 57: 187-95.
50. Chen WN, et al. Induced autocrine signaling through the epidermal growth factor receptor contributes to the response of mammary epithelial cells to tumor necrosis factor alpha. J. Biol. Chem. 2004, 279: 18488-96.
51. Carswell EA, et al. An endotoxin-induced serum factor that causes necrosis of tumors. Proc. Natl. Acad. Sci. U S A 1975, 72: 3666-70.
52. Natoli G, et al. Apoptotic, non-apoptotic, and anti-apoptotic pathways of tumor necrosis factor signalling. Biochem. Pharmacol. 1998, 56: 915-20.
53. Turner T, et al. EGF receptor signaling enhances in vivo invasiveness of DU-145 human prostate carcinoma cells. Clin. Exp. Metastasis. 1996, 14: 409-18.
54. Pilcher BK, et al. Keratinocyte collagenase-1 expression requires an epidermal growth factor receptor autocrine mechanism. J. Biol. Chem. 1999, 274: 10372-81.
55. Angel P, Szabowski A. Function of AP-1 target genes in mesenchymal-epithelial cross-talk in skin. Biochem. Pharmacol. 2002, 64: 949-56.
56. Yates S, Rayner TE, Transcription factor activation in response to cutaneous injury: role of AP-1 in reepithelialization. Wound Repair Regen. 2002, 10: 5-15.

57. Fu S, et al. Heparin-binding epidermal growth factor-like growth factor, a v-Jun target gene, induces oncogenic transformation. Proc. Natl. Acad. Sci. U S A 1999, 96: 5716-21.
58. Johnson AC, et al. Activator protein-1 mediates induced but not basal epidermal growth factor receptor gene expression. Mol. Med. 2000, 6: 17-27.
59. Li G, et al. c-Jun is essential for organization of the epidermal leading edge. Dev. Cell 2003, 4: 865-77.
60. Miettinen PJ, et al. Epithelial immaturity and multiorgan failure in mice lacking epidermal growth factor receptor. Nature 1995, 376: 337-41.
61. Luetteke NC, et al. TGF alpha deficiency results in hair follicle and eye abnormalities in targeted and waved-1 mice. Cell 1993, 73: 263-78.
62. Grose R. Epithelial migration: open your eyes to c-Jun. Curr. Biol. 2003, 13: R678-80.
63. Ornitz DM, Itoh N. Fibroblast growth factors. Genome. Biol. 2001, 2: REVIEWS3005.
64. Cross MJ, Claesson-Welsh L. FGF and VEGF function in angiogenesis: signalling pathways, biological responses and therapeutic inhibition. Trends Pharmacol. Sci. 2001, 22: 201-7.
65. Sato Y, and Rifkin DB. Autocrine activities of basic fibroblast growth factor: regulation of endothelial cell movement, plasminogen activator synthesis, and DNA synthesis. J. Cell Biol. 1988, 107: 1199-205.
66. Mignatti P, Morimoto T, Rifkin DB. Basic fibroblast growth factor released by single, isolated cells stimulates their migration in an autocrine manner. Proc. Natl. Acad. Sci. U S A 1991, 88: 11007-11.
67. Metcalf D. The granulocyte-macrophage colony-stimulating factors. Science 1985, 229: 16-22.
68. Tachibana M, et al. Autocrine growth of transitional cell carcinoma of the bladder induced by granulocyte-colony stimulating factor. Cancer Res. 1995, 55: 3438-43.
69. Mueller MM, et al. Tumor progression of skin carcinoma cells in vivo promoted by clonal selection, mutagenesis, and autocrine growth regulation by granulocyte colony-stimulating factor and granulocyte-macrophage colony-stimulating factor. Am. J. Pathol. 2001, 159: 1567-79.
70. Uemura Y, et al. Effect of serum deprivation on constitutive production of granulocyte-colony stimulating factor and granulocyte macrophage-colony stimulating factor in lung cancer cells. Int. J. Cancer 2004, 109: 826-32.
71. Mueller MM, et al. Autocrine growth regulation by granulocyte colony-stimulating factor and granulocyte macrophage colony-stimulating factor in human gliomas with tumor progression. Am. J. Pathol. 1999, 155: 1557-67.
72. Uehara Y, et al. Placental defect and embryonic lethality in mice lacking hepatocyte growth factor/scatter factor. Nature 1995, 373: 702-5.
73. Borowiak M, et al. Met provides essential signals for liver regeneration. Proc. Natl. Acad. Sci. U S A 2004, 101: 10608-13.
74. Takayama H, et al. Diverse tumorigenesis associated with aberrant development in mice overexpressing hepatocyte growth factor/scatter factor. Proc. Natl. Acad. Sci. U S A 1997, 94: 701-6.
75. Tuck AB, et al. Coexpression of hepatocyte growth factor and receptor (Met) in human breast carcinoma. Am. J. Pathol. 1996, 148: 225-32.
76. Humphrey PA, et al. Hepatocyte growth factor and its receptor (c-MET) in prostatic carcinoma. Am. J. Pathol. 1995, 147: 386-96.

77. Vadnais J, et al. Autocrine activation of the hepatocyte growth factor receptor/met tyrosine kinase induces tumor cell motility by regulating pseudopodial protrusion. J. Biol. Chem. 2002, 277: 48342-50.
78. Shweiki D, et al. Vascular endothelial growth factor induced by hypoxia may mediate hypoxia-initiated angiogenesis. Nature 1992, 359: 843-5.
79. Brown LF, et al. Vascular stroma formation in carcinoma in situ, invasive carcinoma, and metastatic carcinoma of the breast. Clin. Cancer Res. 1999, 5: 1041-56.
80. Salven P, et al. Serum VEGF levels in women with a benign breast tumor or breast cancer. Breast Cancer Res. Treat. 1999, 53: 161-6.
81. Mercurio AM, et al. Autocrine signaling in carcinoma: VEGF and the alpha6beta4 integrin. Semin. Cancer Biol. 2004, 14: 115-22.
82. Bachelder RE, et al. Competing autocrine pathways involving alternative neuropilin-1 ligands regulate chemotaxis of carcinoma cells. Cancer Res. 2003, 63: 5230-3.
83. Forsten KE, Lauffenburger DA. Autocrine ligand binding to cell receptors. Mathematical analysis of competition by solution "decoys". Biophys. J. 1992, 61: 518-29.
84. Shvartsman SY, et al. Spatial range of autocrine signaling: modeling and computational analysis. Biophys. J. 2001, 81: 1854-67.
85. Lauffenburger DA, et al. Real-time quantitative measurement of autocrine ligand binding indicates that autocrine loops are spatially localized. Proc. Natl. Acad. Sci. U S A 1998, 95: 15368-73.
86. Oehrtman GT, Wiley HS, Lauffenburger DA. Escape of autocrine ligands into extracellular medium: experimental test of theoretical model predictions. Biotechnol. Bioeng. 1998, 57: 571-82.
87. DeWitt AE, et al. Quantitative analysis of the EGF receptor autocrine system reveals cryptic regulation of cell response by ligand capture. J. Cell Sci. 2001, 114: 2301-13.
88. DeWitt A, et al. Affinity regulates spatial range of EGF receptor autocrine ligand binding. Dev. Biol. 250: 305-16.
89. Maly IV, Wiley HS, Lauffenburger DA. Self-organization of polarized cell signaling via autocrine circuits: computational model analysis. Biophys. J. 2004, 86: 10-22.
90. Shvartsman SY, Muratov CB, and Lauffenburger DA. Modeling and computational analysis of EGF receptor-mediated cell communication in Drosophila oogenesis. Development 2002, 129: 2577-89.
91. Glading A, et al. Epidermal growth factor receptor activation of calpain is required for fibroblast motility and occurs via an ERK/MAP kinase signaling pathway. J. Biol. Chem. 2000, 275: 2390-8.
92. Glading A, et al. Membrane proximal ERK signaling is required for M-calpain activation downstream of epidermal growth factor receptor signaling. J. Biol. Chem. 2001, 276: 23341-8.
93. Kempiak SJ, et al. Local signaling by the EGF receptor. J. Cell Biol. 2003, 162: 781-7.
94. Haugh JM, et al. Effect of epidermal growth factor receptor internalization on regulation of the phospholipase C-gamma1 signaling pathway. J. Biol. Chem. 1999, 274: 8958-65.
95. Glading A, Lauffenburger DA, Wells A. Cutting to the chase: calpain proteases in cell motility. Trends Cell Biol. 2002, 12: 46-54.
96. Wells A, Grandis JR. Phospholipase C-gamma1 in tumor progression. Clin. Exp. Metastasis. 2003, 20: 285-90.
97. Meinhardt H. Orientation of chemotactic cells and growth cones: models and mechanisms. J. Cell Sci. 1999, 112: 2867-74.

98. Iijima M, Huang YE, Devreotes P. Temporal and spatial regulation of chemotaxis. Dev. Cell 2002, 3: 469-78.
99. Condeelis JS, et al. Lamellipodia in invasion. Semin. Cancer Biol. 2001, 11: 119-28.
100. Sulis ML, Parsons R. PTEN: from pathology to biology. Trends Cell Biol. 2003, 13: 478-83.
101. Wang F, et al. Lipid products of PI(3)Ks maintain persistent cell polarity and directed motility in neutrophils. Nat. Cell Biol. 2002, 4: 513-8.
102. Sawyer C, et al. Regulation of breast cancer cell chemotaxis by the phosphoinositide 3-kinase p110delta. Cancer Res. 2003, 63: 1667-75.
103. Verveer PJ, et al. Quantitative imaging of lateral ErbB1 receptor signal propagation in the plasma membrane. Science 2000, 290: 1567-70.
104. Sawano A, et al. Lateral propagation of EGF signaling after local stimulation is dependent on receptor density. Dev. Cell 2002, 3: 245-57.
105. Segall JE, et al. EGF stimulates lamellipod extension in metastatic mammary adenocarcinoma cells by an actin-dependent mechanism. Clin. Exp. Metastasis. 1996, 14: 61-72.
106. Schlessinger J. All signaling is local? Mol. Cell 2002, 10: 218-9.
107. Reynolds AR, et al. EGFR activation coupled to inhibition of tyrosine phosphatases causes lateral signal propagation. Nat. Cell Biol. 2003, 5: 447-53.
108. Maheshwari G, Wiley HS, and Lauffenburger DA. Autocrine epidermal growth factor signaling stimulates directionally persistent mammary epithelial cell migration. J. Cell Biol. 2001, 155: 1123-8.
109. Wiley HS, et al. Removal of the membrane-anchoring domain of epidermal growth factor leads to intracrine signaling and disruption of mammary epithelial cell organization. J. Cell Biol. 1998, 143: 1317-28.

Chapter 6

SCATTER FACTORS IN TUMOR PROGRESSION

Paolo M. Comoglio and Carla Boccaccio
Institute for Cancer Research and Treatment (IRCC), University of Turin Medical School, Candiolo (TO)-Italy

Abstract: Scatter factors are soluble proteins that activate tyrosine kinase receptors belonging to the *MET* oncogene family. These receptors drive a physiological genetic program known as "invasive growth", which underlies tissue morphogenesis and repair. Aberrant execution of this program in time and localization has been associated to neoplastic transformation, invasion and metastasis. Activation of the *MET* oncogene has been reported in a wide array of human tumors, either as a consequence of germline or sporadic mutations, or, more commonly, of overexpression. Hypoxia has emerged as a key mechanism to induce *MET* transcription and an invasive phenotype. Specific antagonists of *MET* have been shown capable of curing experimental tumors in the mouse, targeting both neoplastic cells and their microenvironment. The Scatter Factor/MET signalling system has thus emerged as a key regulator of tumor growth and progression, whose inhibition offers the possibility to treat metastatic cancer.

Key words: hepatocyte growth factor (HGF), Met, epithelial-mesenchymal transition, de-differentiation

1. THE BIOLOGY OF SCATTER FACTORS: FROM *IN VITRO* CELL DISSOCIATION TO CANCER DISSEMINATION *IN VIVO*

Cancer metastatization recapitulates many aspects of a physiological genetic program, defined as "invasive growth", which is regulated by a family of soluble signals called Scatter Factors. Physiological invasive growth leads to morphogenetic movements and three-dimensional

A. Wells (ed.), Cell Motility in Cancer Invasion and Metastasis, 111-142.

organization of tissues during development and organ regeneration. Aberrant execution of the invasive growth program in time and space is otherwise responsible for anarchic cell spread through tissues and organs, which is the hallmark of cancer malignancy.

Scatter Factor (SF) was first identified in the 80's as a fibroblast secreted molecule, and named after its ability to induce intercellular dissociation of epithelial cultures, within a few hours after administration (1) (Figure 1). Independently, the same protein was isolated either from platelets (2) or from the plasma of patients affected by acute liver failure (2, 3), and characterized as a potent growth factor for hepatocytes in culture (2). After biochemical purification and cDNA cloning, this molecule, since then named Hepatocyte Growth Factor (HGF) (4), and Scatter Factor, were recognized as identical[5]. Then SF/HGF was shown to be the ligand for the receptor encoded by the *MET* proto-oncogene (6, 7). Soon another member joined the family of Scatter Factors: Macrophage Stimulating Protein (MSP, or SF-2) (8), which was shown to be structurally related to SF/HGF, and to induce epithelial cell scatter and proliferation after binding to a *MET* homolgous tyrosine kinase receptor encoded by the oncogene *RON* (9, 10). A third member of the family, *SEA*, turned out to be the avian counterpart of *RON* (11, 12).

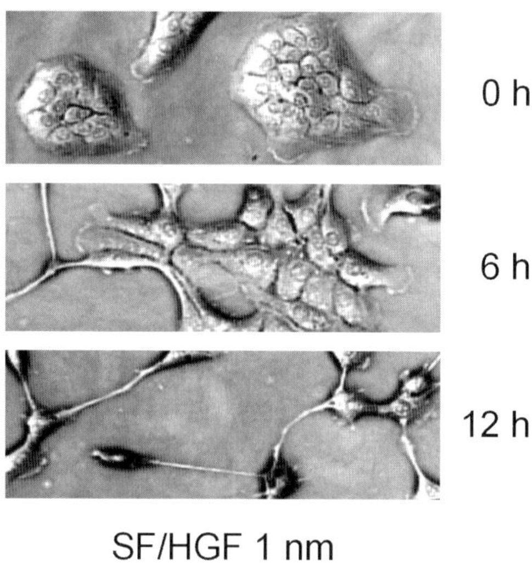

Figure 1. The Scatter effect. Epithelial cells (mouse liver progenitors) grow as compact islands (0 h). Addition of SF/HGF to the culture medium induces cell dissociation within 6 hours (6 h) and acquisition of a mesenchymal phenotype (see text), followed by cell migration (12 h) (micrographs, 400x).

The *MET* oncogene was identified in the 1980s as a transforming gene derived from a chemically transformed human cell line (13), where *MET* was activated by translocation and fusion with the TPR gene (14). The *MET* proto-oncogene counterpart was recognized as the prototype of a new tyrosine kinase receptor family, featuring a membrane-spanning β-chain, covalently linked to a completely extracellular α-chain (15). The corresponding artificial oncogene *TPR-MET* includes most of the intracellular tyrosine kinase domain, which is constitutively dimerized, and thus activated, through a leucine-zipper domain provided by *TPR* (16). The appreciation of the *MET* oncogenic potential culminated with the finding of *MET* missense mutations in hereditary and sporadic papillary renal carcinoma (17). Moreover, *MET* emerged as an "unconventional" oncogene: its activation not only induced transformation, but also a motile and invasive phenotype *in vitro*, and metastatic spread after *in vivo* cell transplantation (18-22). Most importantly, clinical studies pointed at a possible causal role of *MET* in human tumor metastasis. In the case of colorectal carcinoma, amplification of the *MET* gene confers a selective advantage that enables cancer cells to metastasize to the liver (23). Moreover, somatic mutations of *MET* are selected during the metastatic spread of head and neck squamous-cell carcinomas (24).

Although it could be seen as subversion of tissue structure and function, the process of metastasis is rooted in basic properties of undifferentiated cells. In the developing embryo, morphogenic movements require conversion of epithelial cells to a mesenchymal and motile phenotype, which is appropriate for migration in the extracellular environment, and organization in multilayered organs, which eventually encompass several tissues. This is the so called "eptihelial-mesenchymal transition" that takes place since formation of the primitive streak and the first cell movements aimed to transform a flat organism into a three-dimensional one (reviewed in 25). The "scatter effect", induced by the homonymous factors, is a transient epithelial-mesenchymal transition recalling that Scatter Factors are implicated in early developmental processes. This involvement was guessed since Scatter Factors were identified as potent morphogens for epithelial cells resuspended in collagen matrices, which are induced to form branched tubular structures (26, 27). Not surprisingly, the ensuing genetic analysis of SF/HGF and *MET* in mice has shown that this signalling system is essential during development. The *MET* and SF/HGF genes are expressed in dynamic patterns in the embryo, suggesting a paracrine mechanism of action: in general, epithelial cells express the receptor, while cells of mesodermal origin secrete the factor, which has limited capabilities to diffuse *in vivo* (reviewed in 28). In knockout mice, ablation of SF/HGF or *MET* is embyonically lethal, impairing formation of placenta, liver, diaphragm and limb muscles. Interestingly, these muscles are formed by myoblasts that must undertake a long-range migration, from somite-derived axial structures (dermomytomes) to peripheral buds. SF/HGF, which is expressed all along the myoblast pathway, controls both cell delamination from dermomyotomes, and their directional migration (reviewed in 28 and 29).

From elucidation of the role of the SF/*MET* signaling system in embryonic development and cancer progression, the concept of "invasive growth" has emerged (30, 31). This identifies a behavioral pattern, which is evoked by Scatter Factors in many different cell types and in disparate biological contexts. Physiological invasive growth is a highly regulated process, which is typically observed during formation of ramified tubules and papillary outgrowths that compose the parenchymal architecture of epithelial organs such as exocrine glands, liver and kidney, or during the development of the blood circulatory tree (vasculogenesis and angiogenesis). Moreover, a specialized form of invasive growth is seen in the nervous system, when axons elongate through tissues in order to reach their final synaptic target; in the bone marrow, when hemopoietic precursors dissociate from their niches and are released in the blood circulation; in the bone, when osteoclasts proliferate and invade the mineralized matrix to remodel the tissue; and, as mentioned above, invasive growth is observed in the mesoderm, when myoblasts migrate from their primary site of development to colonize limb buds. Pathological invasive growth consists in deregulated cell spread, and inability to organize in ordered tissue structures, as observed in anaplastic and metastatic tumors.

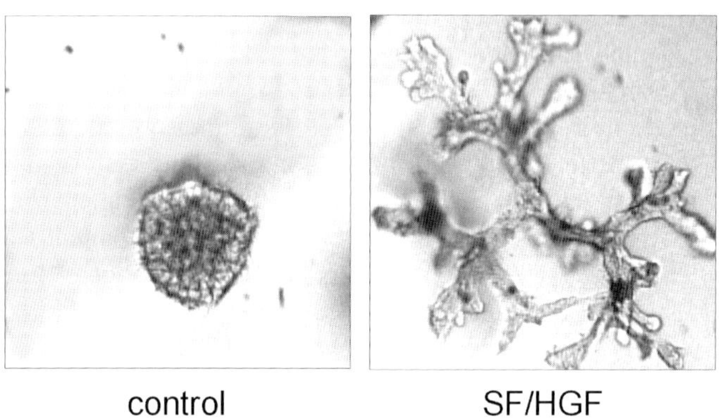

Figure 2. Branching morphogenesis. Epithelial cells (mouse liver progenitors) resuspended in a three-dimensional collagen matrix form spheroidal cysts (control). Growth in the presence of SF/HGF, within a few days induces cell migration and reorganization to form tubules lined by polarized cells (micrographs, 10 x).

In vitro studies has defined the stepwise execution of the invasive growth program elicited by Scatter Factors. Epithelial cells resuspended in a tridimensional collagen matrix, form cysts, that is spherical monolayers of polarized cells that enclose a central lumen. When stimulated by Scatter Factors, cells emit long protrusions in the surrounding matrix. Then, some cells migrate along the path opened by this protrusion, losing polarity and retaining minimal intercellular contacts; then cells dispose in multilayered cords, reestablish junctions and polarization typical of epithelial cells, and form a new lumen, giving rise to a tubular structure (the so-called "branching morphogenesis" or "epithelial tubuligenesis", reviewed in 32) (Figure 2). Constitutive activation of SF/HGF signalling, possibly as a consequence of *MET* oncogenic activation, impairs building of epithelial architecture in three-dimensional cultures, and fosters disordered invasion of the extracellular matrix (ECM, our unpublished observations). A similar, but transient, response is observed when cells are grown on a bi-dimensional plastic support, and undergo the classical "scatter assay": after Scatter Factor stimulation, cells disassemble their junctions and migrate in every direction; moreover, if seeded on the surface of an artificial basement membrane, cells migrate across it (33). This motile/invasive response become constitutive, when cells express an activated *MET* oncogene. During cell scatter or branching morphogenesis, cells may or may not proliferate, but, invariably, they remodel their intercellular junctions and integrin-mediatated cell-matrix contacts. Noteworthy, SF/HGF stimulated cells benefit of active protection from apoptosis, which invariably occurs when cell adhesion is impaired. The invasive growth process imply activation of a complex genetic program, currently under investigation, which likely comprises effectors designed to play a prominent role also in cancer metastasis.

The Scatter Factor Receptor's is now an extended family, including Plexins, the Semaphorin Receptors (34). Both Plexins and their ligands, Semaphorins, contain a structural motif (the Sema domain), which is found also in the *MET* and *RON* extracellular domains. However, the Plexin intracellular region does not contain a tyrosine kinase domain, and has still elusive signalling properties. Recently, it has been shown that Semaphorins can elicit invasive growth: this response is largely mediated by indirect activation of *MET*, which is preassociated to Plexins on the cell surface (35). Although they could not be technically defined as "oncogenes", as they lack transforming properties, Semaphorins and Plexins are now recognized as candidate genes to promote tumor progression towards invasion and metastasis. In fact, alterations in Semaphorin or Plexin expression is reported in human tumors (reviewed in 31and 36).

2. STRUCTURE OF SCATTER FACTORS AND THEIR RECEPTORS

Scatter Factors (SF/HGF and MSP) display an intriguing relationship with members of the blood-clotting cascade (Figure 3). They are phylogenetically related to plasminogen, a circulating proenzyme whose active form (plasmin) is responsible for fibrinolysis (degradation of blood clots). Mature, biologically active Scatter Factors have an unusual large size (94 kDa), and consist of two disulphide-linked chains (α and β). The α chain is characterized by the presence of a N-terminal "hairpin loop", followed by four "kringle" domains (80-amino acid double-looped structure stabilized by internal disulphide bridges), which are a common feature of plasminogen-related proteins (4). The β chain is homologous to serine-proteases (like plasminogen and clotting factors) but lacks proteolytic activity, owing to substitution of three aminoacidic residues critical for catalytic functions[4]. Not only Scatter Factors are related to coagulation proteins in their structure, but also in their activation mechanism. In fact, both SF/HGF and MSP are secreted as single-chain inactive precursors (pro-HGF and pro-MSP), and are activated by a proteolytic cleavage, performed by proteins involved in clotting regulation as well. Urokinase-type plasminogen activator (uPA) was the first enzyme to be shown as a potent activator of pro-HGF (37-39). Then, thrombin, coagulation factor XII and one homologous serine-protease (XII-like factor) were shown to behave ad pro-HGF convertases as well (40,41). SF/HGF binds heparan-sulphate proteoglycans, which provide an extracellular reservoir of the factor *in vivo* and limit its diffusion through extracellular fluids. This favors SF/HGF sequestration in proximity of the synthesis site, and a paracrine modality of action (42). Moreover, proteoglycans, although dispensable for SF/HGF binding to its receptor, couple SF/HGF in symmetrical dimers that simultaneously engage two receptor molecules, inducing receptor trans-activation (43,44). The mechanism of SF/HGF binding to its own receptor is however quite complex, involving both factor α- and β-chain (45).

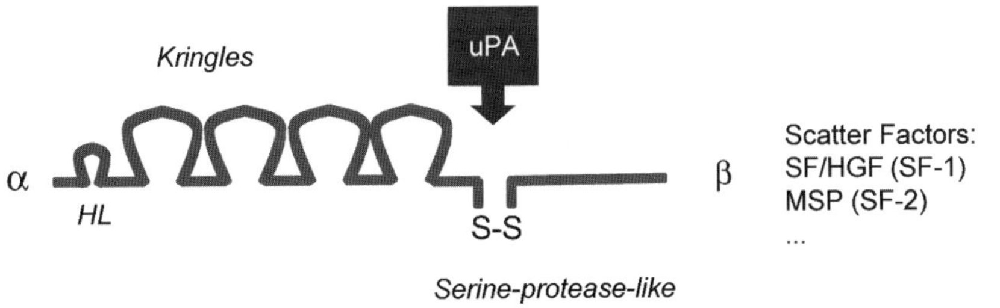

Figure 3. Schematic structure of Scatter Factors. HL: Hairpin Loop; uPA: urokinase-type Plasminogen Activator (for explanation see text).

The SF/HGF Receptor (SF/HGFR), encoded by *MET*, and the MSP Receptor (MSPR), encoded by RON, sharing about 60% homology, are disulphide-linked α/β heterodimers that form by intracellular proteolytic processing of a single-chain precursor (Figure 4). In both receptors, the α subunit is completely extracellular, while the β subunit is a single-pass transmembrane chain encompassing the tyrosine kinase activity. The extracellular region of SF receptors is characterized by the so-called "Sema Domain", common also to the transmembrane receptors Plexins, and to their ligands, Semaphorins (34). In the Scatter Factor Receptors, the Sema Domain spans over 500 aminoacids, including the entire α chain (about 300 aminoacids), and the amino-terminal moiety of the β chain. Recently, mutagenesis studies of SF/HGFR have shown that the Sema Domain is endowed with low-affinity ligand binding (46).

The Scatter Factor Receptors' intracellular domain comprises a tyrosine kinase catalytic site, flanked by juxtamembrane and carboxy-terminal regulatory regions. The juxtamembrane domain contains residues involved in receptor downregulation. These include (a) a serine residue (S985), whose phosporylation inhibits tyrosine kinase activity (47); (b) a tyrosine residue (Y1003), which, upon phosporylation, associates the protein adaptor CBL (48), which is a critical intermediary of *MET* structural downregulation. In fact, CBL activates two distinct receptor-degradation pathways, one mediated by ubiquitination (as CBL is a E3 ubiquitin-ligase) (48), and the other by endocytosis (as CBL recruits regulatory components of endocytic vescicles) (49). The above down-regulatory residues are absent in the oncogenic form of *MET* called *TPR-MET*. Conceivably, their absence contributes, together with constitutive dimerization, to the *TPR-MET* powerful transforming activity (48).

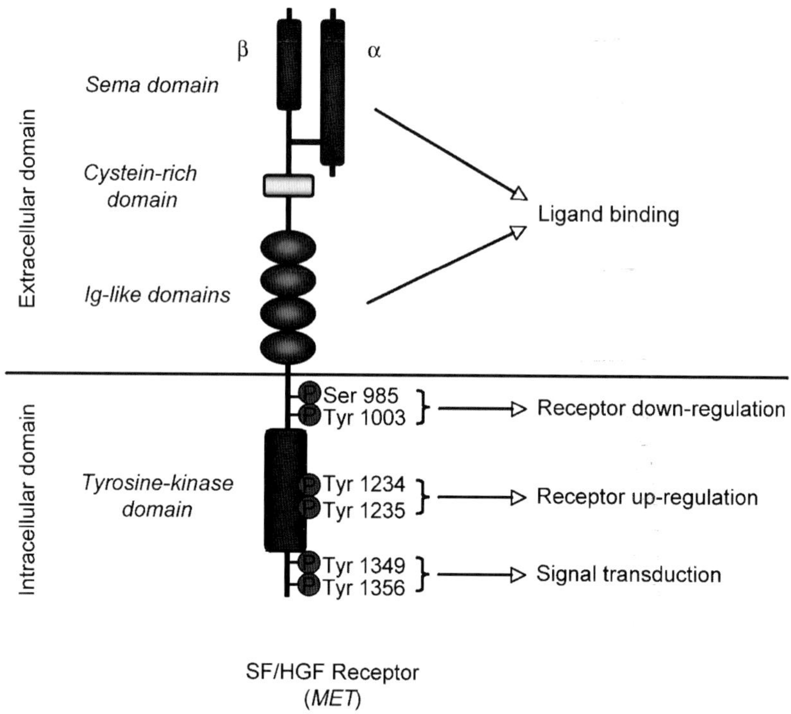

Figure 4. Schematic representation of SF/HGF receptor, the prototype of the Scatter Factor Receptor Family, including also the MSP receptor. Conserved domains and phosphorylated (P) aminoacidic residues (Ser: serine; Tyr: tyrosine) are indicated, which are critical for receptor functions (for explanation see text).

In the catalytic site, it is remarkable the presence of two tyrosines (Y1234 and Y1235) whose phosporylation, occurs early during receptor activation. This results in a conformational change that positively modulate the enzymatic activity exerted both on the receptor carboxy-terminal tail and exogenous substrates ("autocatalytic effect" (50, 51)). It has been shown that *MET* germline mutations (associated to some cancers, see below) can mimic the presence of these phosphorylated tyrosines, and confer to the kinase a sensitized, although not constitutive, activation status (52).

The carboxy-terminal domain of the SF/HGF contains a conserved sequence (Y^{1349}VHV—Y^{1356}VNV) that includes two tyrosines, which, upon phosphorylation, generates the docking sites for signal transducers. While the majority of tyrosine kinase receptors use different phosphotyrosines to recruit distinct SH-2 containing signal effectors, the above sequence, called "multifunctional docking site", is able to recruit the full spectrum of transducers required for invasive growth (53). When, by site-directed mutagenesis, the two tyrosines of the multifunctional docking sites are replaced by phenilalanines, cells become unresponsive to SF/HGF (53). The importance of these residues for receptor biological functions has been confirmed by genetic experiments *in vivo*. A transgenic mouse expressing a receptor in which the two tyrosines of the multifunctional docking site have been mutagenized displays a lethal phenotype, corresponding to that of the *MET*-null mouse (54). Moreover, mutation of the same tyrosines abrogates the oncogenic properties of activated *MET*, without affecting the catalytic function itself (55). The multifunctional docking site is thus the driving force for the biological and transforming properties of the SF/HGF receptor, and possibly it is the key of its specificity. In facts, when inserted in the intracellular domain of other receptor tyrosine kinases, it confers the ability to transduce the invasive growth signals otherwise restricted to the SF/HGF receptor (56).

3. SIGNAL TRANSDUCTION: FROM PRIVATE ADAPTORS TO MULTIPLE CO-RECEPTORS

The multifunctional docking site of the SF/HGF receptor is endowed with the unique property of recruiting – at high affinity – several SH2-containing signal transducer, due to the broad-range specificity of the consensus sequences encompassing its two tyrosines (53). Early studies on signal transduction showed that the SF/HGF receptor could concomitantly activate multiple pathways, including GRB2-RAS, Phosphatidylinositol-3kinase (PI3-K), SRC, and Signal Transducer and Activator of Transcription (STAT) (53, 57-60) (Figure 5).

The above transduction pathways were shown to be critical for the physiological and pathological effects elicited by the SF/HGF receptor-*MET* oncogene. Activation of RAS plays a central role in the entire invasive growth process, affecting both cell scattering and proliferation. When the RAS transduction pathway is impaired, by expression of a dominant-negative RAS (61), or by microinjection of neutralizing antibodies (62), the motile response to SF/HGF is inhibited. Direct activation of the RAS pathway by *MET*, through association of the GRB2 adaptor to the multifunctional docking site, is a key determinant of cell proliferation and thus of the *MET* oncogenic potential: mutagenesis of the *MET* multifunctional docking site that selectively abrogates GRB2 recruitment stops the proliferative signals and abolishes the transforming potential of the activated *MET* oncogene (63). On the contrary, the same *MET* mutant, keeps an unaltered ability to induce cell motility, indicating that the threshold signal to accomplish this branch of the invasive growth response is however reached (63). Another prominent transducer for the motile-invasive response to *MET* is PI3-kinase, which is activated either by direct recruitment through the multifunctional docking site, or as a RAS effector (57, 59). PI3-kinase is indispensable for cell scatter *in vitro*, being required for disassembly of intercellular contacts (64, 65), and remodeling of adhesion to the extracellular substrate (66).

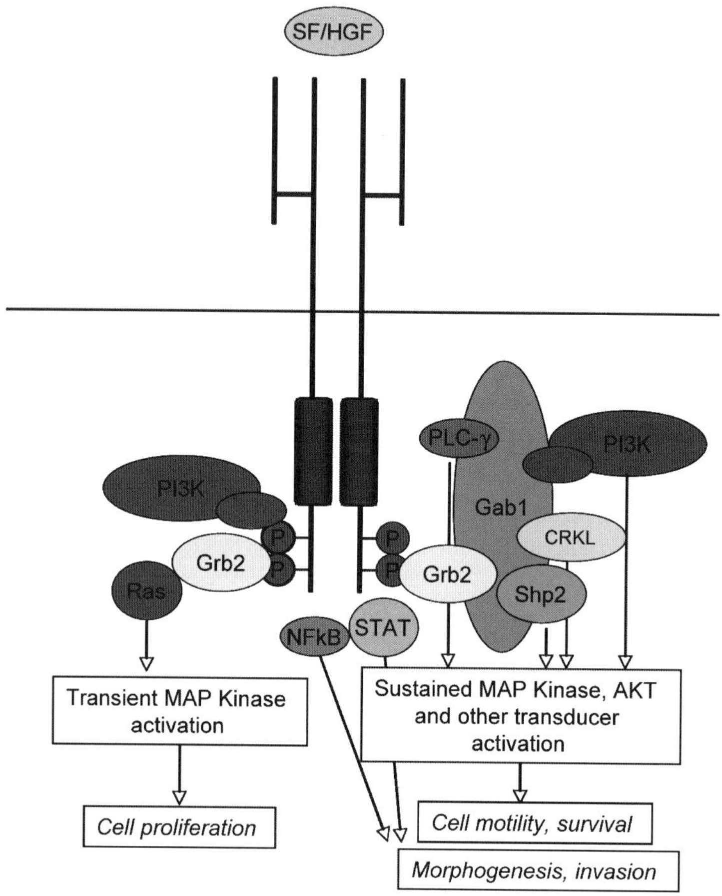

Figure 5. Sig-naling pathways activated by SF/HGF receptor. Through its "multi-functional docking site" including two phosphorylated tyro-sines (P), the recep-tor recruits the full spectrum of trans-ducers (for details see text) required to elicit the invasive growth program, fea-turing cell prolifer-ation, motility, survi-val and morphogen-esis. Deregulated activation of the same signaling path-ways by oncogenic MET leads to cancer invasion and metastasis.

Signaling pathways downstream PI3-K are involved either in cytoskeletal reorganization and motility (such as the small GTPase Rac and the protein kinase PAK (67), or in protection from apoptosis (such as AKT) (68). Anoikis is a form of apoptosis occurring when a cell loses its normal adhesive contacts (69). Protection from anoikis is thus a must when a cell detaches from its primary site and travels through the ECM. Thus, it is not surprising that activation of PI3-kinase by oncogenic *MET* is necessary to accomplish the full process of invasive growth *in vivo*. In fact, mutagenesis of the *MET* docking site that prevents coupling with PI-3 kinase allows cell transformation but not explication of metastatic properties (20). The contribution of RAS and PI3-kinase to the entire invasive growth program is thus equally important, although each of the two pathways preferentially influences different sides of the process (proliferation or invasion). We can conclude that concomitant activation of RAS and PI3-K by *MET* is required to express a fully metastatic phenotype (22).

As well as cancer invasive growth, physiological branching morphogenesis requires the integrity of the full spectrum of signal transducers activated by *MET*, including RAS-MAP kinase (70), PI3-kinase (71), Phospholipase-Cγ (PLC-γ) (72), and transcriptional factors like STAT (60) and NFkB (73). Recently, it has been highlighted the importance of sustained activation of ERK1/2 MAP kinases as a specific feature of SF/HGF signaling, responsible for the induction of cell differentiation and branching morphogenesis (74, 75). The ability to sustain MAP kinase activation sets SF/HGF apart from pure mitogenes that elicit only transient MAP kinase activation (76).

While the role of complex signaling cascades for the physiological and pathological invasive growth process was being elucidated, an intracellular transducer was discovered, named GAB1, which emerged as a key orchestrator of the cell response to *MET* (77, 78) (Figure 5). The importance of GAB1 for SF/HGF signaling is supported by genetic evidence that GAB1 inactivation in mice results in a phenocopy of the *MET* knock-out (79). GAB1 is a scaffolding adaptor (such as the Insulin Receptor Substrate) that lacks obvious SH2 phosphotyrosine interaction domains, but which is recruited directly and indirectly to activated tyrosine kinases. GAB1 association to SF/HGF receptor is mediated directly, through a specific GAB1 phosphotyrosine recognition domain, called MBD (*MET* Binding Domain (77), and, indirectly, through GRB2, bound to the phosphorylated *MET* multifunctional docking site (80, 81). Although GAB-1 is a substrate also for other receptor tyrosine kinases (82), SF/HGF elicits a more prolonged GAB-1 phosphorylation than other growth factors, possibly as a consequence of the low-affinity but high-avidity association between its receptor and GAB1 through the MBD (72, 78). This is possibly the key to understand the role of GAB-1 in inducing invasive growth. In facts, GAB1, phosphorylated by SF/HGF receptor, behaves as a "signal sustainer" enhancing the recruitment and the activation of transducers, including PLC-γ, the protein tyrosine phosphatase SHP2, PI3-K-AKT, and the adaptor CRK-like (CRKL) (72, 83). Shp2 and CRKL lead to the sustained MAP kinase pathway, whose requirement to elicit the invasive growth program is discussed above. Shp2 activates MAPK through complex and still obscure intermolecular interactions (84). CRKL recruits the guanine nucleotide exchanger C3G, which, on its turn, activates the small G protein Rap-1 (85). Most interestingly, it has been reported that, in contrast with Ras, Rap1 is capable of sustaining MAP kinases activation through formation of a stable complex with RafB. This signal has been found responsible for induction of a cell differentiation program (86).

Recent studies have introduced further complexity in the transduction machinery for the invasive growth signal, involving the presence of cell-surface molecules behaving as co-receptors. As mentioned above, Scatter Factors Receptors (*MET* and *RON*) can be activated by Plexins, when the latter are engaged by their ligands Semaphorins (35, 87).

Within the physiological range of ligand concentration, it was shown that the wild-type SF/HGF receptor cannot efficiently recruit signaling effectors and mediate invasive growth, unless it is physically associated with the α6β4 integrin (88). Upon receptor activation by the ligand, the associated integrin is phosphorylated on tyrosines, which provide supplementary docking sites for signal transducers (such as PI3-K and the adaptor SHC). Thus the integrin behaves as an adaptor, which contributes to provide the signaling strength and duration that is the distinctive feature of the invasive growth program. Noticeably, when associated with the SF/HGF receptor, α6β4 integrin elicits intracellular signals independent of cell adhesion, acting as a pure biochemical amplifier of SF/HGF signaling (88). As expected, the integrin requirement can be bypassed forcing the signaling system by non-physiological high concentration of the ligand, or by receptor overexpression (89).

Beside integrins, a variant of the CD44 transmembrane molecule has been found to associate to the SF/HGF receptor, and to cooperate for its activation and signaling (90). CD44 (reviewed in (91)) was known since long-time for its ability to induce cell invasion, but the underlying mechanism remained obscure. CD44 is a receptor for ECM components (such as hyaluronic acid and the protein osteopontin), which, with its intracellular moiety, binds to the ezrin-radixin-moesin (ERM) family of proteins. ERM proteins regulate cell motility, by mediating a dynamic association of the actin cytoskeleton to the plasma membrane. Most interestingly, ezrin has been identified as an effector of the SF/HGF morphogenic program (92). Moreover, deregulated expression of ERM proteins has been correlated to cancer metastasis (93). Thus, CD44, at the same time, structurally and functionally couples SF/HGF receptor (and other tyrosine kinases), to the cytoskeleton and to the extracellular environment, configuring as a molecule capable of localizing the invasive growth process at specific membrane sites.

4. THE GENETIC PROGRAM OF INVASIVE GROWTH

The full accomplishment of the invasive growth program requires transcriptional and post-transcriptional regulation of gene expression, including not only simple induction or repression, but also complex, differential gene modulation over time. The "invasive growth signature" is currently under investigation through genomic and proteomic technologies. The first critical invasive growth effectors have been recognized among the most induced SF/HGF genes, such as the ECM protein osteopontin (94). Noticeably, osteopontin is a ligand for the CD44 molecule (see above), and has been reported, in gene expression profiles, as one of the most transcribed genes in highly metastasizing breast cancer cells (95), and as a predictor of hepatocellular carcinoma metastasis (96).

Meaningful invasive growth genes are expected among effectors that disrupt constraints keeping cells within their physiological location, or that mediate cell migration through foreign districts. During tumor progression as well as morphogenic movements, to initiate ECM invasion, epithelial cells must disassemble adhesive structures that connect them each other, and to the basement membrane. Intercellular adhesion is primarily mediated by adherens junctions, where transmembrane E-cadherins provide omophilic recognition with their extracellular domain, and, with their intracellular moiety, bind to a submembranous scaffold of catenins, that form a structural and functional bridge to the actin cytoskeleton (97). The mechanism of adherens junction destabilization by SF/HGF is still elusive, however it requires cadherin downregulation. Although early reports indicated that the level of E-cadherin/uvomorulin did not change during SF/HGF-induced scatter (33), a decreased expression of the cadherin protein in the same conditions has been reported by other groups, and ascribed to post-translational regulation (98-100). Moreover, it has been shown that SF/HGF induces (a) redistribution of cadherin-catenin complexes from the actin cytoskeleton to the soluble membrane fraction (101, 102), (b) proteolytic cleavage of cadherins from the cell surface (103), and (c) destabilization of adherens juctions through tyrosine phoshorylation of catenins (104, 105). The importance of E-cadherin down-regulation and adherens junction disruption for cancer spreading is emphasized by finding that loss of

E-cadherin, due to genetic or epigenetic mechanisms, is a major determinant of cancer metastasis (reviewed in 106). Interestingly, E-cadherin expression is primarily controlled through the Snail family of transcription factors (107), which regulate epithelial-mesenchymal transition in development and cancer (reviewed in 108and 109). It would be interesting to investigate if SF/HGF signalling modulates Snail transcritpion factors, as suggested by the fact that the latter are a target of the PI3-kinase/AKT pathway (110).

In the invasive growth program, after mutual dissociation, epithelial cells must cross the natural boundary represented by basement membrane to creep into foreign tissues. This implies – one one side- the ability to recognize previously unexplored extracellular substrates, through engagement of appropriate integrins. On the other side, it requires the ability to degrade the ECM through proteases, in order to open a passage through macromolecules and cells, and to expose cryptic adhesion sites. Dynamic remodeling of integrin-mediated extracellular adhesion provides not only a mechanic support for cell motility, but also an indispensable signal that protects cells from anoikis (apoptosis), otherwise occurring when recognition of the extracellular environment fails (69). Finally, local tissue invasion evolves into metastatization, if cells are capable of entering blood and/or lymphatic vessels, survive the circulation, and finally extravasate at a distant site. SF/HGF modulates the expression and the activity of the integrin repertoire, thus allowing successful recognition of the modified extracellular environment during cell invasion. In facts, SF/HGF upregulates integrin transcription, through sustained activation of MAP kinases at least in some cases (111, 112), and stimulates integrin aggregation at specific adhesive sites, increasing their avidity for the substrate (66, 113). SF/HGF is also able to regulate the expression and the activity of primary actors of ECM remodeling, the Matrix Metalloproteases (MMPs) (reviewed in 114). By interacting with integrins and CD44 (see above), MMPs specifically localize matrix digestion at the leading edge of migrating cells. SF/HGF is able to enhance transcription of several MMPs, and to stimulate conversion of their inactive precursors into active enzymes (115-118). However, and surprisingly, it has been recently reported that broad-spectrum MMP inhibition does not impair the initial phases of branching morphogenesis induced by SF/HGF, featuring epithelial-mesenchymal transition and matrix invasion. MMPs are instead required for completion of tubulogenesis and reacquisition of the epithelial, polarized phenotype (75). It is conceivable that MMPs might not be limiting factors for cell invasion, as SF/HGF induces also expression and secretion of other ECM proteases, in particular of urokinase-type plasminogen activator (uPA) (119, 120), which might play a central role both as a regulator and an effector of the invasive growth program. In fact, as mentioned above, uPA is the primary pro-HGF convertase, which binds with high affinity inactive pro-HGF and, by cleavage at a specific site, transforms it into a biologically active molecule (37). Binding of uPA to pro-HGF occurs with an uncommon 1:1 stoichiometry, and results not only in SF/HGF activation, but also in uPA irreversible inactivation (38). Yet, SF/HGF stimulation, by induction of uPA expression, restores the availability of uPA in the microenvironment, possibly sustaining a positive feedback on SF/HGF signaling. As a SF/HGF effector, uPA is responsible for modulation of ECM degradation, through conversion of plasminogen in plasmin, which is active on several extracellular substrates. uPA is known to bind a cell-surface receptor, which localizes uPA proteolytic activity at the cell membrane. Moreover, the uPA receptor is involved in the invasive process through multiple mechanisms, including (a) regulation of cell adhesion, through binding of ECM substrates and modulation of integrin function, and (b) intracellular transduction of chemotactic stimuli (reviewed in 121).

Finally, SF/HGF exerts its pro-invasive activity not only inducing migration and survival of (cancer) cells, but also through "landscaping effects", that is regulation of properties of the tumor microenvironment, which are critical for successful tumor growth and metastatization. Among these, the best characterized is induction of angiogenesis by direct stimulation of endothelial cells (122). Recently, SF/HGF has been found widely implicated in tumor neo-angiogenesis, and shown as an effective target for cancer treatment (see below and 123).

5. THE *MET* ONCOGENE IN TUMOR PROGRESSION

The SF/HGF receptor gene, *MET*, was initially identified as a transforming oncogene derived from a chromosomal translocation in an osteosarcoma cell line treated with a chemical carcinogen (*TPR-MET* (13,14), see above). The same translocation product, which is capable of inducing tumors when expressed in transgenic mice (124), was found only in a small fraction of naturally occurring human gastric cancers (125). However, activation of the *MET* oncogene by different mechanisms has been detected in a large number of human tumors, and its causal role in tumorigenesis has been assessed by *in vitro* and *in vivo* experiments. Constitutive activation of the *MET* tyrosine kinase is usually achieved by one of the following mechanisms: (a) point mutations causing activatory conformational changes in the catalytic site; (b) ligand-receptor autocrine circuits (which free cells from requirement of paracrine SF/HGF supply), or increased paracrine stimulation; (c) *MET* overexpression, which favors either ligand-independent transactivation, or increased sensitivity to the factor.

The most convincing evidence that *MET* plays a causal role in human cancer arose from analysis of patients, affected by a form of hereditary kidney cancer called papillary renal carcinoma (HPRC), who bear germline missense mutations of *MET* (17, 126, 127). The same mutations (and others) have been found also in sporadic tumors, such as sporadic papillary renal cancer (17, 128), childhood hepatocellular carcinoma (129), and gastric cancer (130). Although hereditary and sporadic papillary kidney cancer is usually an "indolent tumor", featuring slow growth and local invasion (131), somatic mutations of *MET* have been associated to increased aggressiveness of hepatocellular carcinoma (129), and to the metastatic spread of head and neck squamous carcinoma (24). Interestingly, in the latter, the population of metastatic cells enriches progressively in *MET* expression, as the tumor stepwise invades lynfonodal stations. Interestingly, some *MET* mutations correspond to those found in the RET oncogene in association to Multiple Endocrine Neoplasia type 2 (132), or the KIT oncogene, in association with gastrointestinal stromal tumors and haematological malignancies (133). Oncogenic *MET* point mutations are concentrated in the intracellular domain, in a few "hotspots" mostly corresponding to regions critical for tyrosine kinase regulation, such as the ATP-binding region or the so-called "activation loop", where tyrosines responsible for kinase up-regulation are located (see above) (131). As a significant example, a *MET* mutation selected during metastatic spread of head and neck carcinoma, causes substitution of one of these tyrosine with a negatively charged aminoacid (aspartate). This mimics the permanent presence of a phosphate group, thus causing a constitutive state of kinase up-regulation. Interestingly, as a general rule, *MET* point mutations do not induce *per se* kinase activation, but a conformational change that highly reduces the threshold for kinase activation (52). These *MET* mutations are thus only weakly tumorigenic, as observed in patients bearing germ-line mutations, who develop cancer late in life. This is not surprising, as these mutations must be compatible with embryonic development and post-natal tissue homeostasis. Thus, it is conceivable that the reported *MET* point mutations are insufficient–alone- to cause cancer onset and that invariably require the occurrence of cooperating genetic lesions, as confirmed by analysis of HPRC patients (17). Intriguingly, a collaborating factor can be also abnormal SF/HGF stimulation (either through paracrine or autocrine mechanisms), as suggested by the finding that, in classical *in vitro* assays, cell transformation by *MET* mutants is possible only in the presence of its ligand, and is impaired in the presence of SF/HGF specific inhibitors (134).

A role for SF/HGF in sustaining *MET*-induced transformation has been reported in human tumors and assessed in mouse models. In patients, SF/HGF autocrine loops and/or enhanced paracrine stimulation are observed in a wide spectrum of cancers, including mesenchymally-derived tumors, such as osteosarcoma (135, 136) and rabdomyosarcoma (136, 137), as well as glioblastoma (138), and breast carcinoma (139, 140), often in association with a highly aggressive tumor behavior. Experimental induction of SF/*MET* autocrine loops in cell lines has proven effective in causing formation of invasive tumors after implantation in mice (18, 21). Transgenic mice expressing SF/HGF under a ubiquitous promoter develop a broad array of neoplasms of both epithelial and mesenchymal origin (141). Among these tumors, melanomas displayed a significant correlation between high metastatic potential and the presence of SF/*MET* autocrine loop potentiated by selection of melanocytes overexpressing *MET* (142). In another mouse model, targeted expression of SF/HGF to the mouse mammary epithelium establishes autocrine and paracrine loops, sustaining formation of metastatic adenocarcinomas (143).

Overexpression – in the absence of any mutation of the coding sequence- is perhaps the most frequent mechanism of *MET* oncogene activation observed in human tumors, again in association with metastatic phenotype and poor prognosis (Table 1). In the case of colorectal carcinoma, *MET* overexpression provides a selective advantage that sustains tumor's capability of forming lymphonode and liver metastasis (23, 144). In an animal model, it is shown that forced expression of wild-type *MET* in hepatocytes is sufficient to cause hepatocarcinomas, which regress following transgene inactivation (145). It is conceivable that increased expression of *MET* favors receptor dimerization and thus ligand-independent activation; however, the availability of SF/HGF could be always mandatory, as suggested by the finding of non-activated, although highly expressed, *MET* in human sarcomas (135). Thus, *in vivo*, the tumor stroma, which physiologically produces, stores and modulates the activation of SF/HGF, could have a critical landscaping role in promoting *MET*-dependent tumor growth, either in the presence of rare *MET* mutations, or the frequent *MET* overexpression.

Table 1. Overexpression of the MET oncogene associated to an aggressive phenotype.

Tumor	Reference
Colorectal carcinoma	23,144
Gastric cancer	179
Thyroid cancer	180-182
Renal cancer	183,184
Ovarian cancer	185,186
Osteosarcoma	135,136,187

6. THE CONTROL OF *MET* EXPRESSION: A KEY TO UNDERSTAND CANCER ONSET AND PROGRESSION

The *MET* promoter has been identified and characterized for positively responding to several mitogenic stimuli, including growth factors (among which SF/HGF itself), tumor promoters (146, 147), and intracellular expression of activated oncogenes (148, 149). Among transcription factors directly controlling *MET* expression, ETS/AP1 is noticeable, as it is activated by *MET* itself through the MAP kinase pathway, thus providing an explanation to why SF/HGF can induce its own receptor (150, 151). Moreover, ETS controls transcription of several genes critical for ECM regulation and thus for invasive growth (152).

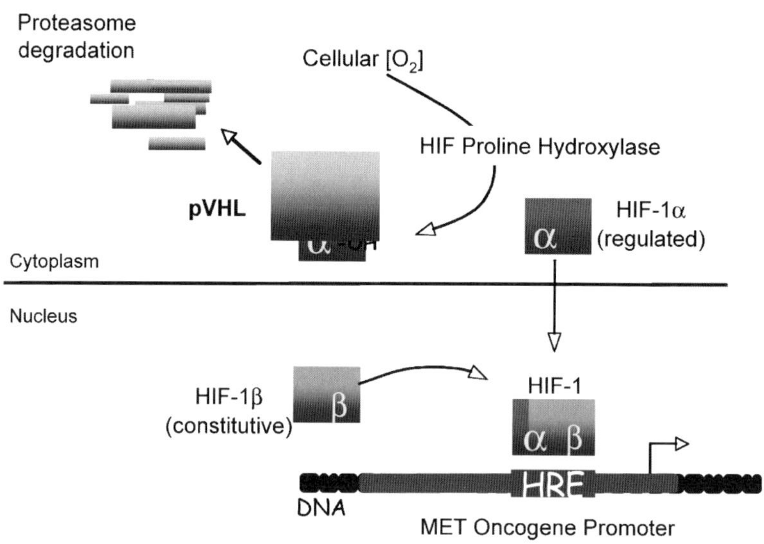

Figure 6. Hypoxia is a major determinant of MET expression. HIF (Hypoxia Inducible Factor)-1α is a transcriptional factor subunit, which, under normal oxygen concentration, is targeted to proteasome degradation by a ubiquitin-ligase, encoded by the VHL tumor suppressor gene. When intracellular oxygen concentration decreases, HIF-1α is prevented from degradation, and migrates to the nucleus, where it couples to the HIF-1β subunit and drives expression of several genes, including the MET oncogene (for details, see text).

Recently, another mechanism of *MET* transcriptional regulation has been discovered, which has profound implications for understanding the metastatic process and for the management of innovative therapies such as the use of anti-angiogenic agents. *MET* expression is highly induced by tissue oxygen deprivation, that is by "hypoxia" (153) (Figure 6). Through inhibition of its degradation, hypoxia activates a transcription factor (Hypoxia Inducible Facor-1α, or HIF-1α (154)), which positively regulates the *MET* promoter. *In vitro* experiments show that hypoxia amplifies SF/HGF signaling and synergizes with SF/HGF in inducing invasive growth. Hypoxia *per se* can induce cell invasion, which is prevented through *MET* inhibition by RNA interference. Most importantly, analysis of human tumors showed that *MET* expression is maximum in hypoxic areas (where expression of HIF-1α is consistently high), while it decreases in proximity of blood vessels (where HIF-1α is absent) (153). An indirect proof of the importance of HIF-1α for *MET* transcriptional control is provided by the observation that *MET* is overexpressed in a renal cell carcinoma subtype, featuring inactivating mutations of the von Hippel-Lindau (VHL) tumor suppressor gene (155). The VHL protein targets HIF proteins for degradation, in the presence of normal oxygen concentrations. Thus, inactivation of VHL prevents HIF-1α downregulation, resulting in increased *MET* transcription. In conclusion, hypoxia emerges as a major determinant of *MET* expression *in vitro* and *in vivo*. Interestingly, hypoxia induces expression not only of *MET*, but also of u-PA receptor (see above) (156), and of the chemokine receptor CXCR4, which is responsible for homing of metastasizing cells into organs producing the CXCR4 specific ligand SDF-1α (157). These findings provide an attractive explanation for the notion that the presence of oxygen-deprived areas in tumours correlates with poor prognosis (reviewed in 158), and challenge the rational employment of anti-angiogenic agents. In fact, deprivation of blood supply, and thus of oxygen, could induce, beside the desirable tumor necrosis, a dangerous tumor "invasive switch". Thus, it is advantageous that anti-angiogenic would be combined with anti-invasive therapy (including *MET* targeting, see below) to maximize the benefits for cancer patients and prevent insidious drawbacks (159).

Another critical feature of *MET* expression is that, in adult tissues, it might be present, or, in some cases even confined, to stem cells. In the hemopoietic system, *MET* expression was found only in a small fraction of bone marrow cells, partly co-expressing the CD34 marker, and thus included in the progenitor or even in the true hemopoietic stem cells population (160). Consistently, the *MET* promoter contains binding elements for the GATA family of transcription factors, which are active in hemopoietic progenitors (147). Although expressed also by mature hepatocytes, *MET* is a hepatocyte stem cell marker, used (together with few other cell-surface proteins) to separate progenitors from differentiated cells with antibody-based cell sorting techniques (161, 162). In the skeletal muscle, *MET* is expressed by myoblasts, but it is downregulated during differentiation in striated fibres, as assessed *in vivo* and *in vitro* (163, 164). Noticeably, in skeletal muscle-derived tumors, rhabdomyosarcomas, *MET* is highly expressed (137), suggesting that either neoplastic cells has re-acquired a feature typical of muscle stem cells (*MET* expression), or, more conceivably, that the tumor directly derives from transformation of myogenic precursors. Thus, *MET* could be simply a marker of "stemness" of the neoplastic population, a feature which is *per se* a negative prognostic factor (165). However, a causal role for *MET* in rabdomyosarcomagenesis is supported by *in vitro* experiments and animal models, where it has been shown that *MET* costitutive activation prevents differention of myoblasts (164), and, in collaboration with other genetic lesions, induces rhabdomyosarcomas (166). Another tumor that provides a connection among *MET*, malignancy and stem cells is colon carcinoma, where *MET* overexpression has been associated with high metastatic ability (see above). Interestingly, *MET* is a transcriptional target for the β-catenin/TCF transcriptional factor (167). The latter is activated by the Wnt signalling pathway, which is physiologically switched on in gut stem cells, and off during enterocyte differentiation (168). The same pathway is constitutively activated in the majority of colon cancers, as a consequence of APC inactivation, thus contributing to explain why *MET* is overexpressed. In conclusion, *MET* is both a gut stem cell gene, and a gene associated to tumor invasion and metastasis. This raises the fascinating hypothesis that invasive growth is a natural genetic program for stem cells. The habit of proliferating and freely circulating through the organism to home in various locations (169), makes stem cells an attractive physiological counterpart of metastatic cells. Thus, the study of inappropriate activation of the invasive program in the stem cell itself will likely provide a keystone to understand tumor onset and progression.

7. *MET* AS A TARGET FOR CANCER THERAPY

Recent years have witnessed a successful employment of inhibitors of receptor tyrosine kinases for cancer therapy: among the most interesting examples are Geftinib©, a small molecule inhibitor of Epidermal Growth Factor Receptor, effective in treating cancers where this receptor is mutated (170), and antibodies against HER-2, which are employed for cancers in which this receptor is overexpressed. Like for other growth factor receptors, many efforts are directed toward the development of low-molecular weight biological inhibitors of *MET* or its ligand, which could provide "leads" for drug development. Small molecules have been isolated, displaying the capability of inhibiting *MET* catalytic activity (171, 172), or *MET* interaction with signal transducers (173), or *MET* expression itself (the antibiotic geldanamycin (174) and small interfering RNAs or ribozymes (175). Immunotherapy, e.g. involving the use of neutralizing antibodies against SF/HGF (176), or antagonistic antibodies against *MET* (177), is in its developmental stage as well. Recently, an innovative approach, combining protein engineering and gene therapy, has proven that *MET* inhibitors are effective anti-cancer agents *in vivo* (123). A sequence encoding a soluble *MET* receptor, including the entire extracellular domain (decoy *MET*), was chosen for its ability to compete with the wild-type receptor for ligand binding and to inhibit receptor dimerization (Figure 7). Decoy *MET* was inserted into lentiviral vectors, and administered locally or systemically to mice bearing tumor xenografts, where it inhibited tumor cell proliferation and survival, and the formation of spontaneous metastasis.

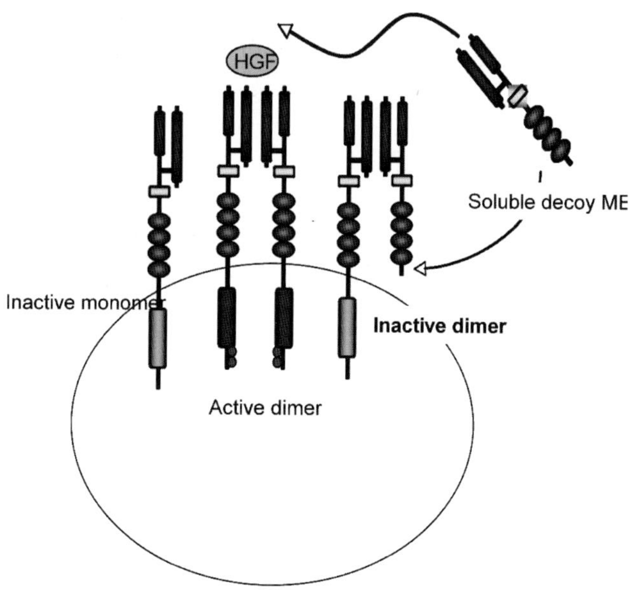

Figure 7. Mechanism of action of decoy MET. This soluble form of the extracellular domain of MET compete both for ligand and receptor binding, inhibiting receptor dimerization and activation.

The target for decoy *MET* includes not only tumor cells themselves, but also the tumor microenvironment, where decoy *MET* is capable of strikingly impairing neo-angiogenesis. Surprisingly, the latter effect can be sufficient to impair growth of tumors that do not express *MET* as well. The legitimate concerns about possible adverse effects of decoy *MET* systemic administration can be apparently neglected, at least in the limits of housekeeping physiological functions, which were preserved in treated animals. The efficacy of decoy *MET* is supported by studies indicating that the isolated Sema domain of the *MET* extracellular region (which is present in decoy *MET*) is a powerful inhibitor of *MET*-mediated signal transduction and invasive growth of tumor cells (178). In another recent study, the same experimental setting employed to investigate decoy *MET* was used to test the efficacy of a mutant of SF/HGF, which is uncleavable by uPA or other pro-HGF convertases, and thus can not be activated. Also in this case, the engineered molecule, administered to xenografted animals, was able to inhibit tumor growth and dissemination, without affecting physiological functions. The mechanisms of action of uncleavable SF/HGF involves competition with wild-type SF/HGF for receptor binding and also inhibition of proteases that converts wild-type proHGF into its mature, biologically active form. We can conclude that these studies provide a strong proof-of-concept for further testing of HGF/*MET* antagonists in clinical trials, and offer hopes to treat the –until now- incurable stage of tumor metastatization.

ACKNOWLEDGEMENTS

We thank G. Sabatino for micrographs and A. Cignetto for excellent secretarial assistance.

REFERENCES

1. Stoker M, Gherardi E, Perryman M, Gray J. Scatter factor is a fibroblast-derived modulator of epithelial cell mobility. Nature 1987, 327: 239-242.
2. Nakamura T, Teramoto H, Ichihara A. Purification and characterization of a growth factor from rat platelets for mature parenchymal hepatocytes in primary cultures. Proc. Natl. Acad. Sci. USA 1986, 83: 6489-6493.
3. Zarnegar R, Michalopoulos G. Purification and biological characterization of human hepatopoietin A, a polypeptide growth factor for hepatocytes. Cancer Res. 1989, 49: 3314-3320.
4. Nakamura T, et al. Molecular cloning and expression of human hepatocyte growth factor. Nature 1989, 342: 440-443.
5. Weidner KM, et al. Evidence for the identity of human scatter factor and human hepatocyte growth factor. Proc. Natl. Acad. Sci. USA 1991, 88: 7001-7005.
6. Naldini L, et al. Scatter factor and hepatocyte growth factor are indistinguishable ligands for the MET receptor. EMBO J. 1991, 10: 2867-2878.
7. Bottaro DP, et al. Identification of the hepatocyte growth factor receptor as the c-met proto-oncogene product. Science, 1991, 251: 802-804.
8. Skeel A, et al. Macrophage stimulating protein: purification, partial amino acid sequence, and cellular activity. J. Exp. Med. 1991, 173: 1227-1234.
9. Gaudino G, et al. RON is a heterodimeric tyrosine kinase receptor activated by the HGF homologue MSP. EMBO Journal 1994, 13: 3524-3532.

10. Wang MH, et al. Identification of the ron gene product as the receptor for the human macrophage stimulating protein. Science 1994, 266: 117-119.
11. Huff JL, Jelinek MA, Borgman CA, Lansing TJ, Parsons JT. The protooncogene c-sea encodes a transmembrane protein-tyrosine kinase related to the Met/hepatocyte growth factor/scatter factor receptor. Proc. Natl. Acad. Sci. USA 1993, 90: 6140-6144.
12. Wahl RC, et al. Chicken macrophage stimulating protein is a ligand of the receptor protein-tyrosine kinase Sea. J. Biol. Chem. 1999, 274: 26361-26368.
13. Cooper CS, et al. Molecular cloning of a new transforming gene from a chemically transformed human cell line. Nature 1984, 311: 29-33.
14. Park M, et al. Mechanism of met oncogene activation. Cell 1986, 45: 895-904.
15. Giordano S, Ponzetto C, Di Renzo, MF, Cooper CS, Comoglio PM. Tyrosine kinase receptor indistinguishable from the c-met protein. Nature 1989, 339: 155-156.
16. Rodrigues GA, Park M. Dimerization mediated through a leucine zipper activates the oncogenic potential of the met receptor tyrosine kinase. Mol. Cell Biol. 1993, 13: 6711-6722.
17. Schmidt L, et al. Germline and somatic mutations in the tyrosine kinase domain of the MET proto-oncogene in papillary renal carcinomas. Nat. Genet. 1997, 16: 68-73.
18. Rong S, Segal S, Anver M, Resau JH, Vande WG. Invasiveness and metastasis of NIH 3T3 cells induced by Met-hepatocyte growth factor/scatter factor autocrine stimulation. Proc. Natl. Acad. Sci. USA 1994, 91: 4731-4735.
19. Jeffers M, Rong S, Anver M, Vande WG. Autocrine hepatocyte growth factor/scatter factor-Met signaling induces transformation and the invasive/metastastic phenotype in C127 cells. Oncogene 1996, 13: 853-856.
20. Giordano S, et al. A point mutation in the MET oncogene abrogates metastasis without affecting transformation. Proc. Natl. Acad. Sci. USA 1997, 94: 13868-13872.
21. Meiners S, Brinkmann V, Naundorf H, Birchmeier W. Role of morphogenetic factors in metastasis of mammary carcinoma cells. Oncogene 1998, 16: 9-20.
22. Bardelli A, et al. Concomitant activation of pathways downstream of Grb2 and PI 3-kinase is required for MET-mediated metastasis. Oncogene 1999, 18: 1139-1146.
23. Di Renzo MF, et al. Overexpression and amplification of the met/HGF receptor gene during the progression of colorectal cancer. Clin. Cancer Res. 1995, 1: 147-154.
24. Di Renzo MF, et al. Somatic mutations of the MET oncogene are selected during metastatic spread of human HNSC carcinomas. Oncogene 2000, 19: 1547-1555.
25. Thiery JP. Epithelial-mesenchymal transitions in tumour progression. Nat. Rev. Cancer 2002, 2: 442-454.
26. Montesano R, Matsumoto K, Nakamura T, Orci L. Identification of a fibroblast-derived epithelial morphogen as hepatocyte growth factor. Cell 1991, 67: 901-908.
27. Medico E, et al. The tyrosine kinase receptors Ron and Sea control "scattering" and morphogenesis of liver progenitor cells in vitro. Mol. Biol. Cell 1996, 7: 495-504.
28. Birchmeier C, Gherardi E. Developmental roles of HGF/SF and its receptor, the c-Met tyrosine kinase. Trends Cell Biol. 1998, 8: 404-410.
29. Birchmeier C, Birchmeier W, Gherardi E, Vande Woude GF. Met, metastasis, motility and more. Nat. Rev. Mol. Cell Biol. 2003, 4: 915-925.
30. Comoglio PM, Trusolino L. Invasive growth: from development to metastasis. J Clin. Invest 2002, 109: 857-862.
31. Trusolino L, Comoglio PM. Scatter-factor and semaphorin receptors: cell signalling for invasive growth. Nat. Rev. Cancer 2002, 4: 289-300.

32. O'Brien LE, Zegers MM, Mostov KE. Opinion: Building epithelial architecture: insights from three-dimensional culture models. Nat. Rev. Mol. Cell Biol. 2002, 3: 531-537.
33. Weidner KM, Behrens J, Vandekerckhove J, Birchmeier W. Scatter factor: molecular characteristics and effect on the invasiveness of epithelial cells. J. Cell Biol. 1990, 111: 2097-2108.
34. Tamagnone L, et al. Plexins are a large family of receptors for transmembrane, secreted, and GPI-anchored semaphorins in vertebrates. Cell 1999, 99: 71-80.
35. Giordano S, et al. The semaphorin 4D receptor controls invasive growth by coupling with Met. Nat. Cell Biol. 2002, 4: 720-724.
36. Tamagnone L, Comoglio PM. To move or not to move? Semaphorin signalling in cell migration. EMBO Rep. 2004, 5: 356-361.
37. Naldini L, et al. Extracellular proteolytic cleavage by urokinase is required for activation of hepatocyte growth factor/scatter factor. EMBO J. 1992, 11: 4825-4833.
38. Naldini L, et al. Biological activation of pro-HGF (hepatocyte growth factor) by urokinase is controlled by a stoichiometric reaction. J. Biol. Chem. 1995, 270: 603-611.
39. Mars WM, Zarnegar R, Michalopoulos GK. Activation of hepatocyte growth factor by the plasminogen activators uPA and tPA. Am. J. Pathol. 1993, 143: 949-958.
40. Shimomura T, et al. Activation of the zymogen of hepatocyte growth factor activator by thrombin. J. Biol. Chem. 1993, 268: 22927-22932.
41. Shimomura T, et al. Activation of hepatocyte growth factor by two homologous proteases, blood-coagulation factor XIIa and hepatocyte growth factor activator. Eur. J. Biochem. 1995, 229: 257-261.
42. Hartmann G, et al. Engineered mutants of HGF/SF with reduced binding to heparan sulphate proteoglycans, decreased clearance and enhanced activity in vivo. Curr. Biol. 1998, 8: 125-134.
43. Chirgadze DY, et al. Crystal structure of the NK1 fragment of HGF/SF suggests a novel mode for growth factor dimerization and receptor binding. Nat. Struct. Biol. 1999, 6: 72-79.
44. Schwall RH, et al. Heparin induces dimerization and confers proliferative activity onto the hepatocyte growth factor antagonists NK1 and NK2. J. Cell Biol. 1996, 133: 709-718.
45. Lokker NA, et al. Structure-function analysis of hepatocyte growth factor: identification of variants that lack mitogenic activity yet retain high affinity receptor binding. EMBO J. 1992, 11: 2503-2510.
46. Gherardi E, et al. Functional map and domain structure of MET, the product of the c-met protooncogene and receptor for hepatocyte growth factor/scatter factor. Proc. Natl. Acad. Sci. USA 2003, 100: 12039-12044.
47. Gandino L, Longati P, Medico E, Prat M, Comoglio PM. Phosphorylation of serine 985 negatively regulates the hepatocyte growth factor receptor kinase. J. Biol. Chem. 1994, 269: 1815-1820.
48. Peschard P, et al. Mutation of the c-Cbl TKB domain binding site on the Met receptor tyrosine kinase converts it into a transforming protein. Mol. Cell 2001, 8: 995-1004.
49. Petrelli A, et al. The endophilin-CIN85-Cbl complex mediates ligand-dependent downregulation of c-Met Nature 2002, 416: 187-190.
50. Naldini L, et al. The tyrosine kinase encoded by the MET proto-oncogene is activated by autophosphorylation. Mol. Cell Biol. 1991, 11: 1793-1803.

51. Longati P, Bardelli A, Ponzetto C, Naldini L, Comoglio PM. Tyrosines1234-1235 are critical for activation of the tyrosine kinase encoded by the MET proto-oncogene (HGF receptor). Oncogene 1994, 9: 49-57.
52. Chiara F, Michieli P, Pugliese L, Comoglio PM. Mutations in the met oncogene unveil a "dual switch" mechanism controlling tyrosine kinase activity. J. Biol. Chem. 2003, 278: 29352-29358.
53. Ponzetto C, et al. A multifunctional docking site mediates signaling and transformation by the hepatocyte growth factor/scatter factor receptor family. Cell 1994, 77: 261-271.
54. Maina F, et al. Uncoupling of Grb2 from the Met receptor in vivo reveals complex roles in muscle development. Cell 1996, 87: 531-542.
55. Bardelli A, et al. Uncoupling signal transducers from oncogenic MET mutants abrogates cell transformation and inhibits invasive growth. Proc. Natl. Acad. Sci. USA 1998, 95: 14379-14383.
56. Sachs M, et al. Motogenic and morphogenic activity of epithelial receptor tyrosine kinases. J. Cell Biol. 1996, 133: 1095-1107.
57. Graziani A, Gramaglia D, Cantley LC, Comoglio PM. The tyrosine-phosphorylated hepatocyte growth factor/scatter factor receptor associates with phosphatidylinositol 3-kinase. J. Biol. Chem. 1991, 266: 22087-22090.
58. Graziani A, Gramaglia D, Dalla ZP, Comoglio PM. Hepatocyte growth factor/scatter factor stimulates the Ras-guanine nucleotide exchanger. J. Biol. Chem. 1993, 268: 9165-9168.
59. Ponzetto C, et al. A novel recognition motif for phosphatidylinositol 3-kinase binding mediates its association with the hepatocyte growth factor/scatter factor receptor. Mol. Cell Biol. 1993, 13: 4600-4608.
60. Boccaccio C, et al. Induction of epithelial tubules by growth factor HGF depends on the STAT pathway. Nature 1998, 391: 285-288.
61. Hartmann G, Weidner KM, Schwarz H, Birchmeier W. The motility signal of scatter factor/hepatocyte growth factor mediated through the receptor tyrosine kinase met requires intracellular action of Ras. J. Biol. Chem. 1994, 269: 21936-21939.
62. Ridley AJ, Comoglio PM, Hall A. Regulation of scatter factor/hepatocyte growth factor responses by Ras, Rac, and Rho in MDCK cells. Mol. Cell Biol. 1995, 15: 1110-1122.
63. Ponzetto C, et al. Specific uncoupling of GRB2 from the Met receptor. Differential effects on transformation and motility. J. Biol. Chem. 1996, 271: 14119-14123.
64. Potempa S, Ridley AJ. Activation of both MAP kinase and phosphatidylinositide 3-kinase by ras is required for hepatocyte growth Factor/Scatter factor-induced adherens junction disassembly [In Process Citation]. Mol. Biol. Cell 1998, 9: 2185-2200.
65. Royal I, Park M. Hepatocyte growth factor-induced scatter of Madin-Darby canine kidney cells requires phosphatidylinositol 3-kinase. J. Biol. Chem. 1995, 270: 27780-27787.
66. Trusolino L, et al. HGF/scatter factor selectively promotes cell invasion by increasing integrin avidity. FASEB J. 2000, 14: 1629-1640.
67. Royal I, Fournier TM, Park M. Differential requirement of Grb2 and PI3-kinase in HGF/SF-induced cell motility and tubulogenesis. J. Cell Physiol. 1997, 173: 196-201.
68. Xiao GH, et al. Anti-apoptotic signaling by hepatocyte growth factor/Met via the phosphatidylinositol 3-kinase/Akt and mitogen-activated protein kinase pathways. Proc. Natl. Acad. Sci. USA 2001, 98: 247-252.
69. Frisch SM, Ruoslahti E. Integrins and anoikis. Curr. Opin. Cell Biol. 1997, 9: 701-706.

70. Fournier TM, Kamikura D, Teng K, Park M. Branching tubulogenesis but not scatter of madin-darby canine kidney cells requires a functional Grb2 binding site in the Met receptor tyrosine kinase. J. Biol. Chem. 1996, 271: 22211-22217.
71. Khwaja A, Lehmann K, Marte BM, Downward J. Phosphoinositide 3-kinase induces scattering and tubulogenesis in epithelial cells through a novel pathway. J. Biol. Chem. 1998, 273: 18793-18801.
72. Gual P, et al. Sustained recruitment of phospholipase C-gamma to Gab1 is required for HGF-induced branching tubulogenesis. Oncogene 2000, 19: 1509-1518.
73. Muller M, Morotti A, Ponzetto C. Activation of NF-kappaB is essential for hepatocyte growth factor-mediated proliferation and tubulogenesis. Mol. Cell Biol. 2002, 22: 1060-1072.
74. Boccaccio C, Ando' M, Comoglio PM. A differentiation switch for genetically modified hepatocytes. FASEB J. 2002, 16: 120-122.
75. O'Brien LE, et al. ERK and MMPs sequentially regulate distinct stages of epithelial tubule development. Dev. Cell 2004, 7: 21-32.
76. Marshall CJ. Specificity of receptor tyrosine kinase signaling: transient versus sustained extracellular signal-regulated kinase activation. Cell 1995, 80: 179-185.
77. Weidner KM, et al. Interaction between Gab1 and the c-Met receptor tyrosine kinase is responsible for epithelial morphogenesis. Nature 1996, 384: 173-176.
78. Maroun CR, et al. The Gab1 PH domain is required for localization of Gab1 at sites of cell-cell contact and epithelial morphogenesis downstream from the met receptor tyrosine kinase. Mol. Cell Biol. 1999, 19: 1784-1799.
79. Sachs M, et al. Essential role of Gab1 for signaling by the c-Met receptor in vivo. J. Cell Biol. 2000, 150: 1375-1384.
80. Bardelli A, Longati P, Gramaglia D, Stella MC, Comoglio PM. Gab1 coupling to the HGF/Met receptor multifunctional docking site requires binding of Grb2 and correlates with the transforming potential. Oncogene 1997, 15: 3103-3111.
81. Nguyen L, et al. Association of the multisubstrate docking protein Gab1 with the hepatocyte growth factor receptor requires a functional Grb2 binding site involving tyrosine 1356. J. Biol. Chem. 1997, 272: 20811-20819.
82. Holgado-Madruga M, Emlet DR, Moscatello DK, Godwin AK, Wong AJ. A Grb2-associated docking protein in EGF- and insulin-receptor signalling. Nature 1996, 379: 560-564.
83. Schaeper U, et al. Coupling of Gab1 to c-Met, Grb2, and Shp2 mediates biological responses. J. Cell Biol. 2000, 149: 1419-1432.
84. Neel BG, Gu H, Pao L. The 'Shp'ing news: SH2 domain-containing tyrosine phosphatases in cell signaling. Trends Biochem. Sci. 2003, 28: 284-293.
85. Sakkab D, et al. Signaling of hepatocyte growth factor/scatter factor (HGF) to the small GTPase Rap1 via the large docking protein Gab1 and the adapter protein CRKL. J. Biol. Chem. 2000, 275: 10772-10778.
86. York RD, et al. Rap1 mediates sustained MAP kinase activation induced by nerve growth factor. Nature 1998, 392: 622-626.
87. Conrotto P, Corso S, Gamberini S, Comoglio PM, Giordano S. Interplay between scatter factor receptors and B plexins controls invasive growth. Oncogene 2004, 23: 5131-5137.
88. Trusolino L, Bertotti A, Comoglio PM. A signaling adapter function for alpha6beta4 integrin in the control of HGF-dependent invasive growth. Cell 2001, 107: 643-654.

89. Chung J, Yoon SO, Lipscomb EA, Mercurio AM. The Met receptor and alpha 6 beta 4 integrin can function independently to promote carcinoma invasion. J. Biol. Chem. 2004, 279: 32287-32293.
90. Orian-Rousseau V, Chen L, Sleeman JP, Herrlich P, Ponta H. CD44 is required for two consecutive steps in HGF/c-Met signaling. Genes Dev. 2002, 16: 3074-3086.
91. Ponta H, Sherman L, Herrlich PA. CD44: from adhesion molecules to signalling regulators. Nat. Rev. Mol. Cell Biol. 2003, 4: 33-45.
92. Crepaldi T, Gautreau A, Comoglio PM, Louvard D, Arpin M. Ezrin is an effector of hepatocyte growth factor-mediated migration and morphogenesis in epithelial cells. J. Cell Biology 1997, 138: 423-434.
93. McClatchey AI. Merlin and ERM proteins: unappreciated roles in cancer development? Nat. Rev. Cancer 2003, 3: 877-883.
94. Medico E, et al. Osteopontin is an autocrine mediator of hepatocyte growth factor-induced invasive growth. Cancer Res. 2001, 61: 5861-5868.
95. Kang Y, et al. A multigenic program mediating breast cancer metastasis to bone. Cancer Cell 2003, 3: 537-549.
96. Ye QH, et al. Predicting hepatitis B virus-positive metastatic hepatocellular carcinomas using gene expression profiling and supervised machine learning. Nat. Med. 2003, 9: 416-423.
97. Nagafuchi A. Molecular architecture of adherens junctions. Curr. Opin. Cell Biol. 2001, 13: 600-603.
98. Tannapfel A, Yasui W, Yokozaki H, Wittekind C, Tahara E. Effect of hepatocyte growth factor on the expression of E- and P-cadherin in gastric carcinoma cell lines. Virchows Arch. 1994, 425: 139-144.
99. Miura H, et al. Effects of hepatocyte growth factor on E-cadherin-mediated cell-cell adhesion in DU145 prostate cancer cells. Urology 2001, 58: 1064-1069.
100. Khoury H, et al. HGF Converts ErbB2/Neu Epithelial Morphogenesis to Cell Invasion. Mol. Biol. Cell 2004,
101. Balkovetz DF, Pollack AL, Mostov KE. Hepatocyte growth factor alters the polarity of Madin-Darby canine kidney cell monolayers. J. Biol. Chem. 1997, 272: 3471-3477.
102. Balkovetz DF, Sambandam V. Dynamics of E-cadherin and gamma-catenin complexes during dedifferentiation of polarized MDCK cells. Kidney Int. 1999, 56: 910-921.
103. Davies G, Jiang WG, Mason MD. Matrilysin mediates extracellular cleavage of E-cadherin from prostate cancer cells: a key mechanism in hepatocyte growth factor/scatter factor-induced cell-cell dissociation and in vitro invasion. Clin. Cancer Res. 2001, 7: 3289-3297.
104. Herynk MH, Tsan R, Radinsky R, Gallick GE. Activation of c-Met in colorectal carcinoma cells leads to constitutive association of tyrosine-phosphorylated beta-catenin. Clin. Exp. Metastasis 2003, 20: 291-300.
105. Shibamoto S, et al. Tyrosine phosphorylation of beta-catenin and plakoglobin enhanced by hepatocyte growth factor and epidermal growth factor in human carcinoma cells. Cell Adhes. Commun. 1994, 1: 295-305.
106. Cavallaro U, Christofori G. Cell adhesion and signalling by cadherins and Ig-CAMs in cancer. Nat. Rev. Cancer 2004, 4: 118-132.
107. Cano A, et al. The transcription factor snail controls epithelial-mesenchymal transitions by repressing E-cadherin expression. Nat. Cell Biol. 2000, 2: 76-83.
108. Nieto MA. The snail superfamily of zinc-finger transcription factors. Nat. Rev. Mol. Cell Biol. 2002, 3: 155-166.
109. Schlessinger K, Hall A. GSK-3beta sets Snail's pace. Nat. Cell Biol. 2004, 6: 913-915.

110. Zhou BP, et al. Dual regulation of Snail by GSK-3beta-mediated phosphorylation in control of epithelial-mesenchymal transition. Nat. Cell Biol. 2004, 6: 931-940.
111. Nebe B, Sanftleben H, Pommerenke H, Peters A, Rychly J. Hepatocyte growth factor enables enhanced integrin-cytoskeleton linkage by affecting integrin expression in subconfluent epithelial cells. Exp. Cell Res. 1998, 243: 263-273.
112. Liang CC, Chen HC. Sustained activation of extracellular signal-regulated kinase stimulated by hepatocyte growth factor leads to integrin alpha 2 expression that is involved in cell scattering. J. Biol. Chem. 2001, 276: 21146-21152.
113. Trusolino L, et al. Growth factor-dependent activation of alphavbeta3 integrin in normal epithelial cells: implications for tumor invasion. J. Cell Biol. 1998, 142: 1145-1156.
114. Egeblad M, Werb Z. New functions for the matrix metalloproteinases in cancer progression. Nat. Rev. Cancer 2002, 2: 161-174.
115. Rosenthal EL, et al. Role of the plasminogen activator and matrix metalloproteinase systems in epidermal growth factor- and scatter factor-stimulated invasion of carcinoma cells. Cancer Res. 1998, 58: 5221-5230.
116. Nabeshima K, et al. Front-cell-specific expression of membrane-type 1 matrix metalloproteinase and gelatinase A during cohort migration of colon carcinoma cells induced by hepatocyte growth factor/scatter factor. Cancer Res. 2000, 60: 3364-3369.
117. Monvoisin A, et al. Involvement of matrix metalloproteinase type-3 in hepatocyte growth factor-induced invasion of human hepatocellular carcinoma cells. Int. J. Cancer 2002, 97: 157-162.
118. Balkovetz DF, et al. Gene expression alterations during HGF-induced dedifferentiation of a renal tubular epithelial cell line (MDCK) using a novel canine DNA microarray. Am. J. Physiol. Renal. Physiol. 2004, 286: F702-F710.
119. Jeffers M, Rong S, Vande Woude GF. Enhanced tumorigenicity and invasion-metastasis by hepatocyte growth factor/scatter factor-met signalling in human cells concomitant with induction of the urokinase proteolysis network. Mol. Cell Biol. 1996, 16: 1115-1125.
120. Pepper MS, Matsumoto K, Nakamura T, Orci L, Montesano R. Hepatocyte growth factor increases urokinase-type plasminogen activator (u-PA) and u-PA receptor expression in Madin-Darby canine kidney epithelial cells. J. Biol. Chem. 1992, 267: 20493-20496.
121. Sidenius N, Blasi F. The urokinase plasminogen activator system in cancer: recent advances and implication for prognosis and therapy. Cancer Metastasis Rev. 2003, 22: 205-222.
122. Bussolino F, et al. Hepatocyte growth factor is a potent angiogenic factor which stimulates endothelial cell motility and growth. J. Cell Biol. 1992, 119: 629-641.
123. Michieli P, et al. Targeting the tumor and its microenvironment by a dual-function decoy Met receptor. Cancer Cell 2004, 6: 61-73.
124. Liang TJ, Reid AE, Xavier R, Cardiff RD, Wang TC. Transgenic expression of tpr-met oncogene leads to development of mammary hyperplasia and tumors. J. Clin. Invest 1996, 97: 2872-2877.
125. Soman NR, Correa P, Ruiz BA, Wogan GN. The TPR-MET oncogenic rearrangement is present and expressed in human gastric carcinoma and precursor lesions. Proc. Natl. Acad. Sci. USA 1991, 88: 4892-4896.
126. Schmidt L, et al. Two North American families with hereditary papillary renal carcinoma and identical novel mutations in the MET proto-oncogene. Cancer Res. 1998, 58: 1719-1722.

127. Olivero M, et al. Novel mutation in the ATP-binding site of the MET oncogene tyrosine kinase in a HPRCC family. Int. J. Cancer 1999, 82: 640-643.
128. Schmidt L, et al. Novel mutations of the MET proto-oncogene in papillary renal carcinomas. Oncogene 1999, 18: 2343-2350.
129. Park WS, et al. Somatic mutations in the kinase domain of the Met/hepatocyte growth factor receptor gene in childhood hepatocellular carcinomas. Cancer Res. 1999, 59: 307-310.
130. Lee JH, et al. A novel germ line juxtamembrane Met mutation in human gastric cancer. Oncogene 2000, 19: 4947-4953.
131. Danilkovitch-Miagkova A, Zbar B. Dysregulation of Met receptor tyrosine kinase activity in invasive tumors. J. Clin. Invest 2002, 109: 863-867.
132. Alberti L, Carniti C, Miranda C, Roccato E, Pierotti MA. RET and NTRK1 proto-oncogenes in human diseases. J. Cell Physiol. 2003, 195: 168-186.
133. Fletcher JA. Role of KIT and platelet-derived growth factor receptors as oncoproteins. Semin. Oncol. 2004, 31: 4-11.
134. Michieli P, et al. Mutant Met-mediated transformation is ligand-dependent and can be inhibited by HGF antagonists. Oncogene 1999, 18: 5221-5231.
135. Ferracini R, et al. The Met/HGF receptor is over-expressed in human osteosarcomas and is activated by either a paracrine or an autocrine circuit. Oncogene 1995, 10: 739-749.
136. Scotlandi K, et al. Expression of Met/hepatocyte growth factor receptor gene and malignant behavior of musculoskeletal tumors. Am. J. Pathol. 1996, 149: 1209-1219.
137. Ferracini R, et al. Retrogenic expression of the MET proto-oncogene correlates with the invasive phenotype of human rhabdomyosarcomas. Oncogene 1996, 12: 1697-1705.
138. Koochekpour S, et al. Met and hepatocyte growth factor/scatter factor expression in human gliomas. Cancer Res. 1997, 57: 5391-5398.
139. Tuck AB, Park M, Sterns EE, Boag A, Elliott BE. Coexpression of hepatocyte growth factor and receptor (Met) in human breast carcinoma. Am. J. Pathol. 1996, 148: 225-232.
140. Yao Y, et al. Scatter factor protein levels in human breast cancers: clinicopathological and biological correlations. Am. J. Pathol. 1996, 149: 1707-1717.
141. Takayama H, et al. Diverse tumorigenesis associated with aberrant development in mice overexpressing hepatocyte growth factor/scatter factor. Proc. Natl. Acad. Sci. USA 1997, 94: 701-706.
142. Otsuka T, et al. c-Met autocrine activation induces development of malignant melanoma and acquisition of the metastatic phenotype. Cancer Res. 1998, 58: 5157-5167.
143. Gallego MI, Bierie B, Hennighausen L. Targeted expression of HGF/SF in mouse mammary epithelium leads to metastatic adenosquamous carcinomas through the activation of multiple signal transduction pathways. Oncogene 2003, 22: 8498-8508.
144. Takeuchi H, et al. c-MET expression level in primary colon cancer: a predictor of tumor invasion and lymph node metastases. Clin. Cancer Res. 2003, 9: 1480-1488.
145. Wang R, Ferrell LD, Faouzi S, Maher JJ, Bishop JM. Activation of the Met receptor by cell attachment induces and sustains hepatocellular carcinomas in transgenic mice. J. Cell Biol. 2001, 153: 1023-1034.
146. Boccaccio C, Gaudino G, Gambarotta G, Galimi F, Comoglio PM. Hepatocyte growth factor (HGF) receptor expression is inducible and is part of the delayed-early response to HGF. J. Biol. Chem. 1994, 269: 12846-12851.

147. Gambarotta G, Pistoi S, Giordano S, Comoglio PM, Santoro C. Structure and inducible regulation of the human MET promoter. J. Biol. Chem. 1994, 269: 12852-12857.
148. Ivan M, Bond JA, Prat M, Comoglio PM, Wynford-Thomas D. Activated ras and ret oncogenes induce over-expression of c-met (hepatocyte growth factor receptor) in human thyroid epithelial cells. Oncogene 1997, 14: 2417-2423.
149. Webb CP, et al. Evidence for a role of Met-HGF/SF during Ras-mediated tumorigenesis/metastasis. Oncogene 1998, 17: 2019-2025.
150. Gambarotta G, et al. Ets up-regulates MET transcription. Oncogene 1996, 13: 1911-1917.
151. Paumelle R, et al. Hepatocyte growth factor/scatter factor activates the ETS1 transcription factor by a RAS-RAF-MEK-ERK signaling pathway. Oncogene 2002, 21: 2309-2319.
152. Trojanowska M. Ets factors and regulation of the extracellular matrix. Oncogene 2000, 19: 6464-6471.
153. Pennacchietti S, et al. Hypoxia promotes invasive growth by transcriptional activation of the met protooncogene. Cancer Cell 2003, 3: 347-361.
154. Semenza GL. Targeting HIF-1 for cancer therapy. Nat. Rev. Cancer 2003, 3: 721-732.
155. Maranchie JK, et al. The contribution of VHL substrate binding and HIF1-alpha to the phenotype of VHL loss in renal cell carcinoma. Cancer Cell 2002, 1: 247-255.
156. Rofstad EK, et al. Hypoxia promotes lymph node metastasis in human melanoma xenografts by up-regulating the urokinase-type plasminogen activator receptor. Cancer Res. 2002, 62: 1847-1853.
157. Muller A, et al. Involvement of chemokine receptors in breast cancer metastasis. Nature 2001, 410: 50-56.
158. Hockel M, Vaupel P. Biological consequences of tumor hypoxia. Semin. Oncol. 2001, 28: 36-41.
159. Steeg PS. Angiogenesis inhibitors: motivators of metastasis? Nat. Med. 2003, 9: 822-823.
160. Galimi F, et al. Hepatocyte growth factor induces proliferation and differentiation of multipotent and erythroid hemopoietic progenitors. J. Cell Biol. 1994, 127: 1743-1754.
161. Suzuki A, et al. Clonal identification and characterization of self-renewing pluripotent stem cells in the developing liver. J. Cell Biol. 2002, 156: 173-184.
162. Zheng YW, Taniguchi H. Diversity of hepatic stem cells in the fetal and adult liver. Semin. Liver Dis. 2003, 23: 337-348.
163. Bladt F, Riethmacher D, Isenmann S, Aguzzi A, Birchmeier C. Essential role for the c-met receptor in the migration of myogenic precursor cells into the limb bud. Nature 1995, 376: 768-771.
164. Anastasi S, et al. A natural hepatocyte growth factor/scatter factor autocrine loop in myoblast cells and the effect of the constitutive Met kinase activation on myogenic differentiation. J. Cell Biol. 1997, 137: 1057-1068.
165. Tu SM, Lin SH, Logothetis CJ. Stem-cell origin of metastasis and heterogeneity in solid tumours. Lancet Oncol. 2002, 3: 508-513.
166. Sharp R, et al. Synergism between INK4a/ARF inactivation and aberrant HGF/SF signaling in rhabdomyosarcomagenesis. Nat. Med. 2002, 8: 1276-1280.
167. Boon EM, van der NR, van de WM, Clevers H, Pals ST. Wnt signaling regulates expression of the receptor tyrosine kinase met in colorectal cancer. Cancer Res. 2002, 62: 5126-5128.
168. van de Wetering M, et al. The beta-catenin/TCF-4 complex imposes a crypt progenitor phenotype on colorectal cancer cells. Cell 2002, 111: 241-250.

169. Wright DE, Wagers AJ, Gulati AP, Johnson FL, Weissman IL. Physiological migration of hematopoietic stem and progenitor cells. Science 2001, 294: 1933-1936.
170. Lynch TJ, et al. Activating mutations in the epidermal growth factor receptor underlying responsiveness of non-small-cell lung cancer to gefitinib. N. Engl. J. Med. 2004, 350: 2129-2139.
171. Morotti A, Mila S, Accornero P, Tagliabue E, Ponzetto C. K252a inhibits the oncogenic properties of Met, the HGF receptor. Oncogene 2002, 21: 4885-4893.
172. Christensen JG, et al. A selective small molecule inhibitor of c-Met kinase inhibits c-Met-dependent phenotypes in vitro and exhibits cytoreductive antitumor activity in vivo. Cancer Res. 2003, 63: 7345-7355.
173. Atabey N, et al. Potent blockade of hepatocyte growth factor-stimulated cell motility, matrix invasion and branching morphogenesis by antagonists of Grb2 Src homology 2 domain interactions. J. Biol. Chem. 2001, 276: 14308-14314.
174. Webb CP, et al. The geldanamycins are potent inhibitors of the hepatocyte growth factor/scatter factor-met-urokinase plasminogen activator-plasmin proteolytic network. Cancer Res. 2000, 60: 342-349.
175. Abounader R, et al. In vivo targeting of SF/HGF and c-met expression via U1snRNA/ribozymes inhibits glioma growth and angiogenesis and promotes apoptosis. FASEB J. 2002, 16: 108-110.
176. Cao B, et al. Neutralizing monoclonal antibodies to hepatocyte growth factor/scatter factor (HGF/SF) display antitumor activity in animal models. Proc. Natl. Acad. Sci. USA 2001, 98: 7443-7448.
177. Prat M, Crepaldi T, Pennacchietti S, Bussolino F, Comoglio PM. Agonistic monoclonal antibodies against the Met receptor dissect the biological responses to HGF. J. Cell Sci. 1998, 111: 237-247.
178. Kong-Beltran M, Stamos J, Wickramasinghe D. The Sema domain of Met is necessary for receptor dimerization and activation. Cancer Cell 2004, 6: 75-84.
179. Nakajima M, et al. The prognostic significance of amplification and overexpression of c-met and c-erb B-2 in human gastric carcinomas. Cancer 1999, 85: 1894-1902.
180. Di Renzo MF, et al. Overexpression of the c-MET/HGF receptor gene in human thyroid carcinomas. Oncogene 1992, 7: 2549-2553.
181. Ramirez R, et al. Over-expression of hepatocyte growth factor/scatter factor (HGF/SF) and the HGF/SF receptor (cMET) are associated with a high risk of metastasis and recurrence for children and young adults with papillary thyroid carcinoma. Clin. Endocrinol. (Oxf) 2000, 53: 635-644.
182. Nardone HC, et al. c-Met expression in tall cell variant papillary carcinoma of the thyroid. Cancer 2003, 98: 1386-1393.
183. Natali PG, et al. Overexpression of the met/HGF receptor in renal cell carcinomas. Int. J. Cancer 1996, 69: 212-217.
184. Sweeney P, El Naggar AK, Lin SH, Pisters LL. Biological significance of c-met over expression in papillary renal cell carcinoma. J. Urol. 2002, 168: 51-55.
185. Di Renzo MF, et al. Overexpression of the Met/HGF receptor in ovarian cancer. Int. J. Cancer 1994, 58: 658-662.
186. Huntsman D, Resau JH, Klineberg E, Auersperg N. Comparison of c-met expression in ovarian epithelial tumors and normal epithelia of the female reproductive tract by quantitative laser scan microscopy. Am. J. Pathol. 1999, 155: 343-348.
187. Wallenius V, et al. Overexpression of the hepatocyte growth factor (HGF) receptor (Met) and presence of a truncated and activated intracellular HGF receptor fragment in

locally aggressive/malignant human musculoskeletal tumors. Am. J. Pathol. 2000, 156: 821-829.

Chapter 7

MATRIX METALLOPROTEINASES IN TUMOR PROGRESSION

William G. Stetler-Stevenson and Dong-Wan Seo
Cell & Cancer Biology Branch, Vascular Biology Faculty, Center for Cancer Research, National Cancer Institute, National Institutes of Health, Bethesda, MD, USA

Abstract: The matrix metalloproteinases (MMPs) are well established as mediators of tumor invasion and metastasis. The classic view that these enzymes simply provide a mechanism for the breakdown of connective tissue barriers led to development of synthetic MMP inhibitors for cancer therapy. The lack of understanding of the complex roles of MMPs in cancer accounts for the failure of this strategy to significantly impact cancer therapy. It is now recognized that members of the MMP family function at all stages of cancer development to both promote and inhibit tumor progression. This overview summarizes recent evidence to support the emerging roles for MMP in all aspects of cancer progression, including tumor cell growth, programmed cell death, and tumor angiogenesis in addition to their classic role in cell invasion and metastasis. The MMP-dependent stimulation of tumor cell growth, stimulation of cell migration, and generation of cryptic sites and signals from within the extracellular matrix (ECM) are reviewed.

Key words: extracellular matrix, migration, proliferation, matrikine signaling, cryptic factors

1. INTRODUCTION

The progression of malignant tumors depends on the successful acquisition of several important phenotypes. These include the continuous (autonomous) growth of the tumor cells, evasion of programmed cell death (apoptosis), overcoming growth inhibitory signals, development of a

A. Wells (ed.), Cell Motility in Cancer Invasion and Metastasis, 143-158.

sustained angiogenic response, local invasion of surrounding tissues and travel to distant organs (metastasize) (1). Tumor invasion and metastasis are an important aspect of tumor progression; the formation of tumor metastasis is the principal factor contributing to cancer morbidity and mortality. Our understanding of the molecular, genetic and biological basis for tumor progression has increased considerably in the last two decades. Numerous studies have demonstrated that successfully invasive tumor cells must coordinate with stromal components (fibroblasts, inflammatory cells and endothelial cells) to orchestrate a precise spatial and temporal co-operation between cell adhesion molecules, cytoskeletal elements, extracellular matrix-degrading proteases and regulatory factors (2, 3).

The MMPs have been implicated in tumor invasion for over twenty years. Early studies focused on the requirement for MMP activity in order for tumor cells to invade extracellular matrix (ECM) barriers such as the sub-epithelial basement membrane. The demonstration that endogenous MMP inhibitors, the tissue inhibitors of metalloproteinases (TIMPs), could block tumor growth and invasion led to the development of synthetic MMP inhibitors for cancer treatment. However, the failure of this strategy to successfully limit tumor progression in cancer patients has been well documented (2, 3). Although the reasons for the failure of this strategy are multiple, it has become apparent that the role of MMPs in cancer progression are complex, and that some components of the MMP proteome may function to inhibit tumor progression (i.e., inhibit tumor-induced angiogenesis) as well as promoting cell invasion. In the past ten years investigators have shown that MMPs are involved in all aspects of tumor progression from regulation of cell growth, release of growth factors sequestered in the ECM, tumor angiogenesis, as well as local invasion and metastasis formation.

2. MATRIX METALLOPROTEINASES

The MMPs, also known as the Matrixins, are members of the metzincin super family of metalloenzymes that also include the ADAM (a metalloproteinase with a disintegrin and metalloproteinase domain), ADAM-TS (ADAM with a thrombospondin-like-domain) and astacin proteases. The mammalian MMP family consists of >20 members that include both soluble secreted proteases as well as cell membrane associated forms, referred to a membrane type- or MT-MMPs (4, 5). These two general classes are further divided into eight distinct sub-classes based on distinct structural features. The domain structure and substrate profiles of MMP family members are the subject of several excellent reviews and as such will not be covered in detail here (4, 5).

All members of the MMP family share a conserved zinc-atom binding site within the catalytic domain and are initially produced in zymogen form requiring proteolytic removal of a pro-domain to achieve enzymatic activity (4, 6). Most MMPs are activated extracellularly by serine- or metalloproteinase-mediated cleavage of their pro-domains, and may also require subsequent auto-proteolytic modification of the pro-domain. These activities disrupt the critical interaction of the sulfhydryl side chain donated by a highly conserved cysteine in the "activation locus" with the zinc atom bound at the catalytic active site. Activation of soluble MMPs is associated with a decrease in molecular mass corresponding to the loss of the pro-fragment. Many "activated" MMP species are thus identified by this shift in molecular weight. However, MT-MMPs, as well as MMP-11 and MMP-28, are activated prior to their appearance on the cell surface by a mechanism that involves furin-like proteases, which again cleave the MMP pro-fragment resulting a decrease in their molecular mass (4, 6).

Under physiologic conditions endogenous protease inhibitors tightly regulate the proteolytic action of MMP family members. These inhibitors include α2-macroglobulin, an abundant plasma protein that is ubiquitous in tissue fluids. MMPs have been shown to cleave the bait region in α2-macroglobulin, which results in sequestration of active MMPs into the "cage" region of this inhibitor (4). The most extensively studied MMP inhibitors are the endogenous inhibitors known as the tissue inhibitors of metalloproteinases or TIMPs. The TIMP family consists of four mammalian genes, *TIMP-1 –2, –3* and *–4* (7-9). The TIMP family members are distinguished by differential expression is various tissues. TIMP-2 is unique in that expression of this inhibitor is constitutive in most tissues, whereas TIMP-1, TIMP-3 and TIMP-4 protein expression is inducible (7-9). Furthermore, recent work suggest that the protease inhibitory activity of TIMP-1 is poor against members of the MT-MMP class, and that TIMP-1 and TIMP-3 inhibit MMPs, as well as members of the related ADAM protease family (7-9). In addition to α2-macroglobulin and TIMPs, a novel membrane-bound MMP inhibitor known as RECK (reversion-inducing cysteine rich protein with kazal motifs) has been identified (10, 11). RECK-deficient mice die during embryogenesis suggesting an important role for this inhibitor in regulation of ECM turnover (11).

The details of a least one cellular mechanism for MMP activation have been worked out. Pro-MMP-2 is activated at the cell surface by a unique pathway that involves MT1-MMP and TIMP-2 (4). This mechanism involves the binding of TIMP-2 to both MT1-MMP and pro-MMP-2 to form a cell surface ternary complex, in which pro-MMP-2 is activated by a second, adjacent molecule of MT1-MMP that is not inhibited by TIMP-2.

Following cleavage of the pro-MMP-2 pro-fragment the active MMP-2 is released from the cell surface. This model implies that inhibitor-free MT1-MMP binds TIMP-2 which then acts as a cell surface receptor for TIMP-2-free pro-MMP-2. However, recent studies have not confirmed that tumor cells or microvascular endothelial cells bind TIMP-2 exclusively via MT1-MMP and in many tumor cell cultures pro-MMP-2 is found exclusively in complex with TIMP-2 (12, 13).

Recent data also challenge the concept that MMP activation is irreversibly associated with the loss of the pro-fragment. Fedarko and colleagues demonstrated that small integrin binding ligand N-linked glycoproteins, also known as the SIBLING family, can bind and activate members of the MMP family, and that this activation can occur reversibly and without proteolytic processing of the MMP profragment (14). These authors demonstrate specificity in the interaction of members of the SIBLING family with MMPs. For example, bone sialoprotein (BSP) co-purified with MMP-2, osteopontin (OPN) with MMP-3 and DMP1 with MMP-9. Interestingly, SIBLING interaction with MMPs could also reverse inhibition of these proteases by the endogenous TIMPs. These findings suggest that members of the SIBLING family can control MMP activity by activation of latent pro-MMPs without pro-domain cleavage, as well as disruption of MMP-TIMP complexes resulting in recovery of proteoltic activity (14). This is an important finding in that it has previously been thought that MMP-TIMP inhibitory complexes were tight binding and essentially irreversible.

3. MMPS IN TUMOR TISSUES

There is now a considerable body of evidence that shows increased MMP expression in most tumors relative to corresponding, adjacent normal tissues (2-5). Enhanced MMP expression in human tumor tissues has been correlated with local invasion, metastasis and poor survival. Early indications of the importance of MMPs in tumor biology came from a study in 1980 of a neutral metalloproteinase secreted from a melanoma cell line that was able to degrade basement membrane collagen (15). The concept derived from this study was that tumor proteases were important in the dissolution of structural components of the ECM. The tumor associated basement-membrane (type IV) collagen degrading activity was identified as the second member of the MMP family (MMP-2). Numerous studies show that tumor cell invasion and metastatic capacity were positively correlated with expression of members of the MMP family (16). Many *in vitro* studies characterized the ability of MMP family members to degrade structural components of the ECM and thus perpetuated the concept that the role of MMPs in tumor progression was the removal of structural barriers to cell migration or invasion (16).

Using human tumor tissues investigators have demonstrated that the increased expression of MMPs associated with tumor progression is due to transcriptional activation of MMPs in both tumor cells and stromal cells (17). Whereas *in vitro* studies reveal the capacity of tumor cells to produce a wide range of MMP activities, studies of tumor tissues reveal that many MMPs are principally expressed by stromal fibroblasts (17). This suggests that tumor cells stimulate stromal MMP production via a paracrine mechanism, consistent with the observation that fibroblast secretion of MMP activity is induced by cytokines and growth factors (18, 19). However, MMPs secreted by stromal cells may be activated and localized by tumor cells (17). For example, *in situ* hybridization studies of *MMP-2* expression in human breast cancer tissues demonstrate that synthesis occurs principally in the stromal fibroblasts, whereas immunohistochemical localization of MMP-2 revealed staining of the cell membranes of both tumor cells and stromal fibroblasts at invasive foci (17, 20).

This paracrine model of enhanced MMP expression in tumors is also consistent with the general observation that MMPs are not up regulated in tumors by gene amplification. Amplification or translocation of *MMP-24 (MT5-MMP)* and *MMP-23,* respectively, has been reported in brain tumors (21-23). However, the role of these genetic alterations in the biology or clinical course of CNS malignancies has not yet been verified. Recent studies have also identified genetic polymorphisms in MMP genes that may contribute to an enhanced susceptibility to invasive carcinoma. These polymorphisms are in the promoter regions of the *MMP-1* and *MMP-3* genes, and result in altered expression of these proteases (24-27). The *MMP-1* polymorphism has been associated with an increased susceptibility to lung cancer and enhanced invasiveness of melanoma cells. Recently, mutations in the *MMP-2* have also been reported in association with a multicentric osteolytic and arthritic syndrome (28), and that a single nucleotide polymorphism in the MMP-2 promoter, abolishing the Sp1-binding site, down regulates MMP-2 expression and is associated with a substantial reduction in risk for breast cancer (29).

4. TUMOR INVASION AND METASTASIS

As described above, the initial model of MMPs function in tumor progression was the disruption of structural components of the ECM. The three-step model of tumor cell invasion, proposed by Liotta and colleagues, described the critical interactions of invasive tumor cells with the subendothelial basement membrane (16). Type IV collagen provides the principal structural component of these basement membranes and as such is a principal barrier to migrating cells. The ability of tumor cells to invade such barriers is critical to their metastatic potential by allowing local invasion as well as entrance to the vascular and lymphatic compartments. The identification of a tumor cell type IV collagenase activity and correlation of this activity with tumor invasion were consistent with this hypothesis. However, at that time it was not known that this type IV collagenase activity was a member of the MMP family (MMP-2), or the number of related protease activities which constitute the MMP family as it is known today, which includes both soluble and membrane-type proteases.

Perhaps a critical turning point in the changing view of the role of MMPs in tumor progression was the demonstration that enhanced expression of a single MMP *in vivo* could promote tumor initiation, as well as the growth of both primary and metastatic tumors (30). Sternlicht and coworkers demonstrated that expression of active stromelysin-1 (MMP-3) in normal epithelial mammary gland structures is sufficient to induce development of invasive tumors and this transition can be blocked by TIMP-1 (30, 31). This compelling evidence suggests that the role of MMPs in tumor progression is not limited to invasion of ECM barriers, but must also include cell growth, migration and angiogenesis.

5. MMPS IN TUMOR GROWTH

The findings described above suggest that MMPs can function in initiation of tumor formation, as well as progression. This is further substantiated in studies using an *MMP-7*-null mouse model of intestinal tumorigenesis (32, 33). The adenomatous polyposis coli (*Apc*) gene functions to regulate cytosolic ß-catenin levels downstream of the tumor suppressor E cadherin (34). Loss of *Apc* function results in increased cytosolic concentration of free ß-catenin that then undergoes nuclear translocation to interact with Tcf (T-cell factor), a transcription repressor. Interaction of ß-catenin with Tcf results in transcriptional activation of genes associated with tumor progression. Studies show that, through interaction with Tcf, ß-catenin significantly upregulates MMP-7 expression in colorectal mucosa. Thus, loss of the tumor suppressor E cadherin can result in upregulation of MMP-7 expression. Interestingly, MMP-7 also cleaves E-cadherin from the cell surface, resulting in increased ß-catenin levels, and the E cadherin cleavage product can stimulate cell invasion (35). These findings implicate MMP-7 in the intial stages of colorectal tumorigenesis (Figure 1A).

The ECM sequesters a variety of growth factors and cytokines, some in latent forms. A second mechanism in which MMPs may contribute to tumor growth is through the release and/or activation of these matrix-associated factors. MMP-9 has been implicated in the release of vascular endothelial growth factor (VEGF) that results in enhanced tumor angiogenesis and growth. This was first demonstrated in the RIP1-Tag2 model of pancreatic islet cell carcinoma in which MMP-9 expression correlated with initiation of tumor angiogenesis, and either chemical or genetic ablation of MMP-9 expression resulted in tumors that failed to undergo vascularization (36). The authors demonstrate that MMP-9, but not MMP-2, specifically enhanced release of VEGF from the ECM. Similarly, MMP-1, MMP-3, as well as plasmin and heparanase degrade endothelial-derived perlecan and release of bound bFGF (37). These findings demonstrate MMP-mediated release of pro-angiogenic growth factors that mediate tumor-induced angiogenesis and enhance tumor growth (Figure 1B).

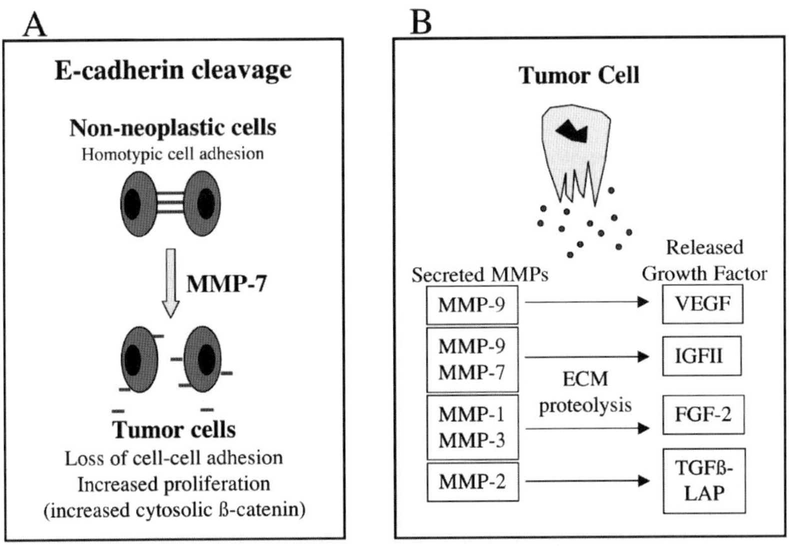

Figure 1. MMPs stimulate tumor growth. A. Increased MMP-7 activity in the tumor microenvironment has been shown to mediate cleavage of the homotypic cell adhesion, tumor suppressor E-cadherin. E-cadherin cleavage releases fragments that stimulate cell migration and increase cytosolic ß-catenin levels, resulting in tumor cell proliferation and MMP-7 secretion. B. Multiple MMP activities shown to enhance release of matrix sequestered growth factors, which can stimulate tumor cell proliferation and/or angiogenesis, indirectly stimulating tumor growth.

MMPs may also regulate growth factor availability in other ways. For example, transgenic animal studies have demonstrated that alteration of the MMP/TIMP balance *in vivo* in favor of TIMP-1 activity can block neoplastic proliferation in the SV40 T antigen-induced model of murine hepatocellular carcinoma (38). The mechanism of this TIMP-1 effect was mediated by direct inhibition of MMP processing of insulin-like growth factor binding protein-3 (IGFBP-3) preventing release of insulin-like growth factor II (IGFII) mitogenic activity. Subsequent studies show that IGFBPs are substrates for both MMP-7 and MMP-9 (38, 39). Thus, cleavage of IGFBP, resulting in the release of IGFs may contribute to enhanced tumor growth. In addition to these two examples, MMPs have also been shown to mediate the release and/or activation of TGFß. TGFß is secreted and maintained in latent form through its interaction with the latency-associated protein (LAP). This latent TGFß complex is sequestered in the ECM through covalent interaction with the fibrillin family protein known as latent TGFß binding protein (LTBP). Several members of the MMP family have been shown to directly cleave LAP resulting in TGFß (40-42). Furthermore, MMP-2, but not the closely related MMP-9, can specifically cleave the ECM bound form of LTBP, suggesting that this protease may function in release of latent TGFß complexes (TGFß-LAP) which are then activated by MMP cleavage of LAP (43).

6. CELL MIGRATION

Cell migration requires transmission of propulsive force from the ECM to the cytoskeleton of the migrating tumor or endothelial cell. Repetitive assembly of cytoskeletal elements to form membrane ruffles, lamellipodia, filopodia and pseudopodia accomplishes cell movement (44, 45). Lamellipodia are broad, flat sheet like structures in comparison to filopodia that are thin, cylindrical projections. Cell movement begins with protrusion of a filopod or lamellipod. These are formed by polymerization of actin to form elongated central filaments in the filopod and a broader cross-weave mesh in the lamellipod (46). At the leading edge of the protruding structures, integrins concentrate in specific regions, and after ligation with ECM ligands form focal adhesions. These focal adhesions are anchored to the actin filaments. Microinjection studies have shown that integrin association with the Rho family of GTPases is critical for organization and assembly of the actin cytoskeleton. It has been demonstrated that there is a hierarchical cascade amongst these GTPases that controls formation of specific cytoskeletal structures. Cdc42 and Rac control formation of filopodia and lamellipodia, respectively, whereas Rho controls stress fiber formation and focal adhesions (47).

The integrin connection to the ECM provides adhesive traction and contraction of the actin filaments that result in forward propulsion of the cell body. As the cell moves new projections occur at the leading edge and are anchored with new focal adhesion. During forward movement the focal adhesions appear to move in a retrograde fashion on the cell surface. This apparent movement of the focal adhesions has been observed using fluorescent tagged beads coated with integrin substrates or anti-integrin antibodies (48-50). Integrins thus play a key role not only in relaying information about the microenvironment by matrix adhesive interactions, but also function to provide the traction necessary for cell migration. Recent data suggest that MMP-integrin interactions may possibly modulate integrin function on the cell surface.

Brooks et al. (51) first demonstrated the interaction of MMPs with members of the integrin family of ECM receptors. The up regulation of $\alpha_v\beta_3$ and $\alpha_v\beta_5$ expression has been observed on "activated" endothelial cells in response to angiogenic stimulation (Figure 2A). Brooks and colleagues showed that the $\alpha_v\beta_3$ integrin binds MMP-2 through interaction with the carboxyl-terminal hemopexin domain of MMP-2 (PEX domain). These findings suggest that MMP-integrin interactions may function to localize MMPs at sites proximal to important ECM ligands that would facilitate cellular invasion. Alternatively, the MMP-integrin interaction could function to alter outside-in signaling events associated with receptor ligation. However, it appears that the cellular processing of MMP-2 results in generation of an intact C-terminal PEX domain that can compete for binding of intact MMP-2 to the $\alpha_v\beta_3$ receptor and inhibit the angiogenic response *in vivo* (52). Thus it appears that while binding of active MMP-2 to $\alpha_v\beta_3$ facilitates the angiogenic response, generation of the catalytically inactive PEX domain signals completion of angiogenesis by competing for binding and displacement of active MMP-2 from the endothelial cell surface. MMP-2-deficient mice have reduced angiogenesis and tumor growth (53), however, the critical MMP-2 substrate regulating these events *in vivo* has not been identified.

The requirement for MMP activity during cell movement through reconstituted ECM is not absolute. Friedl and co-workers have shown that when proteolysis of the ECM is completely inhibited tumor cells can adopt an amoeboid type movement for trafficking through ECM (54, 55). This suggests that tumor cells can negotiate ECM by mechanical means alone.

However, tissue-invading tumor cells appear to assume a mesenchymal cell-like phenotype, not an amoeboid shape, and it has been suggested that invading tumor cells resemble fibroblasts (54, 55) (Figure 2B). One critical point in tumor cell invasion has recently been clarified by the systematic investigations of Weiss and colleagues (56-58). These investigators have focused on the relative contribution of the soluble and membrane type MMPs to cellular invasion. Their results clearly show that MT1-MMP activity is essential for the ability of both tumor and endothelial cells to survive in and invade provisional as well as interstitial matricies (56-58). Expression of soluble MMPs, such as MMP-2, MMP-3 or MMP-9 did not facilitate cell invasion or cell survival in a three-dimensional type I collagen matrix. Similar findings are also reported for fibroblasts, suggesting that MT1-MMP serves as the major cell-associated proteinase necessary to confer both normal (fibroblasts and endothelial cells) and neoplastic cells with invasive capacity (59).

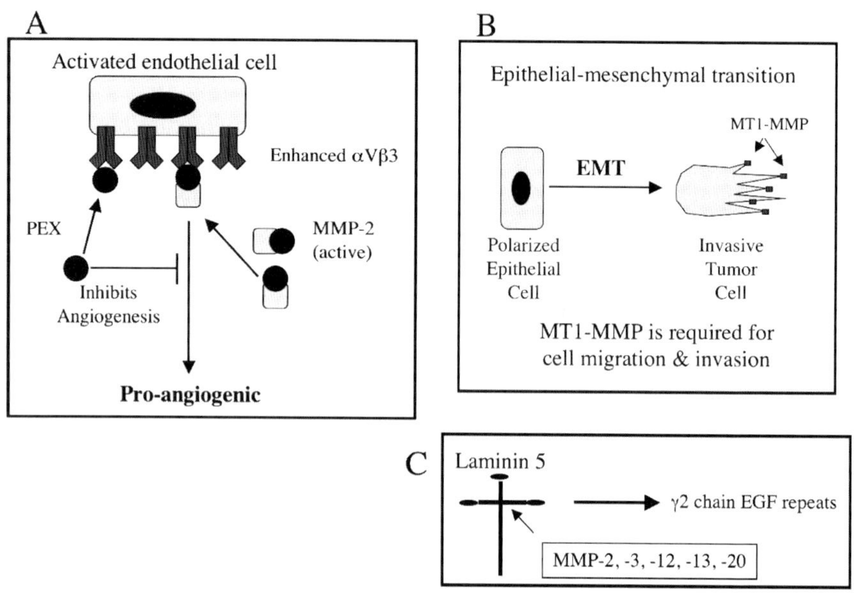

Figure 2. MMPs promote cell migration. A. MMPs may interact with integrin receptors to focus MMP activity at the cell surface and enhance cell migration. The binding of MMP-2 to $\alpha v\beta 3$ is observed on activated endothelial cells and is mediated by the C-terminal hemopexin-like domain (PEX). Cleavage of MMP-2 results in generation of a proteolytically inactive fragment (PEX) that is anti-angiogenic in vivo. B. Enhanced MT1-MMP expression is observed in migrating fibroblasts, endothelial cells and tumor cells. Only reconstitution of MT1-MMP expression facilitates invasion of these cells. Soluble, secreted MMPs are required for invasion of 3 dimensional collagen matricies. C. MMP-mediated cleavage of laminin 5 γ2 chain releases a cryptic fragment containing EGF-like repeats which stimulate cell migration.

In 1997 Quaranta and co-workers demonstrated that MMP-2 cleavage of laminin 5 resulted in exposure of a cryptic site that induces mammary epithelial cell migration (60). This was the first demonstration that MMP activity could modify the ECM and influence cell behavior. Later studies demonstrated that several MMPs were capable of degrading laminin 5 to generate soluble fragment of the $\gamma 2$ chain that is comprised of epidermal growth factor (EGF)-type repeats. Binding of this fragment to the EGF receptor activates this receptor and downstream signaling that results in enhanced cell migration (Figure 2C). Although MMP-2, MMP-3, MMP-12, MMP-13 and MMP-20 also reportedly cleave the laminin 5 $\gamma 2$ chain the *in vivo* significance of these mechanisms have not been reported. MT1-MMP-deficient mice have reduced tissue levels of the laminin 5 $\gamma 2$ chain fragment, suggesting that MT1-MMP is the critical activity for generation of this fragment *in vivo*. This also highlights the problems in relating *in vitro* substrate analysis to identification of critical substrate-degrading activity *in vivo*.

7. ANGIOGENESIS

An emerging concept in the field of angiogenesis research is that many endogenous angiogenesis inhibitors are generated by proteolytic modification of ECM components already present in the matrix. Protease activity on these substrates uncovers neoepitopes or releases crytpic sites that function to modulate a cellular response.

Interestingly, it has been demonstrated that MMPs play an important role in promoting the angiogenic response. As mentioned previously, MMP-2 and MMP-9-deficient mice have impaired angiogenic responses (36, 53), and both of these proteases have been shown to expose cryptic sites in type IV collagen that promotes angiogenesis (61, 62). However, MMP activity is also critical for generating the potent angiogenesis inhibitors *in vivo* (63, 64). MMPs degrade plasminogen to generate an N-terminal fragment that inhibits endothelial cell proliferation, and is known as angiostatin. These include MMP-3, MMP-7, MMP-9 and MMP-12. MMP-12 is the most potent in generating angiostatin that in turm inhibits microvascular endothelial cell proliferation *in vitro* (64, 65). Angiostatin production also correlates with MMP-12 synthesis in macrophages cultured from Lewis lung tumors grown in mice (64). Endostatin, a proteolytic fragment of collagen XVIII, may also be generated by MMP-7 activity directly, or through multistep processing involving elastase. Perhaps the best example of MMP-mediated generation of angio-inhibitory peptides is the release of tumstatin from the NC1 domain of type IV collagen (66). MMP-9-null mice have decreased circulating tumstatin and accelerated tumor growth (66). Reconstitution of MMP-9 expression restored circulating tumstatin levels and overexpression of tumstatin inhibited tumor growth. The exact role of MMP activity in generation of other angiogenesis inhibitors, such as canstatin, arrestin and endorepellin, a fragment of the proteoglycan perlecan, are suspected but not definitively demonstrated.

8. SUMMARY

Recent studies have demonstrated that MMP activity contributes to tumor progression through a variety of mechanisms. In addition to their previously identified role in tumor invasion, MMP activities have now been shown to promote tumor cell growth, migration and invasion. The mechanisms involved require MMP-mediated cleavage of both macromolecular components of the ECM, as well as selective cleavage of cell surface molecules mediating both cell-cell (E cadherin) and/or cell-matrix adhesion sites (integrins). MMP-mediated cleavage of ECM components can result in liberation of sequestered growth factors or cytokines, as well as bioactive components of ECM macromolecules. In addition, it is now appreciated that MMP activities may both promote and/or inhibit tumor progression. These recent findings help explain the failure of broad-spectrum synthetic MMP inhibitors to attenuate tumor progression, and suggest that selective targeting of specific members of the MMP family may still hold promise as therapeutic cancer modality. The findings presented in this review further suggest that there is a specific contribution of membrane-associated MMP activity, in particular MT1-MMP, to the invasive capacity of both normal and neoplastic cells. In addition, MMPs, previously thought important for tumor progression (e.g., MMP-2 and MMP-9), may actually prevent tumor progression by generating bioactive fragments of ECM that inhibit tumor-associated angiogenesis and tumor growth. The challenge for the future will be to selectively target MMPs that promote tumor progression, while leaving those protease activities that inhibit angiogenesis intact. However, this task may be further complicated by the fact that the role of a specific MMP in tumor progression may be dictated by the specific stage or temporal progression of different tumors. Finally, just as the paradigm for the role of MMPs in tumor progression is changing, the role of endogenous MMP inhibitors (TIMPs) is also changing and new inhibitors are being discovered

(RECK). TIMPs have been shown to possess growth-modulating activities that independent of their MMP inhibitory role. Understanding these novel biological effects of TIMPs, and their role in tumor progression are also essential to development of new cancer therapies.

REFERENCES

1. Hanahan D, Weinberg RA. The hallmarks of cancer. Cell 2000, 100: 57-70.
2. Coussens LM, Fingleton B, Matrisian LM. Matrix metalloproteinase inhibitors and cancer: trials and tribulations. Science 2002, 295: 2387-92.
3. Egeblad M, Werb Z. New functions for the matrix metalloproteinases in cancer progression. Nat. Rev. Cancer 2002, 2: 161-74.
4. Visse R, Nagase H. Matrix metalloproteinases and tissue inhibitors of metalloproteinases: structure, function, and biochemistry. Circ. Res. 2003, 92: 827-39.
5. Overall CM, Lopez-Otin C. Strategies for MMP inhibition in cancer: innovations for the post-trial era. Nature Reviews. Cancer 2002, 2: 657-72.
6. Bode, W. Structural basis of matrix metalloproteinase function. Biochemical Society Symposia 2003, 1-14.
7. Crocker SJ, Pagenstecher A, Campbell LL. The TIMPs tango with MMPs and more in the central nervous system. J. Neurosci. Res. 2004, 75: 1-11.
8. Brew K, Dinakarpandian D, Nagase H. Tissue inhibitors of metalloproteinases: evolution, structure and function. Biochim. Biophys. Acta. 2000, 1477: 267-83.
9. Lambert E, et al. TIMPs as multifacial proteins. Critical Reviews in Oncology Hematology 2004, 49: 187-198.
10. Takahashi C, et al. Regulation of matrix metalloproteinase-9 and inhibition of tumor invasion by the membrane-anchored glycoprotein RECK. Proc. Natl. Acad. Sci. USA 1998, 95: 13221-6.
11. Oh J, et al. The membrane-anchored MMP inhibitor RECK is a key regulator of extracellular matrix integrity and angiogenesis. Cell 2001, 107: 789-800.
12. Hoegy SE, et al. Tissue inhibitor of metalloproteinases-2 (TIMP-2) suppresses TKR-growth factor signaling independent of metalloproteinase inhibition. J. Biol. Chem. 2001, 276: 3203-14.
13. Seo DW, et al. TIMP-2 mediated inhibition of angiogenesis: an MMP-independent mechanism. Cell 2003, 114: 171-80.
14. Fedarko NS, et al. Three small integrin binding ligand N-linked glycoproteins (SIBLINGs) bind and activate specific matrix metalloproteinases. FASEB Journal 2004, 18: 734-6.
15. Liotta LA, et al. Preferential digestion of basement membrane collagen by an enzyme derived from a metastatic murine tumor. Proc. Natl. Acad. Sci. USA 1979, 76: 2268-72.
16. Liotta LA, Steeg PS, Stetler-Stevenson WG. Cancer metastasis and angiogenesis: an imbalance of positive and negative regulation. Cell 1991, 64: 327-36.
17. Polette M, et al. Tumour invasion and matrix metalloproteinases. Critical Reviews in Oncology-Hematology 2004, 49: 179-86.
18. Sternlicht MD, Werb Z. How matrix metalloproteinases regulate cell behavior. Annual Review of Cell & Developmental Biology 2001, 17: 463-516.

19. Toole BP, Emmprin (CD147), a cell surface regulator of matrix metalloproteinase production and function. Current Topics in Developmental Biology 2003, 54: 371-89.
20. Monteagudo C, et al. Immunohistochemical distribution of type IV collagenase in normal, benign, and malignant breast tissue. Am. J. Pathol. 1990, 136: 585-92.
21. Gururajan R, et al. Duplication of a genomic region containing the Cdc2L1-2 and MMP21-22 genes on human chromosome 1p36.3 and their linkage to D1Z2. Genome Research 1998, 8: 929-39.
22. Gururajan R, et al. Isolation and characterization of two novel metalloproteinase genes linked to the Cdc2L locus on human chromosome 1p36.3. Genomics 1998, 52: 101-6.
23. Llano E, et al. Identification and characterization of human MT5-MMP, a new membrane-bound activator of progelatinase a overexpressed in brain tumors. Cancer Res. 1999, 59: 2570-6.
24. Rutter JL, et al. A single nucleotide polymorphism in the matrix metalloproteinase-1 promoter creates an Ets binding site and augments transcription. Cancer Res. 1998, 58: 5321-5.
25. Tower GB, et al. Erk 1/2 differentially regulates the expression from the 1G/2G single nucleotide polymorphism in the MMP-1 promoter in melanoma cells. Biochimica et Biophysica Acta 2002, 1586: 265-74.
26. Tower GB, Coon CI, Brinckerhoff CE. The 2G single nucleotide polymorphism (SNP) in the MMP-1 promoter contributes to high levels of MMP-1 transcription in MCF-7/ADR breast cancer cells. Breast Cancer Res. Treat. 2003, 82: 75-82.
27. Tower GB, et al. Fra-1 targets the AP-1 site/2G single nucleotide polymorphism (ETS site) in the MMP-1 promoter. Eur. J. Biochem. 2003, 270: 4216-25.
28. Martignetti JA, et al. Mutation of the matrix metalloproteinase 2 gene (MMP2) causes a multicentric osteolysis and arthritis syndrome. Nat. Genet. 2001, 28: 261-5.
29. Zhou Y, et al. Substantial reduction in risk of breast cancer associated with genetic polymorphisms in the promoters of the matrix metalloproteinase-2 and tissue inhibitor of metalloproteinase-2 genes. Carcinogenesis 2004, 25: 399-404.
30. Sternlicht MD, et al. The stromal proteinase MMP3/stromelysin-1 promotes mammary carcinogenesis. Cell 1999, 98: 137-46.
31. Sternlicht MD, Bissell MJ, Werb Z. The matrix metalloproteinase stromelysin-1 acts as a natural mammary tumor promoter. Oncogene 2000, 19: 1102-13.
32. Wilson CL, et al. Intertinal tumorigenesis is suppressed in mice lacking the metalloproteinase matrilysin. Proc. Natl. Acad. Sci. USA 1997, 94: 1402-1407.
33. Crawford HC, et al. The metalloproteinase matrilysin is a target of ß-catenin transactivation in intestinal tumors. Oncogene 1999, 18: 2883-2891.
34. George SJ, Dwivedi A. MMPs, Cadherins, and Cell Proliferation. Trends in Cardiovascular Medicine 2004, 14: 100-105.
35. Noe V, et al. Release of an invasion promoter E-cadherin fragment by matrilysin and stromelysin-1. J. Cell Science 2001, 114: 111-118.
36. Bergers G, et al. Matrix metalloproteinase-9 triggers the angiogenic switch during carcinogenesis. Nature Cell Biology 2000, 2: 737-744.
37. Whitelock JM, et al. The degradation of human endothelial cell-derived perlecan and release of bound basic fibroblast growth factor by stromelysin, collagenase, plasmin, and heparanases. J. Biol. Chem. 1996, 271: 10079-10086.
38. Manes S, et al. The matrix metalloproteinase-9 regulates the insulin-like growth factor-triggered autocrine response in DU-145 carcinoma cells. J. Biol. Chem. 1999, 274: 6935-45.

39. Manes S, et al. Identification of insulin-like growth factor-binding protein-1 as a potential physiological substrate for human stromelysin-3. J. Biol. Chem. 1997, 272: 25706-12.
40. Mu D, et al. The integrin alpha(v)beta8 mediates epithelial homeostasis through MT1-MMP-dependent activation of TGF-beta1. J. Cell Biol. 2002, 157: 493-507.
41. Yu Q, Stamenkovic I. Localization of matrix metalloproteinase 9 to the cell surface provides a mechanism for CD44-mediated tumor invasion. Genes. Dev. 1999, 13: 35-48.
42. Yu Q, Stamenkovic I. Cell surface-localized matrix metalloproteinase-9 proteolytically activates TGF-beta and promotes tumor invasion and angiogenesis. Genes & Development 2000, 14: 163-76.
43. Dallas SL, et al. Proteolysis of latent transforming growth factor-beta (TGF-beta)-binding protein-1 by osteoclasts. A cellular mechanism for release of TGF-beta from bone matrix. J. Biol. Chem. 2002, 277: 21352-60.
44. Lauffenburger DA, Horowitz AF. Cell Migration: A Physically Integrated Molecular Process. Cell 1996, 84: 359-369.
45. Gumbiner BM. Cell Adhesion: The Molecular Basis of Tissue Architecture and Morphogenesis. Cell 1996, 84: 345-357.
46. Mitchison TJ, Cramer LP. Actin-Base Cell Motility and Cell Locomotion. Cell 1996, 84: 371-379.
47. Tapon N, Hall A. Rho, Rac and Cdc42 GTPases regulate the organization of the actin cytoskeleton. Curr. Opin. Cell Biol. 1997, 9: 86-92.
48. Felsenfeld DP, et al. Selective regulation of integrin--cytoskeleton interactions by the tyrosine kinase Src. Nat. Cell Biol. 1999, 1: 200-6.
49. Choquet D, Felsenfeld DP, Sheetz MP. Extracellular matrix rigidity causes strengthening of integrin- cytoskeleton linkages. Cell 1997, 88: 39-48.
50. Sheetz MP, Felsenfeld DP, Galbraith CG. Cell migration: regulation of force on extracellular-matrix-integrin complexes. Trends Cell Biol. 1998, 8: 51-4.
51. Brooks PC, et al. Localization of matrix metalloproteinase MMP-2 to the surface of invasive cells by interaction with integrin alpha v beta 3. Cell 1996, 85: 683-693.
52. Brooks PC, et al. Disruption of angiogenesis by PEX, a noncatalytic metalloproteinase fragment with integrin binding activity. Cell 1998, 92: 391-400.
53. Itoh T, et al. Reduced angiogenesis and tumor progression in gelatinase A- deficient mice. Cancer Res. 1998, 58: 1048-1051.
54. Wolf K, et al. Compensation mechanism in tumor cell migration: mesenchymal-amoeboid transition after blocking of pericellular proteolysis. J. Cell Biol. 2003, 160: 267-77.
55. Friedl P, Wolf K. Tumour-cell invasion and migration: diversity and escape mechanisms. Nature Reviews. Cancer 2003, 3: 362-74.
56. Hiraoka N, et al. Matrix metalloproteinases regulate neovascularization by acting as pericellular fibrinolysins. Cell 1998, 95: 365-377.
57. Hotary KB, et al. Matrix metalloproteinases (MMPs) regulate fibrin-invasive activity via MT1-MMP-dependent and -independent processes. Journal of Experimental Medicine 2002, 195: 295-308.
58. Hotary KB, et al. Membrane type I matrix metalloproteinase usurps tumor growth control imposed by the three-dimensional extracellular matrix. Cell 2003, 114: 33-45.
59. Sabeh F, et al. Tumor cell traffic through the extracellular matrix is controlled by the membrane-anchored collagenase MT1-MMP. J. Cell Biol. 2004, 167: 769-781.

60. Giannelli G, et al. Induction of cell migration by matrix metalloprotease-2 cleavage of laminin-5. Science 1997, 277: 225-8.
61. Hangai M, et al. Matrix metalloproteinase-9-dependent exposure of a cryptic migratory control site in collagen is required before retinal angiogenesis. Am. J. Pathol. 2002, 161: 1429-37.
62. Xu J, et al. Proteolytic exposure of a cryptic site within collagen type IV is required for angiogenesis and tumor growth in vivo. J. Cell Biol. 2001, 154: 1069-79.
63. O'Reilly MS, et al. Regulation of angiostatin production by matrix metalloproteinase-2 in a model of concomitant resistance. J. Biol. Chem. 1999, 274: 29568-71.
64. Dong ZY, et al. Macrophage-derived metalloelastase is responsible for the generation of angiostatin in Lewis lung carcinoma. Cell 1997, 88: 801-810.
65. Cornelius LA, et al. Matrix metalloproteinases generate angiostatin: Effects on neovascularization. Journal of Immunology 1998, 161: 6845-6852.
66. Hamano Y, et al. Physiological levels of tumstatin, a fragment of collagen IV alpha 3 chain, are generated by MMP-9 proteolysis and suppress angiogenesis via alpha V beta 3 integrin. Cancer Cell 2003, 3: 589-601.

Chapter 8

DYNAMIC FUNCTIONS OF THE α6β4 INTEGRIN IN CARCINOMA

Isaac Rabinovitz and Arthur M. Mercurio
Division of Cancer Biology and Angiogenesis, Department of Pathology, Beth Israel Deaconess Medical Center and Harvard Medical School, Boston, MA, USA

Abstract: The α6β4 integrin plays pivotal but distinct roles in the biology of epithelial and carcinoma cells. In healthy epithelia, its major function is to anchor the epithelium to the basement membrane as a component of either Type I or Type II hemidesmosomes. The signaling capacity of this integrin in the hemidesmosome appears to be minimal. Epithelial wounds or, more importantly, factors linked to malignant transformation and progression can induce dramatic changes in the function of α6β4. In fact, a scenario is emerging of how the function of α6β4 is altered in carcinoma. Factors in the host-tumor microenvironment have the potential to mobilize α6β4 from hemidesmosomes and promote its association with F-actin. This association with F-actin enables this integrin to function in cell migration and to harness traction forces on laminin-containing matrices such as basement membranes, a process that could contribute to the remodeling of basement membranes during tumor invasion. Importantly, this altered localization of α6β4 appears to be coupled to an activation of its signaling potential. The primal signaling event triggered by α6β4 appears to be activation of PI3-K. Although the mechanism by which this occurs needs to be deciphered in more detail, especially with respect to the involvement of growth factor receptors, α6β4-mediated activation of PI3-K and its effectors such as Akt, mTOR and Rac has profound consequences on the biology of carcinoma cells. Arguably, the ability of α6β4 to stimulate the translation of VEGF and possibly other growth factors may be the most significant contribution of this integrin to cancer because of the potential autocrine and paracrine effects of these factors.

Key words: extracellular matrix (ECM), epithelial-mesenchymal transition (EMT), migration, invasion

A. Wells (ed.), Cell Motility in Cancer Invasion and Metastasis, 159-187.

1. INTRODUCTION

Carcinoma progression denotes stages in the evolution of a cancer where malignant cells of epithelial origin leave their normal confines to invade the adjacent stroma and eventually metastasize to distant tissues. Progression is an area of intensive research primarily because it is the reason for most of the morbidity and mortality associated with cancer. A major goal of such research is to understand the biology that underlies the behavior of carcinoma cells and to use such information as the basis for clinical intervention. In this direction, it is widely assumed that invasion and metastasis are dependent on the migration of cells within the extracellular matrix (ECM), a fundamental process that is also important in embryonic development, wound healing and inflammation. Most of the evidence to implicate cell migration in the invasion of primary carcinomas derives from inferences made from the pathological analysis of tumors. Such analyses have revealed that carcinomas can invade as cell sheets or as single cells, the latter morphology indicative of an epithelial to mesenchymal transition (EMT) and associated with the most aggressive cancers (1-3). In addition, numerous experimental observations indicate that the translocation of tumor cells from one tissue compartment to another (e.g., epithelium to stroma) involves active cell movement (4-7).

Migration is mediated largely by the integrin adhesion receptors, which comprise a family of homologous, transmembrane heterodimers built from a combination of one of 12 α and one of 9 β subunits, providing the variety and specificity necessary to recognize most of the ECM molecules (8). Integrins link the ECM and the actin cytoskeleton in a transmembrane bridge, which plays an important role in the transmission of forces required for migration (9-11). In addition, signaling through these receptors can influence migration independently of their adhesive function (12-14).

Integrins also function in cell growth and survival, processes that are essential for tumor formation and progression. With regard to survival, it has been known for some time that the attachment of primary epithelial cells to extracellular matrix (ECM) proteins is essential for their survival (15). More specifically, integrin-mediated interactions with ECM proteins initiate signals that sustain survival (16). Survival is enhanced significantly by growth factor stimulation of attached cells providing evidence that integrins and growth factor receptors cooperate to promote survival (16). The environment within the epithelium that supports survival, however, is progressively lost during malignant transformation, and especially as malignant cells become invasive and metastatic. Positional cues such as basement membrane anchoring and cell-cell adhesive contacts that provide survival signals in the normal epithelium are often absent in invasive and metastatic cancer (17, 18). Moreover, the environment encountered by invasive and metastatic cells is foreign and often pro-apoptotic. The conclusion can be drawn from these considerations that invasive and metastatic cells must acquire mechanisms that maintain their survival outside the confines of the epithelium. Indeed, cancer progression can be considered an evolutionary process that selects for cells that exhibit the capacity for survival, among other properties (19, 20). One important implication of this hypothesis is that those cells that do survive will be the most aggressive and have the greatest propensity to metastasize. In other terms, a strong selective pressure promotes the growth of metastatic cells that have evolved mechanisms for surviving in hostile environments.

The involvement of integrins in many aspects of tumor biology raises the key issue of how malignant transformation and progression influence either the expression or function of specific integrins. To this end, most studies have focused on changes in integrin expression as a function of transformation and progression. Indeed, the literature abounds with studies that employ immunohistochemistry to compare the expression of a specific integrin in normal and malignant tissues. Although data obtained from such studies can be informative, they are often limited by technical considerations (e.g., fixation, antibodies) and fail to account for possible changes in integrin localization. For example, a diffuse staining pattern for a given integrin may be mistaken for decreased expression. This staining approach is optimal for those few integrins that exhibit clear-cut differences in expression in tumors. A prime example is the integrin $\alpha\varpi\beta6$ that is not expressed in most adult epithelia but whose expression is induced in some carcinomas. For most other integrins, it would appear that more sophisticated methods such as laser capture microscopy coupled with real-time PCR are needed to obtain an accurate assessment of changes in integrin expression as a function of transformation and progression.

Numerous studies using tumor cell lines have substantiated the hypothesis that alterations in integrin expression can affect the behavior of tumor cells. For example, an increase in the expression of the $\alpha5\beta1$ integrin (fibronectin receptor) in CHO ovary carcinoma cells can reduce their proliferative ability and tumorigenicity (21), while the *de novo* expression of the $\alpha v\beta3$ integrin (vitronectin receptor) in melanoma cells increases their survival and correlates with the vertical phase of this cancer (22). Integrins can also affect the tumor environment, which plays a critical role in the nourishment of the tumor and the determination of escape routes. In this respect, integrins such as $\alpha v\beta3$ have been shown to provide the signaling necessary for tumor angiogenesis (23). Protease activity, which is important for tumor invasion, can also be influenced by integrin expression (24).

A mechanism of how integrins contribute to cancer that has been considered much less than gross alterations in their expression is that malignant transformation alters the functions and signaling properties of specific integrins. This mechanism is exemplified best by the integrin α6β4, which has been implicated in the biology of carcinoma and is the focus of this review. What will emerge from the analysis presented is that the functions of this integrin in carcinoma cells are distinct from its established role in normal epithelial cells and that this transition of function is intrinsic to the nature of malignant carcinoma. Moreover, it is becoming apparent that α6β4 functions at multiple levels to influence the migration, invasion and survival of carcinoma cells.

2. FUNCTIONS OF THE α6β4 INTEGRIN IN CARCINOMA PROGRESSION AND FORMATION

2.1 Overview

The α6β4 integrin recognizes a variety of laminins, a family of ECM molecules that are primary components of the basement membrane (25). The basement membrane separates normal epithelia from the surrounding stroma and plays an important role in the confinement of newly transformed cells (carcinoma *in situ*) prior to their penetration into stroma (invasive carcinoma) (26). The α6β4 integrin provides a strong attachment of epithelia to the basement membrane by connecting the basement membrane with the cytokeratin network in structures termed hemidesmosomes (27, 28). Mutations that affect the function of the α6β4 integrin, as found in patients with epidermolysis bullosa, produce characteristic skin blistering that results from deficient attachment of the epidermis to the basement membrane (29), a finding substantiated by analysis of β4 null mice (30). The link between the cytokeratins and laminin bridged by the transmembrane α6β4 is critical for maintaining cohesive tensile strength necessary for an epithelial cell to resist mechanical loads (30). The interaction of α6β4 with the cytokeratin network is indirect and uses an additional group of molecules that together build the hemidesmosome. This group of molecules includes hemidesmosome1/plectin, BPAG1(Bullous Pemphigoid Antigen), BPAG2 and the tetraspanin CD151 (28).

There is also evidence that α6β4 on carcinoma cells may adhere to ligands other than laminin. Specifically, it has been suggested that α6β4-

containing carcinoma cells bind to CLCA (chloride channel, calcium-activated) proteins that are present on endothelial cells (31, 32). This interaction may facilitate vascular arrest of these carcinoma cells and promote their intravascular metastatic growth. Although this scenario is intriguing, more studies are needed to validate the interaction between α6β4 and CLCA proteins and to assess its importance in carcinoma biology.

The α6 subunit is the only known partner of the β4 subunit, although truncated forms of α6 have been described (33). In contrast, α6 can partner with both β4 and β1 subunits (34, 35). Distinct from β1 subunits that have a short cytoplasmic tail, the β4 subunit has a 1,000-amino-acid-long cytoplasmic tail containing several structural and functional domains that are critical for its association with the other hemidesmosomal proteins (Figure 1) (36-38). These regions include two pairs of FN type III repeats separated by a connecting segment. The β4 subunit is known to bind hemidesmosome1/plectin within two large regions: part of the FN type III repeat 2 along with a region of the connecting segment and the last FN repeat plus the carboxy terminal residues (39-42). The FN type III repeats 3 and 4 are also important for binding BPAG-1 and -2 (43, 44). There are two types of hemidesmosomes: type I hemidesmosomes, which are present in skin and several types of epithelia, and contain α6β4, hemidesmosome1/plectin, BPAG1 and BPAG2 (45); type II hemidesmosomes, which are present in intestinal epithelia and contain only α6β4 and hemidesmosome1/plectin (45-47).

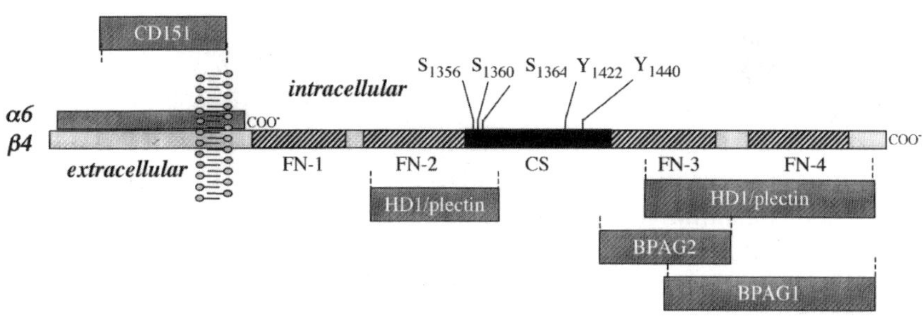

Figure 1. Schematic representation of the structure and domain organization of the α6β4 integrin. FN1-4 fibronectin III repeats; CS connecting segment. Boxes below α6β4 represent the regions in the α6β4 integrin involved in the interaction with other hemidesmosomal components (BPAG1, BPAG2, HD1/Plectin, and CD151).

2.2 Expression of the α6β4 Integrin In Carcinomas

Given its role in hemidesmosome-mediated adhesion, it was thought by some that α6β4 expression had to be suppressed to enable tumor cells to progress to invasive carcinoma (48). The catalyst for studying the expressionand function of α6β4 in cancer was the report by Falcioni et al. (49), which identified a tumor antigen associated with metastasis (TS180) that was shown to be identical to the β4 integrin subunit (50). Subsequently, it became apparent that expression of α6β4 is maintained or even increased in several types of invasive and metastatic carcinomas, including squamous, gastric, colon and bladder, and that α6β4 expression actually correlates with the progression of these carcinomas (Table 1).
In some tumors such as thyroid carcinoma, α6β4 is not expressed in normal cells but it is expressed as a consequence of malignant transformation and the level of expression appears to correlate with the aggressiveness of the tumor. Such examples provide the strongest correlative evidence that α6β4 is linked to carcinoma progression. At the same time, other studies reported that expression of this integrin is reduced in some carcinomas (reviewed in [48]). A common observation in many of these latter studies is that the localization of the α6β4 integrin in invasive and metastatic carcinoma cells is diffuse in comparison to its polarized, basal localization in normal epithelia. This altered localization of α6β4 could account for the apparent decrease in α6β4 expression. Moreover, the loss of polarization of the α6β4 integrin could be related to the invasive phenotype of carcinomas. In murine squamous papillomas, for example, the loss of polarization and diffuse suprabasal expression of α6β4 integrin has been associated with a high risk of malignant progression (51), and in human head and neck carcinomas a similar phenotype has been related to poor prognosis (52). It is also interesting to note that studies that have examined α6β4 integrin expression using more quantitative methods such as immunoblots and Northern blots have observed an increase in the expression of this subunit in some carcinomas in comparison to normal tissue (53, 54). An intriguing cancer with respect to α6β4 expression is prostate carcinoma. Current evidence indicates that the expression of this integrin, as well as several basement membrane components such as laminin 5 and collagen VII, is decreased in primary tumors compared to normal prostatic epithelium (55). Interestingly, there appears to be continued or even increased expression of the β4 integrin mRNA in these tumors (56). These data contrast with the fact that most prostate carcinoma cell lines express this integrin *in vitro* (57), a finding that suggests the possibility of epigenetic control of β4 expression. An interesting hypothesis in this regard is that the absence of laminin 5 in tumors might

impede the translation of β4 or increase its internalization and degradation. This hypothesis suggests that α6β4 might be re-expressed during the metastatic progression of prostate carcinoma if the tumor microenvironment is conducive to such expression. Clearly, studies on the expression of α6β4 in metastatic prostate lesions should be a high priority.

Table 1. Expression of α6β4 Integrin In Human Carcinomas.

carcinoma	α6β4 pattern of expression	Ref.
Skin- squamous cell	High expression and pericellular distribution of α6β4.	(111, 112)
Skin- basal cell	Loss of α6β4, laminin 5 and collagen VII. Laminin 1 and collagen IV preserved. In other cases α6β4 integrin was strongly stained in the superficial portions of the basal cell carcinoma.	(113)
Other squamous cell	Loss of basal polarization and increased intensity of expression of α6β4 in cancer. Expression associated with progression and reduced survival.	(52)
Cervical	In normal squamous epithelium, α6β4 is detected in basal and parabasal cells. In high-grade CIN (2-3) there is increased α6β4 to the whole epithelial thickness. In invasive squamous cell carcinomas, α6β4 chain is diffusely expressed.	(114)
Ovarian	Staining for α6β4 was irregular. When present, the basement membrane `stained positively for laminin.	(115)
Thyroid	α6β4 and laminin 5 are absent in normal cells. In thyroid carcinomas, neoexpression of α6β4 is correlated to aggressiveness.	(54)
Breast	α6β4 expression is maintained in tumors, co-expression with laminin correlates with poor prognosis.	(116)
Lung	α6β4 subunit was detectable in squamous cell carcinoma and adenocarcinomas but not in small cell cancers.	(117)
Prostate	Prostatic intraepithelial neoplasia (PIN) lesions preserve α6β4 expression, while protein expression is lost in carcinoma *in situ* and invasive carcinoma.	(55, 118, 119)
Bladder	Expression increased in tumors, linked to reduced survival.	(20)
Pancreas	α6β4 is expressed in carcinomas. Microarray analysis reports increased expression.	(121-123)
Gastric	Increased α6β4 at the interface between malignant cells and stroma; coincident with laminin-1 and -5.	(124)
Colorectal	α6β4 expressed in all carcinomas; inversely correlated to degree of differentiation. Diffuse and intense in less differentiated forms.	(53)

2.3 Invasion and Migration

Although the expression studies described above suggested that α6β4 may contribute to the functions of carcinoma cells, the nature of this contribution was elusive. The first breakthrough in this area was the unexpected observation that α6β4 functions in the invasion and migration of carcinoma cells. More specifically, initial evidence to implicate α6β4 in invasion was provided by the finding that transfection of this integrin into β4-deficient colon and breast carcinoma cell lines dramatically increased their ability to invade *in vitro*. More recently, studies that employ RNA interference to reduce β4 expression in highly invasive carcinoma cells have shown a corresponding reduction in the ability of these cells to invade (58). Other studies demonstrated that the expression of α6β4 stimulates the migration of carcinoma cells. Interestingly, this integrin stimulates chemotaxis and not haptotaxis (59), an observation consistent with its purported ability to influence the function of specific growth factor receptors that can mediate chemotaxis (see below).

An unexpected finding from the studies mentioned above was that the ability of α6β4 to stimulate migration and invasion can be independent of ligand binding, i.e., engagement of α6β4 by laminin substrata is not essential. This conclusion is based on the observation that in some model systems migration and invasion are not blocked by antibodies that inhibit α6β4 adhesive interactions, an observation that discounts the possibility of adhesion to laminins deposited on the substrate by the cells (59). In addition, expression of a β4 deletion mutant that lacks the extracellular, ligand-binding domain was shown to be sufficient to enhance the migration and invasion of some breast carcinoma cells (60). Furthermore, reduced expression of the β4 subunit results in decreased migration of breast carcinoma cells on non-laminin substrata (58). This ‘ligand-independent’ function of α6β4 has significant implications for the behavior of carcinoma cells *in vivo* because it implies that the ability of α6β4 to stimulate migration and invasion is not limited to specific matrix environments. As will be discussed below, however, there are situations in which the contribution of α6β4 to cell migration is clearly dependent on its interactions with the laminins.

2.4 Survival

In addition to its involvement in migration and invasion, a role for α6β4 in promoting cell survival was discovered using breast carcinoma cells that had been deprived of matrix attachment and serum (61). The ability of such cells to resist apoptosis under such ‘stress’ conditions is a function of α6β4 expression and it relates to the ability of α6β4 to activate the PI3-K/Akt pathway. A surprising twist to α6β4-mediated survival signaling is that it can be regulated by p53. Specifically, α6β4 activates Akt and promotes survival in carcinoma cells that express mutant forms of p53 (61). These events are not observed, however, in carcinoma cells that express wild-type p53 because α6β4 stimulates p53-dependent caspase-3 activity that results in the cleavage and inactivation of Akt (61, 62).

A fruitful extension of this work was the observation that α6β4-mediated survival in breast carcinoma cells is dependent on vascular endothelial growth factor (VEGF) (63). In cells in which VEGF expression had been reduced using antisense oligonucleotides, α6β4 lost its ability to prevent apoptosis in stress conditions. This finding was consistent with other data indicating that autocrine VEGF is important for the survival of breast carcinoma cells (64). In addition, it raised the possibility that α6β4 can regulate VEGF expression. Indeed, subsequent studies revealed that α6β4 can regulate VEGF expression at the translational level via the mTOR/eIF-4E pathway (63) (see below).

The ability of α6β4 to regulate the expression of growth factors such as VEGF that are critical for carcinoma biology has significant ramifications for our understanding of how integrins contribute to cancer. The widely accepted notion is that integrins, often in concert with growth factor receptors, activate specific signaling pathways that sustain survival, e.g., (65). The data on α6β4 and VEGF indicate that the survival function of integrins may not only be mediated by activation of a key survival kinase such as Akt and the consequent effects of Akt on apoptotic signaling (66) but also by the integrin-dependent translation and expression of growth factors such as VEGF that promote survival in an autocrine and, possibly paracrine, fashion. In this direction, it is becoming more apparent that tumor cells can acquire a certain degree of self-sufficiency by elaborating autocrine signaling pathways that facilitate key functions of growth, survival and invasion (20). Autocrine pathways are more significant as tumors progress towards invasive and metastatic disease because the environment of such tumors is increasingly hostile. As such, autocrine signaling pathways represent a prime target for therapy aimed at impeding tumor dissemination. In this regard, it is tempting to speculate that the expression of α6β4 and VEGF are linked at certain stages of progression and that this linkage provides a survival advantage to the tumor cells.

Other studies employing three-dimensional cultures have substantiated the importance of α6β4 for the survival of breast carcinoma cells. Specifically, it was observed that three-dimensional polarity confers resistance to apoptosis and that such polarity is maintained by α6β4 (67). Subsequent studies suggested that α6β4 can mediate the anchorage-independent survival of breast carcinoma cells by a mechanism that involves autocrine laminin-5, which ligates α6β4 and activates a Rac GTPase/NF-□B signaling pathway that sustains survival (68). It will be informative to determine if VEGF or other autocrine growth factors mediate α6β4-dependent survival in these model systems.

2.5 The α6β4 Integrin in Carcinoma Formation

Recent studies indicate that α6β4 may not only have an impact in tumor progression but also on the general susceptibility of epidermal cells to form tumors. Although it was known that expression of α6β4 increases markedly in squamous cell carcinomas (SCC) but not in basal cell carcinomas (69), which also originate from epidermal cells but are relatively non-metastatic, the report that α6β4 may be involved in the formation of SCCs was rather surprising (70). Using a model system that involves the retroviral expression of specific genes in primary, human keratinocytes and the subsequent use of these cells for grafting to regenerate human skin on immune-deficient mice, Khavari's group discovered that expression of oncogenic Ras alone induces growth arrest and graft failure but that the co-expression of oncogenic Ras with molecules that impede Ras-induced growth arrest results in invasive epidermal carcinoma resembling SCC (70). Of interest, antibodies specific for either α6β4 or its ligand laminin-5 prevent the genesis of invasive SCC in this model. In a more definitive experiment, it was observed that keratinocytes deficient in the expression of either the β4 subunit or laminin-5 (isolated from patients with blistering skin diseases) were unable to form tumors but expression of the respective genes restored their ability to form invasive tumors. In a related study from Fiona Watt's lab that may shed more insight into mechanism, targeted expression of α6β4 in the suprabasal layer of the epidermis in transgenic mice increased significantly the frequency of papillomas, carcinomas and metastases induced by chemical carcinogenesis (71). The mechanism appears to involve the suppression of TGF-β signaling by α6β4 and the resulting inhibition of TGF-β from suppressing the clonal expansion of initiated cells in the epidermal basal layer (71). Clearly, these studies on skin tumors add a new dimension to our understanding of the contribution of α6β4 to cancer and it should be quite fruitful to assess the possible role of α6β4 in the promotion of other carcinomas and to elucidate the mechanisms involved.

3. MECHANISTIC INSIGHTS INTO THE ALTERED FUNCTIONS OF α6β4 IN CARCINOMA

3.1 Overview

The studies described above implicate α6β4 in the invasive process, but they do not provide a mechanism to explain how an integrin, which associates with intermediate filaments and forms stable adhesive contacts, can promote the dynamic process of invasion. The presence of α6β4-containing hemidesmosomes would be expected to impede invasion. In fact, hemidesmosomes in carcinoma cells usually disappear, although α6β4 continues to be expressed. This section will discuss existing data on the functional changes in α6β4 that occur in carcinoma and how such changes are linked to carcinoma migration and invasion.

3.2 Dynamic Interactions with the Cytoskeleton

One possible explanation for the ability of α6β4 to promote invasion is that α6β4, upon dissociation from the hemidesmosome, might interact with the actin cytoskeleton and participate in the migration process. Supporting this idea was our finding that α6β4 can associate with F-actin and is localized at the leading edges of invasive carcinoma cells (72). Moreover, we demonstrated that α6β4 actually mediates the migration of such cells through its ability to associate with the actin cytoskeleton and promote the formation and stabilization of filopodia and lamellae (72). This finding implied that the function and cytoskeletal association of α6β4 in invasive carcinoma cells are distinct from its established role of anchoring epithelial cells to the basement membrane through its association with cytokeratins.

In normal cells, the interaction of α6β4 with the actin cytoskeleton might occur when cells are induced to migrate such as during wound healing. Indeed, we and others have shown that α6β4 is mobilized from hemidesmosomes and redistributes to actin-rich protrusions in response to chemotactic factors (73, 74). Moreover, in epithelial wound models such as cornea and intestinal mucosa, hemidesmosome disassembly occurs in the wound margins, and translocation of α6β4 to lamellipodia is observed (75-77). Several studies have implicated the involvement of α6β4 as well as α3β1 in wound healing based on the use of function-blocking antibodies, and there is evidence that α6β4 may contribute to this process by stimulating α3β1 function (78). An association of α6β4 with F-actin may also occur in the early stages of hemidesmosome formation. In this regard, Marchisio and colleagues using 804G bladder carcinoma cells, which form hemidesmosomes *in vitro*, observed that hemidesmosomes organize around a core of actin filaments and that the assembly of hemidesmosomes is preceded by the recruitment of α6β4 to the actin core (79). Moreover, when 804G cells are induced to migrate, hemidesmosomes disassemble and α6β4 becomes co-localized with F-actin at the cell margins.

Several lines of evidence suggest an association of α6β4 with actin filaments, although the molecular basis of this association has not been elucidated. Specifically, the actin-severing protein gelsolin is able to liberate α6β4 along with actin from invasive colon carcinoma cells (72)]. Also, immunostaining studies reveal that both filopodia and the edges of lamellae contain α6β4 distributed in the form of discrete clusters that co-localize with F-actin. An association between α6β4 and F-actin is also supported by the fact that α6β4 immunostaining of the marginal edges of the cell (which do not contain cytokeratins) is resistant to solubilization with Triton-X-100, a treatment that preserves the actin cytoskeleton (72). As mentioned before, a functional interaction between a6b4 and F-actin is suggested by the fact that α6β4 function-blocking antibodies destabilize the lamella and filopodia resulting in the reduced migration of colon carcinoma cells. Most likely, the association between α6β4 and F-actin is indirect because the β4 cytoplasmic domain lacks an actin-binding motif. One potential 'linker' molecule is HD-1/plectin, which also connects the β4 subunit to intermediate filaments, because it contains an actin-binding domain (80). Although β4 binds to HD-1/plectin at the actin-binding domain (81), this linker molecule exists as a homotetramer, which provides three additional actin-binding domains that could link β4 to actin.

The presence of α6β4 in filopodia may be important in determining the directionality during cell migration on laminin-1. This integrin has an extremely high adhesive strength for laminin-1 (82). Specifically, laminin-1 adhesion mediated by α6β4 is able to resist shear forces up to 100 dynes/cm^2, in contrast to the β1 integrins that have a much lower resistance. The high adhesive strength of this integrin for laminin-1 supports its involvement in filopodial stabilization. The stabilization of filopodia correlates with lamellar protrusion in the direction of the filopodia in several models of migration including colon carcinoma cells. For example, when fibroblasts are plated on surfaces coated with gold, lamellipodia are extended primarily where exploratory filopodia find and contact a gold-coated area and their extension follows the direction of the stabilized filopodia (83).

The importance of filopodial contacts in directing cell migration is also underscored by the pioneer growth cone model in grasshoppers where a single filopodial contact can re-orient an entire growth cone (84). An important implication of these studies is that carcinoma cells may follow invasive paths in which α6β4 ligands are detected by exploratory filopodia.

3.3 The Epithelial-Mesenchymal Transition and α6β4 Mobilization

The foregoing discussion on the mobilization of α6β4 and disassembly of hemidesmosomes in response to chemotactic factors raises the important and timely issue of α6β4 dynamics and the epithelial-mesenchymal transition (EMT). During this process, which is essential for embryonic development and also can occur during the progression of carcinomas, epithelium is converted into mesenchyme as a result of E-cadherin loss or dysfunction and the consequent expression of mesenchymal proteins such as vimentin. Of relevance to cancer, the EMT results in the induction of cell migration and the ability of epithelial cells to survive autonomously. An obvious consideration in this regard is the impact of the EMT on hemidesmosomes and α6β4. *A priori*, the assumption can be made that the EMT results in hemidesmosome disassembly and the mobilization of α6β4 to F-actin containing structures. Surprisingly, no studies have addressed this issue directly. Such studies have the potential to increase our understanding of how hemidesmosomes and α6β4 are regulated during the progression of a carcinoma, and they should provide insight into the more general issue of the relationship between cell-cell and cell-matrix adhesion.

3.4 α6β4 and Basement Membrane Remodeling

The association of α6β4 with the actin cytoskeleton suggests a mechanical contribution of α6β4 to migration. Using a traction-detection system we were able to show that α6β4 is capable of transmitting forces generated by the actin cytoskeleton onto a laminin substrate (85). The forces mediated by α6β4 could provide a clue as to how this integrin functions in carcinoma invasion. Although it is commonly assumed that invading cells break through the basement membrane and migrate into stroma, it is recognized that several invasive tumors can produce basement membrane components and form structures that range from sparse patches of basement membrane to an apparently intact membrane. These tumor-produced basement membranes could facilitate invasion by providing a temporary substrate for the invading cells to move using integrins such as α6β4. The remodeling of basement membranes and other ECMs to generate 'paths' for migration has been reported for several types of cells *in vitro*. Endothelial cells, for example, can reorganize reconstituted basement membranes into tubular paths through which they move. In this regard, we observed that colon carcinoma cells that express the α6β4 integrin remodel or 'compress' reconstituted basement membranes into very compact structures, and form paths connecting distant cell aggregates. The importance of α6β4 in this compression is evidenced by the inhibition of the remodeling using function-blocking antibodies (85). The ability of α6β4 to remodel basement membranes may derive from its ability to generate traction on such matrices. This α6β4-dependent remodeling is dependent on an active actin cytoskeleton and it appears that most of the compression occurs under the lamellae (85).

The above observations lead to the hypothesis that α6β4-dependent remodeling of basement membranes might be a mechanism by which tumor cells breach basement membranes. For example, this process could result in the compression of the basement membrane in discrete regions, a process that would deplete basement membrane components from surrounding regions and create 'gaps' through which carcinoma cells could escape. In this regard, the ectopic expression of α6β4 in β4-deficient breast carcinoma cells, which dramatically promotes their invasive ability, correlates with a substantial increase in their ability to compress basement membranes. Other models support as well the idea of remodeling as means of dissemination. For example, a correlation between remodeling activity and invasive capability has also been observed for melanoma cells, in which their aggressiveness correlates with their ability to remodel ECM (86, 87). Similar studies observed that highly invasive melanoma cells can remodel three-dimensional matrices, although partial matrix degradation was also observed (87, 88). Based on these findings, it appears reasonable to postulate that invasion involves both matrix remodeling and proteolytic degradation. Interestingly, however, no studies to date have linked α6β4 to either the expression, localization or activity of specific proteases.

3.5 Mechanisms of Hemidesmosome Disassembly

As mentioned above, hemidesmosomes provide epithelial cells with a strong and stable anchor to the basal lamina and, as a consequence, hinder migration. During wound healing or in carcinoma progression, hemidesmosomes disassemble to facilitate migration. Therefore, deciphering the mechanisms involved in hemidesmosome disassembly is critical to understanding how epithelial cells migrate and invade. Initial insight into this problem was provided by the finding made by our group and others that EGF, acting as a chemotactic factor, induces mobilization of the α6β4 integrin and promotes hemidesmosome disassembly (Figure 2) (73, 74). EGF also stimulates a PKCα dependent pathway that results in the phosphorylation of the β4 integrin subunit and its re-distribution to actin-rich structures (74). More recently, we extended these findings and determined that in keratinocytes and carcinoma cells EGF induces the phosphorylation of the β4 subunit on a cluster of serines (S_{1356}, S_{1360} and S_{1364}), located within the connecting segment of the cytoplasmic tail (89). Using a mutational analysis strategy on cos7 cells that can assemble hemidesmosomes type II upon transfection of the β4 cDNA, we observed that the ability of EGF to disrupt hemidesmosomes is reduced significantly by mutations that prevent the phosphorylation of S_{1356}, S_{1360} and S_{1364}. The phosphorylation of these serines is dependent on PKCα, which can directly phosphorylate some of these residues (89). Moreover, PKCα is sufficient to induce both the phosphorylation of these residues and the disruption of hemidesmosomes.

Although ~95 % of the stimulus-induced phosphorylation of the β4 subunit occurs on serine residues (89), EGF-induced phosphorylation of β4 on specific tyrosine residues has been reported. Moreover, preventing the phosphorylation of two tyrosines in the connecting segment of β4 (Y_{1422} and Y_{1440}) results in a partial inhibition of hemidesmosome disassembly induced by EGF (90). These tyrosines are phosphorylated through a Fyn-dependent pathway that involves the binding of Shc to those residues (90, 91). Together, these studies indicate an important role for serine and tyrosine phosphorylation of the β4 integrin subunit in regulating hemidesmosome disassembly. Moreover, the phosphorylation of other hemidesmosome omponents, such as hemidesmosome1/plectin (92), has been documented and is likely to play a regulatory role in the destabilization of this structure.

The destabilization of the hemidesmosome by the phosphorylation of β4 could result from conformational changes caused by the electrostatic charges of the added phosphate groups. Phosphorylation of β4 might inhibit β4 association with plectin and other hemidesmosomal components, or destabilize the hemidesmosome by interfering with β4 self-association. Both intermolecular and intramolecular interactions of β4 involving the connecting segment have been reported, and are likely to be influenced by phosphorylation (42, 93). Alternatively, serine phosphorylation may provide the binding site for other destabilizing molecules. In this regard, it has been reported that PKCα mediates serine phosphorylation on unknown residues in the connecting segment that induces the association of 14-3-3 proteins and, in concert with signals from the Ron receptor, may contribute to the destabilization of the hemidesmosome (94)]. Ron may also induce tyrosine phosphorylation of β4 through src-fyn, a pathway also implicated in EGF-mediated disassembly of hemidesmosomes (91).

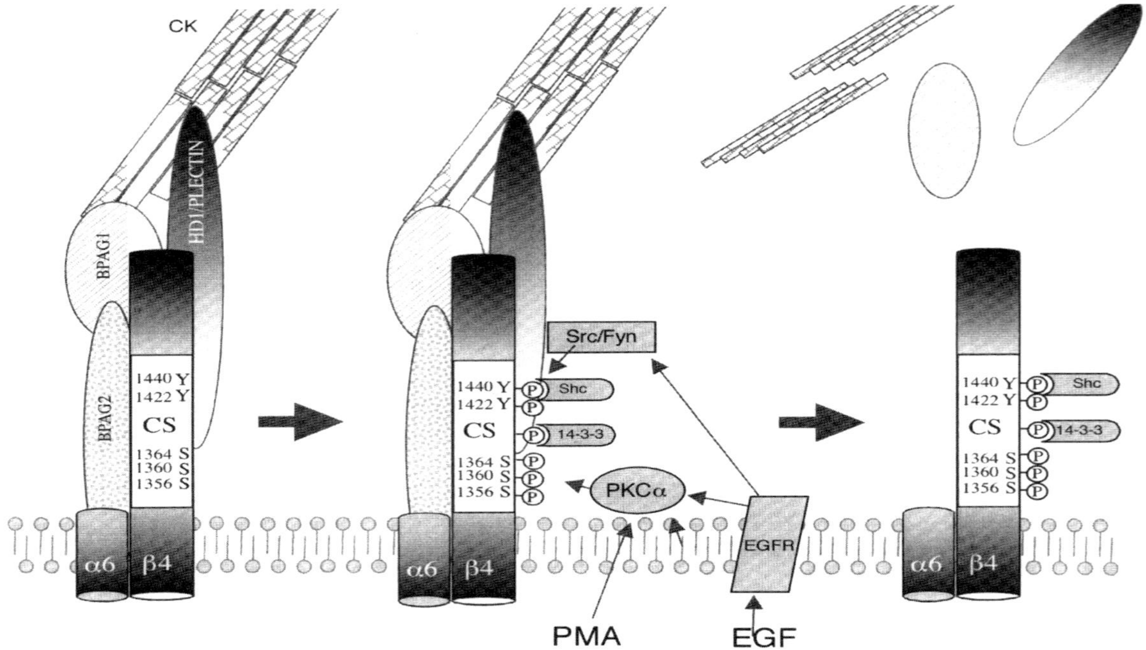

Figure 2. **Dynamics of hemidesmosome disassembly.** Growth factor stimulation results in the disassembly of hemidesmosomes as a consequence of the phosphorylation of β4 integrin on specific residues. A number of phosphorylation sites have been reported. One group includes a cluster of serines in the CS (connecting segment) that are phosphorylated by PKCα. A number of tyrosines are phosphorylated by Fyn, which allows them to recruit Shc. The PKCα-dependent phosphorylation of β4 on unidentified serine residues in the connecting segment provides a docking site for 14-3-3 proteins, facilitating hemidesmosome disassembly. CK cytokeratins

Although data from several studies strongly suggest that the presence of hemidesmosomes impedes migration, the notion of hemidesmosome as an unchanging and non-dynamic structure has been challenged. Using photobleaching experiments, it has been shown recently that the turnover of α6β4 in hemidesmosomes is surprisingly fast in contrast to other hemidesmosomal components, such as BPAG2 (95, 96). It will be important to define whether these observations made *in vitro* reflect the behavior of hemidesmosomes *in vivo*. Nonetheless, these studies suggest that the stability provided by the hemidesmosome should be viewed in terms of a dynamic equilibrium, where structure and function are maintained but components are subject to rapid turnover.

4. ACTIVATION OF A SIGNALING COMPETENT α6β4 INTEGRIN IN CARCINOMA

4.1 Overview

A hypothesis that emerges from the foregoing discussion is that stimuli that induce epithelial migration or the progression to invasive carcinoma trigger alterations in the localization of α6β4 that influence its signaling capacity. Several studies support this hypothesis. Initial studies from our group demonstrated that the localization of α6β4 in lamellipodia and filopodia was associated with its ability to signal migration and invasion (97). More recently, the Ron receptor studies (see above) provided evidence that the mobilization of α6β4 from hemidesmosomes to lamellipodia triggers the signaling capacity of this integrin and concluded that in keratinocytes the 'function of α6β4 switches from a mechanical adhesive device into a signaling competent (receptor)' that drives epidermal wound healing (94). Additional evidence to support this notion derives from the observation that the β4 subunit can be palmitoylated and, although palmitoylation is not required for α6β4 mediated adhesion or assembly into hemidesmosomes, it does promote the association of α6β4 with lipid rafts that are signaling competent (98). Thus, there appears to be a strong correlation between α6β4 localization and its signaling capacity. This section will summarize the data on the signaling properties of α6β4 in carcinoma cells from the perspective that such signaling is linked to the mobilization of α6β4 from hemidesmosomes.

4.2 Activation of PI3-K

A discussion of the functions of carcinoma cells that are influenced by α6β4 cannot be had without mentioning PI3-K, the lipid kinase that is a nexus of intracellular signaling. A seminal finding was that α6β4 has the ability to activate PI3-K potently in carcinoma cells (97), a finding that has been substantiated by several laboratories (60, 78, 94, 98-100). Indeed, this activation underlies the contribution of α6β4 to migration, invasion and survival. Moreover, a number of downstream targets of PI3-K including Akt, Rac and mTOR are central to the functions of carcinoma cells and their regulation by α6β4 has been established (63, 97, 101). Activation of PI3-K by α6β4 may also stimulate the function of other integrins important for the biology of carcinoma cells, such as α3β1 (78, 94, 100). A critical issue that arises from these findings is the mechanism by which α6β4 activates PI3-K. This mechanism is probably not a direct activation of PI3-K by α6β4 because the β4 cytoplasmic domain lacks a consensus sequence for binding the p85 regulatory subunit of PI3-K (97). Possible mechanisms that have been proposed include the involvement of insulin receptor substrate proteins (IRS-1 and IRS-2), which contain multiple PI3-K binding motifs and are tyrosine phosphorylated and bind to PI3-K upon ligation of α6β4 (102). In addition, the compartmentalization of α6β4 into lipid rafts may facilitate interactions with tyrosine kinases that can activate PI3-K (98). Finally, the reported ability of α6β4 to associate with specific growth factor receptors, which are known to activate PI3-K, and to facilitate the function of these receptors may account for the ability of α6β4 to activate PI3-K (see below). These mechanisms are not mutually exclusive and they may function in concert to mediate the α6β4-dependent activation of PI3-K.

4.3 The α6β4-Growth Factor Receptor Connection

An emerging consensus is that α6β4 has the potential to form complexes with specific growth factor receptors and that such macromolecular complexes act synergistically to activate key signaling pathways and mediate the functional contributions of α6β4 to carcinoma cells. Although this paradigm has been proposed for other integrins, studies by Falcioni and colleagues provided the first evidence that α6β4 may associate with erbB2, an orphan receptor of the EGFR family, on the surface of some breast carcinoma cell lines (103). Subsequent studies using a 3T3 cell model system demonstrated that both α6β4 and erbB2 are required for PI3-K activation and the stimulation of chemoinvasion (99). These data are provocative because erbB2 is thought to function only when it heterodimerizes with other members of the EGF receptor family (104).

Thus, the possibility that $\alpha6\beta4$ actually associates with an erbB2 heterodimer needs to be evaluated.

An intriguing example of the functional significance of $\alpha6\beta4$ growth factor receptor interactions is the report that a constitutive, physical association of the Met receptor and $\alpha6\beta4$ exists on the surface of invasive carcinoma cells and that $\alpha6\beta4$ functions as an essential 'adaptor' protein for Met that enables hepatocyte growth factor (HGF)-mediated invasion (60). In this scenario, $\alpha6\beta4$ is needed to facilitate the signaling functions of Met that promote invasion such as activation of PI3-K. The impact of this finding is amplified by the fact that substantial evidence exists for the importance of Met in the scattering, invasion and metastasis of tumor cells (105). Indeed, if $\alpha6\beta4$ were an essential, specific adaptor for Met function in these events, the consequences for carcinoma biology and therapy would be profound. Recently, however, our group was unable to detect a 'constitutive, physical association' between $\alpha6\beta4$ and Met on the surface of invasive carcinoma cells by co-immunoprecipitation (106). The discrepancy between these studies may reflect the fact that the association between these receptors, if it does occur, is tenuous. Along these lines, it should be noted that all of the evidence to support an association of $\alpha6\beta4$ with specific growth factor receptors is based on co-immunoprecipitation experiments that have the potential to yield spurious results and that are not rigorous enough to establish direct protein-protein interactions. A positive co-immunoprecipitation result, for example, may reflect the fact that $\alpha6\beta4$ and a growth factor receptor do not associate directly but co-exist in a membrane micro-domain such as a lipid raft.

A significant observation made by Chung et al. (106) is that the invasive function of Met can be independent of $\alpha6\beta4$ Cells that lack expression of $\alpha6\beta4$ exhibit Met-dependent invasion towards HGF and increasing Met expression can enhance their rate of invasion. Moreover, expression of exogenous $\alpha6\beta4$ in such cells increases their ability to invade towards several growth factors including HGF. Thus, it appears that Met is one of several growth factor receptors whose function may be enhanced by $\alpha6\beta4$ expression but that can signal and promote invasion in the absence of $\alpha6\beta4$ expression.

The most conclusive evidence to date for the existence of an $\alpha6\beta4$/growth factor receptor complex is the work on the Ron receptor in keratinocytes described above (94). These data indicate that the association between Ron and $\alpha6\beta4$ is dependent upon activation of Ron by its ligand, macrophage-stimulating protein. Specifically, Ron triggers the PKC-dependent phosphorylation of both itself and $\alpha6\beta4$ that generates 14-3-3 binding sites. Presumably, 14-3-3 molecules bridge these binding sites and enable a complex to form between these two receptors. Although more data are needed to validate this hypothesis, this is the first study to provide a mechanism to account for a direct interaction between α6β4 and a specific growth factor receptor. Moreover, it supports the hypothesis that associations between α6β4 and specific growth factor receptors may occur in response to specific physiological or oncogenic stimuli. This hypothesis meshes with the notion that the signaling capacity of α6β4 is manifested upon the mobilization of α6β4 from hemidesmosomes because such stimuli also induce its mobilization.

Clearly, an intimate relationship between α6β4 and specific growth factor receptors exists. Whether such relationships are manifested as direct physical associations between these receptors, their association in membrane micro-domains or simply cooperation at the functional level remains to be determined. Regardless, a better understanding of the mechanisms by which these two classes of receptors cooperate to influence signaling and function is of paramount importance.

4.4 α6β4 and Protein Translation

Many mRNAs that encode proteins essential for cell growth and survival are characterized by significant 5'UTR secondary structure and their translation is dependent on the concentration of translation initiation factor eIF-4E (107). eIF-4E interacts with a large scaffolding protein termed eIF-4G that directs the translational machinery to the 5' end of the mRNA. Of note, activation of the PI3-K/Akt pathway results in the phosphorylation and activation of mammalian target of rapamycin (mTOR), a kinase related to PI3-K (107). One target of mTOR is the 4E-binding protein (4E-BP) family, a family of three small polypeptides that complex with eIF-4E and prevents its interaction with eIF-4G. Upon phosphorylation of 4E-BP1 by mTOR, its binding to eIF-4E is disrupted, enabling eIF-4E to interact with eIF-4G and initiate translation (107). Thus, phosphorylation of 4E-BP is a critical event that links signaling to protein translation. As mentioned above, we demonstrated that α6β4 stimulates 4E-BP phosphorylation in breast carcinoma cells and, consequently, regulates the translation of VEGF, a finding confirmed by polysome analysis (63).

Obviously, numerous factors such as growth factors and other nutrients, as well as mutations in tumor suppressor genes such as PTEN and the tuberous sclerosis complex (TSC), have the potential to activate the mTOR pathway and influence protein translation (108). Our belief is that the contribution of α6β4 to this pathway is manifested in those environments in which the presence of other factors or such mutations is reduced. Our finding that the α6β4-dependent stimulation of 4E-BP1 phosphorylation and increase in VEGF translation is detected most readily under conditions of serum-deprivation substantiates this belief (63). Interestingly, it is under these same conditions that the survival function of α6β4 is most apparent and this function is linked to VEGF expression (63). Given these observations, one can speculate that this function of α6β4 may be important in the tumor microenvironment where the supply of nutrients and other factors may be limiting.

Another important point that emerges from the work on α6β4 and VEGF translation is that the α6β4 mediated activation of PI3-K and the consequent increase in VEGF expression, coupled with the ability of VEGF to activate PI3-K, results in an amplification of PI3-K activation. In fact, this pathway could provide one mechanism for the high basal activity of PI3-kinase that has been noted in invasive carcinomas and that may be essential for the behavior of these cells. In this direction, it is more than likely that the translation of other growth factors and cytokines is influenced by α6β4 and that these factors impact the signaling capacity and functions of carcinoma cells. Future studies comparing the profiles of mRNAs associated with actively translating polysomes as a function of α6β4 expression in cell lines and carcinomas will undoubtedly yield a wealth of information and likely some unexpected surprises.

4.5 A Possible Link between α6β4 and Transcription

Although α6β4 appears to have a significant impact on protein translation, there is also evidence that it may influence the transcription of genes important for invasion. This possibility is based on the report that expression of the transcription factor NFAT (nuclear factor of activated T cells) is linked to α6β4 is some breast and colon carcinoma cell lines and that NFAT transcriptional activity can be induced by ligation of α6β4 (109). The identification of specific NFAT target genes that are important for α6β4-mediated invasion would strengthen the link between this integrin and transcription. Also, it would be informative to determine whether the stimulation of NFAT transcriptional activity by α6β4 is PI3-K-dependent, given the central role of this kinase in α6β4 signaling.

5. CONCLUSIONS

The α6β4 integrin plays pivotal but distinct roles in the biology of epithelial and carcinoma cells. In healthy epithelia, its major function is to anchor the epithelium to the basement membrane as a component of either Type I or Type II hemidesmosomes. The signaling capacity of this integrin in the hemidesmosome appears to be minimal. Epithelial wounds or, more importantly, factors linked to malignant transformation and progression can induce dramatic changes in the function of α6β4. In fact, a scenario is emerging of how the function of α6β4 is altered in carcinoma. Factors in the host-tumor microenvironment have the potential to mobilize α6β4 from hemidesmosomes and promote its association with F-actin. This association with F-actin enables this integrin to function in cell migration and to harness traction forces on laminin-containing matrices such as basement membranes, a process that could contribute to the remodeling of basement membranes during tumor invasion. Importantly, this altered localization of α6β4 appears to be coupled to an activation of its signaling potential, which may occur through its association with growth factor receptors or lipid rafts, possibilities that are not mutually exclusive. The primal signaling event triggered by α6β4 appears to be activation of PI3-K. Although the mechanism by which this occurs needs to be deciphered in more detail, especially with respect to the involvement of growth factor receptors, α6β4-mediated activation of PI3-K and its effectors such as Akt, mTOR and Rac has profound consequences on the biology of carcinoma cells. Arguably, the ability of α6β4 to stimulate the translation of VEGF and possibly other growth factors may be the most significant contribution of this integrin to cancer because of the potential autocrine and paracrine effects of these factors.

The recent reports that α6β4 may contribute to the promotion and formation of epidermal tumors is both intriguing and revealing (70, 71), especially given the assumption that this integrin contributes to later stages of progression such as invasion and metastasis (110). One hypothesis that arises from the existing data is that a critical function of α6β4 in cancer is tumor cell survival. This finding implies that this survival function is necessary for transformed keratinocytes to form squamous cell carcinomas and that it is necessary for the subsequent survival of invasive and metastatic carcinoma cells, especially in the 'stress' conditions present within the tumor microenvironment. This proposed survival function for α6β4 is likely to involve its ability to regulate the expression of growth factors such as VEGF that have profound effects on both tumor and stromal cells. Also, the observation that α6β4 may inhibit the growth suppressive effects of TGF-β needs to be considered. These observations do not discount the contribution of α6β4 to the migration and invasion of carcinoma cells. Clearly, more data are needed to assess these possibilities, especially data from mouse tumor models in which the expression of α6β4 can be regulated at specific stages in the progression from normal epithelia to metastatic carcinoma.

ACKNOWLEDGEMENTS

The work in the authors' laboratories that contributed to this review was supported by NIH Grants CA88979 (IR) and CA80789 (AMM).

REFERENCES

1. Behrens J. Cadherins and catenins: role in signal transduction and tumor progression. Cancer Metastasis Rev. 1999, 18: 15-30.
2. Birchmeier W, Hulsken J, Behrens J. E-cadherin as an invasion suppressor. Ciba Found Symp. 1995, 189: 124-36; discussion 136-41, 174-6.
3. Wijnhoven BP, Dinjens WN, Pignatelli M. E-cadherin-catenin cell-cell adhesion complex and human cancer. Br. J. Surg. 2000, 87: 992-1005.
4. McCoy JP, Jr., et al. Identification of a laminin-like substance on the surface of high-malignant murine fibrosarcoma cells. J. Cell Sci. 1984, 65: 139-51.
5. Barsky SH, et al. Laminin molecular domains which alter metastasis in a murine model. J. Clin. Invest. 1984, 74: 843-8.
6. Wyckoff JB, et al. A critical step in metastasis: in vivo analysis of intravasation at the primary tumor. Cancer Res. 2000, 60: 2504-11.
7. Kan Z, Liu TJ. Video microscopy of tumor metastasis: using the green fluorescent protein (GFP) gene as a cancer-cell-labeling system. Clinical & Experimental Metastasis 1999, 17: 49-55.
8. Hynes RO. Integrins: versatility, modulation, and signaling in cell adhesion. Cell 1992, 69: 11-25.
9. Burridge K, et al. Focal adhesions: transmembrane junctions between the extracellular matrix and the cytoskeleton. Annual Review of Cell Biology 1988, 4: 487-525.
10. DeMali KA, Wennerberg K, Burridge K, Integrin signaling to the actin cytoskeleton. Curr. Opin. Cell Biol. 2003, 15: 572-82.
11. Lee J, Jacobson K. The composition and dynamics of cell-substratum adhesions in locomoting fish keratocytes. J. Cell Sci. 1997, 110: 2833-44.
12. Schwartz MA, Schaller MD, Ginsberg MH. Integrins: emerging paradigms of signal transduction. Annual Review of Cell & Developmental Biology 1995, 11: 549-99.
13. Alahari SK, Reddig PJ, Juliano RL. Biological aspects of signal transduction by cell adhesion receptors. Int. Rev. Cytol. 2002, 220: 145-84.
14. Miranti CK, Brugge JS. Sensing the environment: a historical perspective on integrin signal transduction. Nature Cell Biology 2002, 4.
15. Frisch SM, Francis H. Disruption of epithelial cell-matrix interactions induces apoptosis. J. Cell Biol. 1994, 124: 619-26.
16. Schwartz MA. Integrins, oncogenes, and anchorage independence. J. Cell Biol. 1997, 139: 575-8.
17. Hagios C, Lochter A, Bissell MJ, Tissue architecture: the ultimate regulator of epithelial function? Philos. Trans R. Soc. Lond. B. Biol. Sci. 1998, 353: 857-70.
18. Boudreau N, Bissell MJ. Extracellular matrix signaling: integration of form and function in normal and malignant cells. Curr. Opin. Cell. Biol. 1998, 10: 640-6.
19. Fearon ER. Cancer progression. Curr. Biol. 1999, 9: R873-5.
20. Hanahan D, Weinberg R.A. The hallmarks of cancer. Cell 2000, 100: 57-70.

21. Giancotti FG, Ruoslahti E. Elevated levels of the alpha 5 beta 1 fibronectin receptor suppress the transformed phenotype of Chinese hamster ovary cells. Cell 1990, 60: 849-59.
22. Nip J, et al. Human melanoma cells derived from lymphatic metastases use integrin alpha v beta 3 to adhere to lymph node vitronectin. J. Clin. Invest. 1992, 90: 1406-13.
23. Brooks PC, Clark RA, Cheresh DA. Requirement of vascular integrin alpha v beta 3 for angiogenesis. Science 1994, 264: 569-71.
24. Brooks PC, et al. Localization of matrix metalloproteinase MMP-2 to the surface of invasive cells by interaction with integrin alpha v beta 3. Cell 1996, 85: 683-93.
25. Mercurio AM. Laminin receptors - achieving specificity through cooperation. Trends Cell Biol. 1995, 5: 419-423.
26. Liotta LA, Rao CN, Wewer UM. Biochemical interactions of tumor cells with the basement membrane. Ann. Rev. Biochem. 1986, 55: 1037-57.
27. Green KJ, Jones JCR. Desmosomes and hemidesmosomes - structure and function of molecular components. FASEB Journal 1996, 10: 871-881.
28. Borradori L, Sonnenberg A. Hemidesmosomes - Roles in adhesion, signaling and human diseases. Curr. Opin. Cell Biol. 1996, 8: 647-656.
29. Vidal F, et al. Integrin beta 4 mutations associated with junctional epidermolysis bullosa with pyloric atresia. Nat. Genet. 1995, 10: 229-34.
30. Dowling J, Yu QC, Fuchs E. Beta-4 Integrin is required for hemidesmosome formation, cell adhesion and cell survival. J. Cell Biol. 1996, 134: 559-572.
31. Abdel-Ghany M, et al. The breast cancer beta 4 integrin and endothelial human CLCA2 mediate lung metastasis. J. Biol. Chem. 2001, 276: 25438-46.
32. Abdel-Ghany M, et al. Focal adhesion kinase activated by beta(4) integrin ligation to mCLCA1 mediates early metastatic growth. J. Biol. Chem. 2002, 277: 34391-400.
33. Davis TL, et al. Identification of a novel structural variant of the alpha 6 integrin. J. Biol. Chem. 2001, 276: 26099-106.
34. Sonnenberg A, Modderman PW, Hogervorst F. Laminin receptor on platelets is the integrin VLA-6. Nature 1988, 336: 487-9.
35. Hemler ME, Crouse C, Sonnenberg A. Association of the VLA alpha 6 subunit with a novel protein. A possible alternative to the common VLA beta 1 subunit on certain cell lines. J. Biol. Chem. 1989, 264: 6529-35.
36. Suzuki S, Naitoh Y. Amino acid sequence of a novel integrin beta 4 subunit and primary expression of the mRNA in epithelial cells. Embo. J. 1990, 9: 757-63.
37. Hogervorst F, et al. Cloning and sequence analysis of beta-4 cDNA: an integrin subunit that contains a unique 118 kd cytoplasmic domain. Embo. J. 1990, 9: 765-70.
38. Nievers MG, et al. Ligand-independent role of the beta 4 integrin subunit in the formation of hemidesmosomes. J. Cell Sci. 1998, 111: 1659-72.
39. Niessen CM, et al. A minimal region on the integrin beta4 subunit that is critical to its localization in hemidesmosomes regulates the distribution of HD1/plectin in COS-7 cells. J. Cell Sci. 1997, 110: 1705-16.
40. Niessen CM, et al. Integrin alpha 6 beta 4 forms a complex with the cytoskeletal protein HD1 and induces its redistribution in transfected COS-7 cells. Mol. Biol. Cell 1997, 8: 555-66.
41. Nievers MG, et al. Formation of hemidesmosome-like structures in the absence of ligand binding by the (alpha)6(beta)4 integrin requires binding of HD1/plectin to the cytoplasmic domain of the (beta)4 integrin subunit. J. Cell Sci. 2000, 113: 963-73.

42. Rezniczek GA, et al. Linking integrin alpha6beta4-based cell adhesion to the intermediate filament cytoskeleton: direct interaction between the beta4 subunit and plectin at multiple molecular sites. J. Cell Biol. 1998, 141: 209-25.
43. Borradori L, et al. The localization of bullous pemphigoid antigen 180 (BP180) in hemidesmosomes is mediated by its cytoplasmic domain and seems to be regulated by the beta4 integrin subunit. J. Cell Biol. 1997, 136: 1333-47.
44. Koster J, et al. Analysis of the interactions between BP180, BP230, plectin and the integrin alpha6beta4 important for hemidesmosome assembly. J. Cell Sci. 2003, 116: 387-99.
45. Borradori L, Sonnenberg A. Structure and function of hemidesmosomes: more than simple adhesion complexes. J. Invest Dermatol. 1999, 112: 411-8.
46. Hieda Y, et al. Identification of a new hemidesmosomal protein, HD1: a major, high molecular mass component of isolated hemidesmosomes. J. Cell Biol. 1992, 116: 1497-506.
47. Uematsu J, et al. Demonstration of type II hemidesmosomes in a mammary gland epithelial cell line, BMGE-H. J. Biochem. 1994, 115: 469-76.
48. Rabinovitz I, Mercurio AM. The integrin alpha 6 beta 4 and the biology of carcinoma. Biochem. Cell Biol. 1996, 74: 811-21.
49. Falcioni R, et al. Expression of tumor antigen correlated with metastatic potential of Lewis lung carcinoma and B16 melanoma clones in mice. Cancer Res. 1986, 46: 5772-8.
50. Kennel SJ, et al. Analysis of the tumor-associated antigen TSP-180. Identity with alpha 6-beta 4 in the integrin superfamily. J. Biol. Chem. 1989, 264: 15515-21.
51. Tennenbaum T, et al. The suprabasal expression of alpha 6 beta 4 integrin is associated with a high risk for malignant progression in mouse skin carcinogenesis. Cancer Res. 1993, 53: 4803-10.
52. Van Waes C, et al. The A9 antigen associated with aggressive human squamous carcinoma is structurally and functionally similar to the newly defined integrin alpha 6 beta 4. Cancer Res. 1991, 51: 2395-402.
53. Falcioni, R., et al. Integrin beta 4 expression in colorectal cancer. Int. J. Oncol. 1994, 5: 573-578.
54. Serini G, et al. Changes In Integrin and E-Cadherin Expression In Neoplastic Versus Normal Thyroid Tissue. J. Natl. Cancer Inst. 1996, 88: 442-449.
55. Knox JD, et al. Differential expression of extracellular matrix molecules and the alpha 6-integrins in the normal and neoplastic prostate. Am. J. Pathol. 1994, 145: 167-74.
56. Hao J, et al. Investigation into the mechanism of the loss of laminin 5 (alpha3beta3gamma2) expression in prostate cancer. Am. J. Pathol. 2001, 158: 1129-35.
57. Witkowski CM, et al. Characterization of integrin subunits, cellular adhesion and tumorgenicity of four human prostate cell lines. J. Cancer Res. Clin. Oncol. 1993, 119: 637-44.
58. Lipscomb EA, et al. Use of RNA interference to inhibit integrin (alpha6beta4)-mediated invasion and migration of breast carcinoma cells. Clin. Exp. Metastasis 2003, 20: 569-76.
59. O'Connor KL, Shaw LM, Mercurio AM. Release of cAMP gating by the alpha 6 beta 4 integrin stimulates lamellae formation and the chemotactic migration of invasive carcinoma cells. J. Cell Biol. 1998, 143: 1749-1760.
60. Trusolino L, Bertotti A, Comoglio PM. A signaling adapter function for alpha6beta4 integrin in the control of HGF-dependent invasive growth. Cell 2001, 107: 643-54.

61. Bachelder RE, et al. p53 inhibits alpha 6 beta 4 integrin survival signaling by promoting the caspase 3-dependent cleavage of AKT/PKB. J. Cell Biol. 1999, 147: 1063-72.
62. Bachelder RE, et al. The cleavage of Akt/protein kinase B by death receptor signaling is an important event in detachment-induced apoptosis. J. Biol. Chem. 2001, 276: 34702-7.
63. Chung J, et al. Integrin (alpha 6 beta 4) regulation of eIF-4E activity and VEGF translation: a survival mechanism for carcinoma cells. J. Cell Biol. 2002, 158: 165-74.
64. Bachelder RE, et al. Vascular endothelial growth factor is an autocrine survival factor for neuropilin-expressing breast carcinoma cells. Cancer Res. 2001, 61: 5736-40.
65. Colognato H, et al. CNS integrins switch growth factor signalling to promote target-dependent survival. Nat. Cell Biol. 2002, 4: 833-41.
66. Datta SR, Brunet A, Greenberg ME, Cellular survival: a play in three Akts. Genes. Dev. 1999, 13: 2905-27.
67. Weaver VM, et al. beta4 integrin-dependent formation of polarized three-dimensional architecture confers resistance to apoptosis in normal and malignant mammary epithelium. Cancer Cell 2002, 2: 205-16.
68. Zahir N, et al. Autocrine laminin-5 ligates alpha6beta4 integrin and activates RAC and NFkappaB to mediate anchorage-independent survival of mammary tumors. J. Cell Biol. 2003, 163: 1397-407.
69. Rossen K, et al. Expression of the alpha 6 beta 4 integrin by squamous cell carcinomas and basal cell carcinomas: possible relation to invasive potential? Acta. Derm Venereol. 1994, 74: 101-5.
70. Dajee M, et al. NF-kappaB blockade and oncogenic Ras trigger invasive human epidermal neoplasia. Nature 2003, 421: 639-43.
71. Owens DM, et al. Suprabasal alpha6beta4 integrin expression in epidermis results in enhanced tumourigenesis and disruption of TGFbeta signalling. J. Cell Sci. 2003, 116: 3783-91.
72. Rabinovitz I, Mercurio AM. The integrin alpha6beta4 functions in carcinoma cell migration on laminin-1 by mediating the formation and stabilization of actin-containing motility structures. J. Cell Biol. 1997, 139: 1873-84.
73. Mainiero F, et al. The intracellular functions of alpha6beta4 integrin are regulated by EGF. J. Cell Biol. 1996, 134: 241-53.
74. Rabinovitz I, Toker A, Mercurio AM. Protein kinase C-dependent mobilization of the alpha6beta4 integrin from hemidesmosomes and its association with actin-rich cell protrusions drive the chemotactic migration of carcinoma cells. J. Cell Biol. 1999, 146: 1147-60.
75. Larjava H, et al. Expression of integrins and basement membrane components by wound keratinocytes. J. Clin. Invest. 1993, 92: 1425-35.
76. Gipson IK, et al. Redistribution of the hemidesmosome components alpha 6 beta 4 integrin and bullous pemphigoid antigens during epithelial wound healing. Exp. Cell Res. 1993, 207: 86-98.
77. Lotz MM, Rabinovitz I, Mercurio AM. Intestinal restitution: progression of actin cytoskeleton rearrangements and integrin function in a model of epithelial wound healing. Am. J. Pathol. 2000, 156: 985-96.
78. Nguyen BP, Gil SG, Carter WG. Deposition of laminin 5 by keratinocytes regulates integrin adhesion and signaling. J. Biol. Chem. 2000, 275: 31896-907.
79. Spinardi L, et al. A dynamic podosome-like structure of epithelial cells. Exp. Cell Res. 2004, 295: 360-74.

80. Wiche G. Role of plectin in cytoskeleton organization and dynamics. J. Cell Sci. 1998, 111: 2477-86.
81. Geerts D, et al. Binding of integrin alpha6beta4 to plectin prevents plectin association with F-actin but does not interfere with intermediate filament binding. J. Cell Biol. 1999, 147: 417-34.
82. Tozeren A, et al. Integrin alpha 6 beta 4 mediates dynamic interactions with laminin. J. Cell Sci. 1994, 107: 3153-63.
83. Albrecht-Buehler G. Filopodia of spreading 3T3 cells. Do they have a substrate-exploring function? J. Cell Biol. 1976, 69: 275-86.
84. O'Connor TP, Duerr JS, Bentley D. Pioneer growth cone steering decisions mediated by single filopodial contacts in situ. J. Neurosci. 1990, 10: 3935-46.
85. Rabinovitz I, Gipson IK, Mercurio AM. Traction forces mediated by alpha6beta4 integrin: implications for basement membrane organization and tumor invasion. Mol. Biol. Cell 2001, 12: 4030-43.
86. Klein CE, et al. Integrin alpha 2 beta 1 is upregulated in fibroblasts and highly aggressive melanoma cells in three-dimensional collagen lattices and mediates the reorganization of collagen I fibrils. J. Cell Biol. 1991, 115: 1427-36.
87. Silletti S, Paku S, Raz A. Autocrine motility factor and the extracellular matrix. II. Degradation or remodeling of substratum components directs the motile response of tumor cells. Int. J. Cancer 1998, 76: 129-35.
88. Friedl P, et al. Migration of highly aggressive MV3 melanoma cells in 3-dimensional collagen lattices results in local matrix reorganization and shedding of alpha2 and beta1 integrins and CD44. Cancer Res. 1997, 57: 2061-70.
89. Rabinovitz I, Tsomo L, Mercurio AM. Protein kinase C-alpha phosphorylation of specific serines in the connecting segment of the beta 4 integrin regulates the dynamics of type II hemidesmosomes. Mol. Cell Biol. 2004, 24: 4351-60.
90. Dans M, et al. Tyrosine phosphorylation of the beta 4 integrin cytoplasmic domain mediates Shc signaling to extracellular signal-regulated kinase and antagonizes formation of hemidesmosomes. J. Biol. Chem. 2001, 276: 1494-502.
91. Mariotti A, et al. EGF-R signaling through Fyn kinase disrupts the function of integrin alpha6beta4 at hemidesmosomes: role in epithelial cell migration and carcinoma invasion. J. Cell Biol. 2001, 155: 447-58.
92. Foisner R, Traub P, Wiche G. Protein kinase A- and protein kinase C-regulated interaction of plectin with lamin B and vimentin. Proc. Natl. Acad. Sci. USA 1991, 88: 3812-6.
93. Schaapveld RQ, et al. Hemidesmosome formation is initiated by the beta4 integrin subunit, requires complex formation of beta4 and HD1/plectin, and involves a direct interaction between beta4 and the bullous pemphigoid antigen 180. J. Cell Biol. 1998, 142: 271-84.
94. Santoro MM, Gaudino G, Marchisio PC. The MSP receptor regulates alpha 6 beta 4 and alpha 3 beta 1 integrins via 14-3-3 proteins in keratinocyte migration. Dev. Cell 2003, 5: 257-271.
95. Tsuruta D, Hopkinson SB, Jones JC. Hemidesmosome protein dynamics in live epithelial cells. Cell Motil Cytoskeleton. 2003, 54: 122-34.
96. Geuijen CA, Sonnenberg A. Dynamics of the alpha6beta4 integrin in keratinocytes. Mol. Biol. Cell 2002, 13: 3845-58.
97. Shaw LM, et al. Activation of phosphoinositide 3-OH kinase by the alpha6beta4 integrin promotes carcinoma invasion. Cell 1997, 91: 949-60.

98. Gagnoux-Palacios L, et al. Compartmentalization of integrin alpha6beta4 signaling in lipid rafts. J. Cell Biol. 2003, 162: 1189-96.
99. Gambaletta D, et al. Cooperative signaling between alpha(6)beta(4) integrin and ErbB-2 receptor is required to promote phosphatidylinositol 3-kinase-dependent invasion. J. Biol. Chem. 2000, 275: 10604-10.
100. Hintermann E, et al. Inhibitory Role of alpha6beta4-associated erbB-2 and Phosphoinositide 3- Kinase in Keratinocyte Haptotactic Migration Dependent on alpha3beta1 Integrin. J. Cell Biol. 2001, 153: 465-78.
101. Russell AJ, et al. Alpha 6 beta 4 integrin regulates keratinocyte chemotaxis through differential GTPase activation and antagonism of alpha 3 beta 1 integrin. J. Cell Sci. 2003, 116: 3543-56.
102. Shaw LM. Identification of insulin receptor substrate 1 (IRS-1) and IRS-2 as signaling intermediates in the alpha6beta4 integrin-dependent activation of phosphoinositide 3-OH kinase and promotion of invasion. Mol. Cell Biol. 2001, 21: 5082-93.
103. Falcioni R, et al. Alpha 6 beta 4 and alpha 6 beta 1 integrins associate with ErbB-2 in human carcinoma cell lines. Exp. Cell Res. 1997, 236: 76-85.
104. Riese DJ. 2nd and Stern DF. Specificity within the EGF family/ErbB receptor family signaling network. Bioessays. 1998, 20: 41-8.
105. Comoglio PM, Trusolino L. Invasive growth: from development to metastasis. J. Clin. Invest. 2002, 109: 857-62.
106. Chung J, et al. The Met receptor and alpha 6 beta 4 integrin can function independently to promote carcinoma invasion. J. Biol. Chem. 2004, 279: 32287-93.
107. Gingras AC, Raught B, Sonenberg N. Regulation of translation initiation by FRAP/mTOR. Genes. Dev. 2001, 15: 807-26.
108. Shaw RJ, et al. The LKB1 tumor suppressor negatively regulates mTOR signaling. Cancer Cell 2004, 6: 91-9.
109. Jauliac S, et al. The role of NFAT transcription factors in integrin-mediated carcinoma invasion. Nat. Cell Biol. 2002, 4: 540-4.
110. Mercurio AM, Rabinovitz I. Towards a mechanistic understanding of tumor invasion--lessons from the alpha6beta 4 integrin. Semin. Cancer Biol. 2001, 11: 129-41.
111. Savoia P, et al. Expression and topography of integrins and basement membrane proteins in epidermal carcinomas: basal but not squamous cell carcinomas display loss of alpha 6 beta 4 and BM-600/nicein. J. Invest. Dermatol. 1993, 101: 352-8.
112. Rossen K, et al. Expression of the alpha 6 beta 4 integrin by squamous cell carcinomas and basal cell carcinomas: possible relation to invasive potential? Acta. Derm. Venereol. 1994, 74: 101-5.
113. Korman NJ, Hrabovsky SL. Basal cell carcinomas display extensive abnormalities in the hemidesmosome anchoring fibril complex. Exp. Dermatol. 1993, 2: 139-44.
114. Carico E, et al. Integrin beta 4 expression in the neoplastic progression of cervical epithelium. Gynecol. Oncol. 1993, 49: 61-6.
115. Skubitz AP, et al. Expression of alpha 6 and beta 4 integrins in serous ovarian carcinoma correlates with expression of the basement membrane protein laminin. Am. J. Pathol. 1996, 148: 1445-61.
116. Gui GP, et al. Integrin expression in primary breast cancer and its relation to axillary nodal status. Surgery 1995, 117: 102-8.
117. Costantini RM, et al. Integrin (alpha 6/beta 4) expression in human lung cancer as monitored by specific monoclonal antibodies. Cancer Res. 1990, 50: 6107-12.
118. Cress AE, et al. The alpha 6 beta 1 and alpha 6 beta 4 integrins in human prostate cancer progression. Cancer Metastasis Rev. 1995, 14: 219-28.

119. Nagle RB, et al. Expression of hemidesmosomal and extracellular matrix proteins by normal and malignant human prostate tissue. Am. J. Pathol. 1995, 146: 1498-507.
120. Grossman HB, et al. Expression of the alpha6beta4 integrin provides prognostic information in bladder cancer. Oncol. Rep. 2000, 7: 13-6.
121. Hall PA, et al. Characterization of integrin chains in normal and neoplastic human pancreas. J. of Pathol. 1991, 165: 33-41.
122. Weinel RJ, et al. The alpha 6-integrin receptor in pancreatic carcinoma. Gastroenterology 1995, 108: 523-32.
123. Logsdon CD, et al. Molecular profiling of pancreatic adenocarcinoma and chronic pancreatitis identifies multiple genes differentially regulated in pancreatic cancer.[erratum appears in Cancer Res. 2003 Jun 15;63(12):3445]. Cancer Res. 2003, 63: 2649-57.
124. Tani T, et al. Alpha(6)Beta(4) Integrin and Newly Deposited Laminin-1 and Laminin-5 Form the Adhesion Mechanism Of Gastric Carcinoma - Continuous Expression Of Laminins But Not That Of Collagen Vii Is Preserved In Invasive Parts Of the Carcinomas - Implications For Acquisition Of the Invading Phenotype. Am. J. Pathol. 1996, 149: 781-793.

Chapter 9

RHO GTPASES IN CELL MOTILITY AND TUMORIGENESIS

Amra Hajdo-Milašinović, Alexander E. Mertens, Irene H. L. Hamelers, John G. Collard
The Netherlands Cancer Institute, Department of Cell Biology, Amsterdam, The Netherlands

Abstract: Rho proteins are small regulatory molecules that belong to the Ras superfamily of proteins. They act as molecular switches, shuffling between an active, GTP-bound state, and an inactive, GDP-bound state. Upon activation, they interact with a multitude of downstream effectors. In this way Rho proteins regulate a broad range of cellular processes, including cell motility, cell growth, apoptosis, and gene transcription. Therefore, it is not surprising that Rho proteins are also involved in different aspects of tumorigenesis. In particular, as key regulators of cell motility, Rho GTPases are implicated in invasion and metastasis of a tumor. In this review we will focus on the involvement of Rho proteins in cell migration and the different steps of tumorigenesis.

Key words: Rho, Rac, cdc42, GTP, actin cytoskeleton, adhesion, protrusion

1. INTRODUCTION

Similar to Ras proteins, Rho GTPases bind guanine nucleotides and act as molecular binary switches. In contrast to the mutated Ras oncogenes, Rho proteins have never been found mutated in human tumors; however, their involvement in cancer development has been clearly established and is an active area of research.

Rho GTPases are ubiquitously expressed and so far 20 Rho proteins have been described in humans (1). Based on primary sequence and known functions, they can roughly be divided into 5 groups, being the Rho-like,

A. Wells (ed.), Cell Motility in Cancer Invasion and Metastasis, 189-220.

Rac-like, Cdc42-like, Rnd, and RhoBTB subfamilies (Figure 1). Of these 20 members, RhoA/B, Rac1/2 and Cdc42 are the most widely studied. Rho proteins are involved in the regulation of many cellular processes, including cytoskeletal organization, cell motility, cell cycle progression and growth, apoptosis, gene transcription and vesicle transport. Consequently, it is not surprising that Rho proteins are involved in different steps of tumor development and progression. In this chapter, we describe in more detail how the Rho proteins are regulated, what their role is in cell migration and maintenance of cell-cell adhesions. In addition, we summarize the data that implicate this protein family in cancer development.

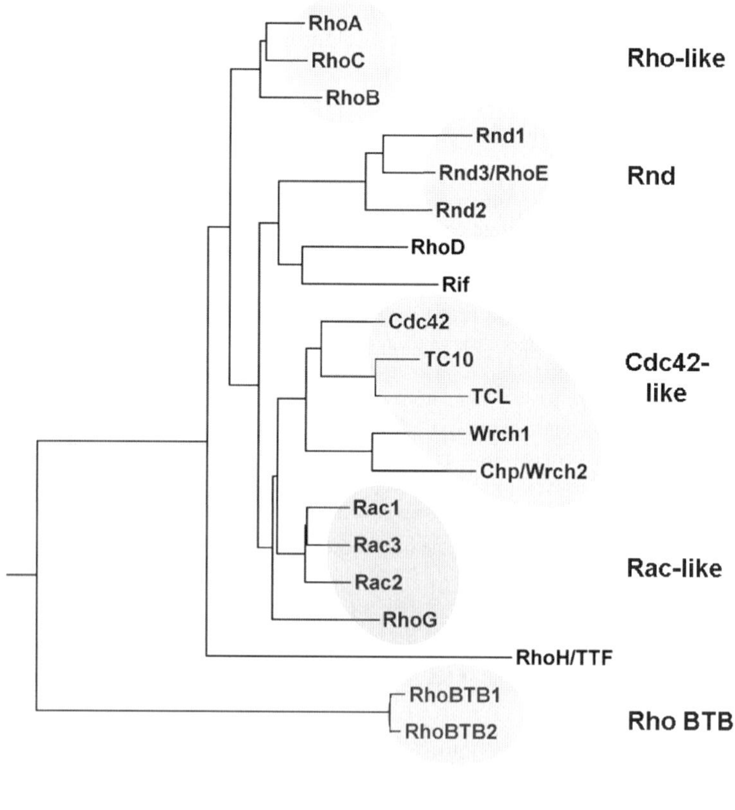

Figure 1. Rho protein family tree. Based on sequence homology and function, the 20 Rho protein family members are divided into 5 groups: Rho-like, Rac-like, Cdc42-like, Rnd and RhoBTB.

2. REGULATION OF RHO GTPASE ACTIVITY

Rho proteins cycle between an active, GTP-bound conformation and an inactive, GDP-bound conformation. In the GTP-bound form, they interact with downstream target proteins to induce cellular responses. Rho proteins exchange nucleotide and hydrolyse GTP at slow rates. These reactions are catalyzed by guanine nucleotide- exchange factors (GEFs) and GTPase-activating proteins (GAPs), respectively (Figure 2).

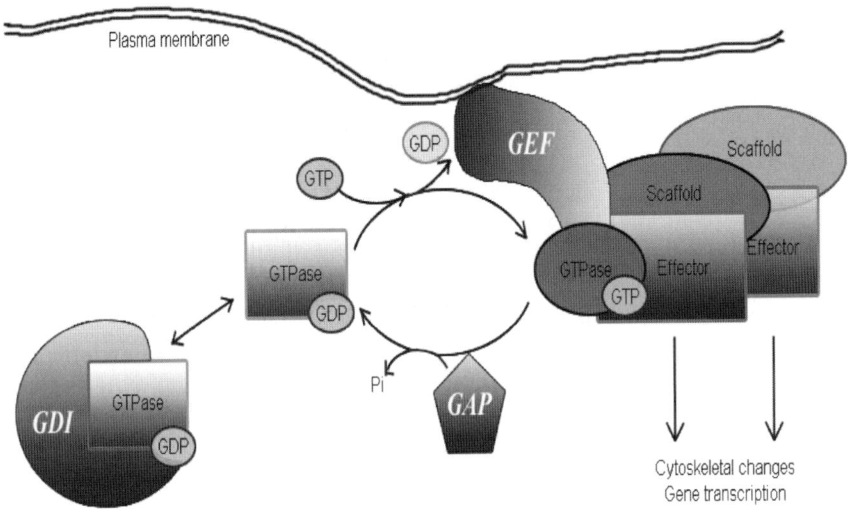

Figure 2. : Regulation of RhoGTPase function. Rho GTPases shuttle between an active and an inactive state. RhoGDIs keep Rho proteins in a GDP-bound inactive state. GEFs and GAPs regulate the GDP release and the GTP hydrolysis, respectively.

The guanine nucleotide exchange factors (GEFs) promote the exchange of GDP for GTP, in this fashion activating the GTPase. RhoGEFs are generally large, multidomain proteins that typically contain a Dbl homology (DH) domain (2) which catalyses GDP-GTP exchange. The DH domain is often flanked by pleckstrin homology (PH) domain. The DH-PH domain units show varying specificities for subsets of Rho family proteins (3). There are many more GEF proteins than there are Rho proteins; so far, about 60 GEFs have been identified in the mammalian genomes (4). This diversity probably represents the engagement in a wide variety of signaling pathways in different tissues. In addition, RhoGEFs are not just simply activators of Rho proteins, but they also serve as docking sites for adaptor proteins and downstream effector proteins of Rho-like GTPases (reviewed in 5). This additional feature promotes RhoGEFs into tools for fine-tuned spatial and temporal regulation of Rho GTPases and their downstream signaling.

GTPase-activating proteins (GAPs) stimulate the rate of GTP hydrolysis, thereby converting GTPases to their inactive, GDP-bound form. So far, there are about 70 proteins identified that contain a RhoGAP domain (6). Apart from their function to inactivate Rho GTPases there is evidence that some GAPs also act as Rho effectors and transmit signals downstream from Rho proteins.

The third group of Rho family regulators is known as the guanine-nucleotide dissociation inhibitors (GDI). GDIs bind a subset of Rho proteins, inhibit nucleotide exchange and prevent the binding of these proteins to the membranes, where they are activated. To date, three RhoGDIs have been identified, and a few other molecules have been suggested to have RhoGDI activity. A large fraction of Rho proteins in the cell is likely to be bound to GDIs (7). How RhoGDIs interact with Rho proteins is known from structural and biochemical studies (8, 9), but it is still unclear how this interaction is regulated. GDI phosphorylation and the binding of other proteins to GDI can stimulate release of Rho GTPases from GDIs. In this way, Rho proteins can be delivered by the GDIs to a particular site of action in the cell.

The localization of Rho proteins is regulated by lipid modifications. Rho GTPases are generally post-translationally modified at their C-termini by prenylation of a conserved cysteine, and this is a determinant for targeting these proteins to the plasma membranes (10).

Various tools have been employed to analyse Rho protein function. Two types of point mutants have been used extensively: activated mutants, which are constitutively GTP-bound (the GAP proteins can not bind anymore) and dominant-negative mutants, which generally have reduced affinity for nucleotides (11) and are titrating out GEFs (12). Bacterial toxins that inactivate Rho proteins have also been used to study Rho function (reviewed in 13). A prototypical agent is C3 exotransferase, produced by *Clostridium botulinum*. C3 (ADP-)ribosylates RhoA, B, and C leading to the inactivation of these proteins (14). More recently, the downregulation of various Rho proteins by short interference RNA sequences have become a powerful tool to study the specific function of Rho proteins.

3. FUNCTION OF RHO GTPASES

The most intensively investigated function of Rho proteins is the regulation of the cytoskeletal organization in response to extracellular signals. However, over the past few years it has been shown that Rho GTPases also play crucial roles in many other cellular events, such as membrane trafficking, transcriptional regulation, apoptosis, cell polarization, cell growth control, and cell differentiation. Within the scope of this chapter we will discuss in particular the involvement of Rho proteins in cell migration and cell adhesion, processes associated with the formation and progression of tumors.

3.1 Rho GTPases in migration

Cell migration is essential to normal development of multicellular organisms and plays a crucial role in morphogenesis throughout embryonic development. Cells migrate from epithelial layers to end up in locations where they start to differentiate and form specialized tissues and organs. Migration is also a prominent component of tissue repair and immune protection. It is not surprising that migration is a crucial factor in many pathological processes as well. Vascular diseases, osteoporosis, chronic inflammatory diseases, mental retardation and cancer correlate tightly with impaired, deregulated or intensified migratory capacity of cells. So, understanding the mechanisms underlying cell migration might potentially lead to more effective therapeutic approaches for treating disease.

Cell migration can be seen as a cyclic process, consisted of four mechanistically separated steps (15). Firstly, in response to migration-promoting agent, cells start to polarize and extend a leading edge protrusions or lamellipodium in the direction of migration. These protrusions are dependent on actin polymerisation. Secondly, the new adhesions are being established at the front of the cell, serving also as traction sites for migration as the cell move over them. Subsequently, the contraction of the cell body occurs, and finally, the adhesions at the cell rear become detached.

Cells start to migrate in response to different extracellular cues (e.g., growth factors, signals on neighboring cells, or signals from the extracellular matrix) that act either as attractants or repellents. This evokes signaling cascades within the cell that lead to cytoskeletal changes, cell-substrate adhesions, loss of cell-cell adhesions and other processes that are required for forward movement.

During the migratory processes, Rho proteins regulate the changes in cell adhesion and the actin cytoskeletal changes during cell migration (Figure 3). In fibroblasts, Rho can be activated by the addition of extracellular ligands and activated Rho leads to the assembly of contractile actin-myosin filaments (stress fibers) and of associated focal adhesion complexes. Rac is activated by a distinct set of agonists (different growth factors, insulin), and activated Rac induces the assembly of a meshwork of actin filaments at the cell periphery to produce lamellipodia and membrane ruffles. Furthermore, activation of Cdc42 induces actin-rich surface protrusions called filopodia.

Other processes regulated by Rho GTPases that are important for cell migration, are focal adhesion complex formation and turn-over, the establishment of cell polarity, microtubule dynamics, vesicular transport pathways and the signaling up- and downstream of integrins. Rather then exploring these functions one by one, we will discuss the various steps of migration together with the Rho protein function in every particular step.

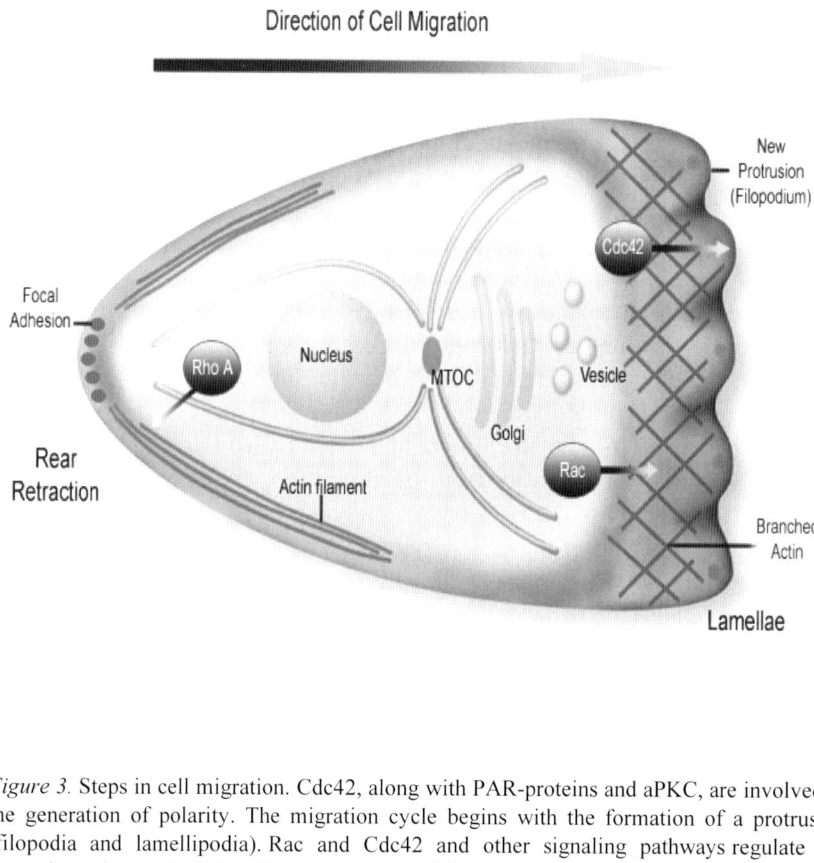

Figure 3. Steps in cell migration. Cdc42, along with PAR-proteins and aPKC, are involved in the generation of polarity. The migration cycle begins with the formation of a protrusion (filopodia and lamellipodia). Rac and Cdc42 and other signaling pathways regulate the formation of actin branches Protrusions are stabilized by the formation of adhesions. This process requires integrin activation, clustering, and the recruitment of structural and signaling components to nascent adhesions. At the cell rear, adhesions disassemble as the rear retracts. This process is mediated by Rho.

3.1.1 Cell polarization

In order to migrate, a cell must be polarized, meaning that different molecular processes must define the front and the rear of the cell. A key player in the regulation of cell polarity is Cdc42. The GTPase Cdc42 is active toward the front of migrating cells (16), and both inhibition and over-activation of Cdc42 can disrupt directionality of migration (17). Cdc42 restricts the area where the lamellipodia can be formed (18), and it also affects polarity by localizing the microtubule-organizing center (MTOC) in front of the nucleus, oriented toward the leading edge (Figure 2) (17, 19). The effects of Cdc42 on MTOC position appear to involve the Cdc42 effector PAR6, which forms a complex with PAR3 and an atypical protein kinase C (aPKC) (20, 21). An additional role of Cdc42 in cell migration is its role in filopodia formation, by initiating the actin polymerisation. These protrusions are formed in the process of cell polarization and are required for the direction-sensing in many different cell types (22, 23). Cdc42 is able to stimulate actin polymerization via its interaction with WASp and N-WASp, leading to activation of the Arp2/3 complex that mediates actin filament assembly. For proper cell polarization, a tight regulation of the microtubule dynamics is needed, which involve not only Cdc42 (24, 25) but also both Rac (24, 25), and Rho (26-29).

3.1.2 Lamellipodia

Lamellipodia are protrusive structures generated at the leading edge of migrating cells. They consist of branched actin filament networks formed through the actin-nucleating activity of the Arp2/3 complex (30). Through localized activation of the Arp2/3 complex, the lamellipodia are induced to grow in a particular direction, providing the basis for directional migration. Activation of the Arp2/3 complex occurs by WASp/WAVE family members, which are downstream effectors of Cdc42 and Rac respectively (31).

Rac is required for lamellipodium extension induced by growth factors, cytokines and extracellular matrix components and inhibition of Rac activity impairs cell migration (22, 23, 32). Activated Rac is localized preferentially towards the front of migrating cells (33). A major protein that can activate Rac at the leading edge is phosphoinositide 3-kinase (PI 3-kinase) that is activated either via tyrosine kinases or G-protein-coupled receptors (34-36). The products of PI 3-kinase, PtdIns (3, 4, 5) P3 / PtdIns (3, 4) P2 (PIP3/PIP2), appear to be enriched at the leading edge of migrating cells (37-39). Tiam1 and Vav2 are PIP3-responsive GEF proteins localised at the leading edge of cells. They are likely candidates for regulating local Rac activity during directed cell migration (34, 40, 41).

A downstream effector of Rac important for cytoskeletal rearrange-ments and membrane ruffling is p21-activated kinase (PAK). Both Rac and Cdc42 activate PAK; its activation promotes formation of lamellipodia (42) and leads to the loss of stress fibers and focal adhesions (43). Via PAK, Rac has been reported to stimulate the activity of LIM-kinase (44, 45), which phosphorylates and inactivates cofilin, a protein that promotes actin depolymerization (45). Thus Rac is able to inhibit cofilin-induced depolymerization.

In addition, myosins have been implicated in cell migration (46), and Rac can affect the phosphorylation of both myosin II heavy chain (MHC; [47]) and myosin light chain (MLC) via PAK (48-50). Phosphorylation of the myosin heavy chain by PAK inhibits myosin function and causes the disassembly of actomyosin structures (47). In addition to myosins, several other targets downstream of Rac and PAK have been implicated in actin reorganisation, including the actin binding protein filamin, the paxillin/Pix/PKL complex, formins and the adaptor protein Nck (reviewed in 13).

3.1.3 Cell-matrix adhesions

Prior to cell migration, a protrusion must extend, which subsequently needs to be stabilized by adhesion to a substrate. Small focal complex structures are localized in the lamellipodia of most migrating cells, and are important in mediating the attachment of the extending lamellipodium to the extracellular matrix (ECM; 15). Although many different receptors are involved in the migration of different cell types, the integrins are the major migration-promoting receptors. Integrins act as the "feet" of a migrating cell by supporting adhesion to the ECM or other cells and by linking via adapters with the actin filaments inside of the cell.

Rac is required for focal complex assembly (51-53). In the case of Rho, integrin clustering is very pronounced and results from tension aggregating dispersed integrins, such that they align through their attachment with the ends of stress fibres in focal adhesions (54). In some cell lines, the inhibition of Rho decreases adhesion, causing a retraction of the lamellae and rounding of the cell body. The speed of cell migration is dependent on the substrate composition. Indeed, the relative levels of Rho, Rac and Cdc42 activation vary with extracellular matrix composition (55-57). There is continuous crosstalk between integrins and Rac to allow cells to respond to changes in extracellular matrix composition of growth factor receptors (58).

The turnover of focal complexes/adhesions is important for migration of cells. If they are persistently sustained, the cell cannot migrate because it cannot detach from the ECM. On the other hand, if they are properly disassembled but their assembly is deregulated, migration is again impaired because the cell does not have sufficient grip on EMC to move forward. Both Rac and Rho are directly and indirectly involved in the regulation of turnover of focal adhesions/complexes (48, 59, 60).

3.1.4 Cell body contraction

In order to migrate, the cell needs to contract and retract the rear. Cell body contraction is dependent on actomyosin contractility (61), a process regulated by Rho. Rho acts via ROCKs (also known as Rho-kinases) to affect MLC phosphorylation, both by inhibiting MLC phosphatase and by direct phosphorylation of MLC (62, 63). MLC phosphorylation is also regulated by MLC kinase (MLCK), which is activated by calcium, and stimulated by the ERK MAPKs (64). It is likely that ROCKs and MLCK act in concert to regulate different aspects of cell contractility. ROCKs are required for MLC phosphorylation associated with actin filaments in the cell body, whereas MLCK is required at the cell periphery (65). Rho promotes myosin contractility and the resulting tension drives the formation of stress fibers and focal adhesions. Thus, reducing Rho activity has two opposing effects: it promotes cell migration by lowering adhesion, but decreases it on the other hand by inhibiting cell body contraction.

3.2 Rho-GTPases as regulators of cell-cell adhesion

Invasion of surrounding tissue is one of the most important steps in tumor metastasis. It requires the release of tumor cells from the primary tumor. The major intercellular adhesion molecules expressed by epithelial cells belong to the family of the classical cadherins (E/P/N-cadherins), where E-cadherin is the most abundantly expressed isoform. In addition to cadherins, which are concentrated in the so-called adherens junction (AJ), epithelial cells form intercellular contacts through tight junctions (TJ) and desmosomes (Figure 4). Cadherin complexes and tight junctional components associate with the cortical F-actin cytoskeleton, whereas desmosomal cadherins are linked to intermediate filaments called cytokeratins.

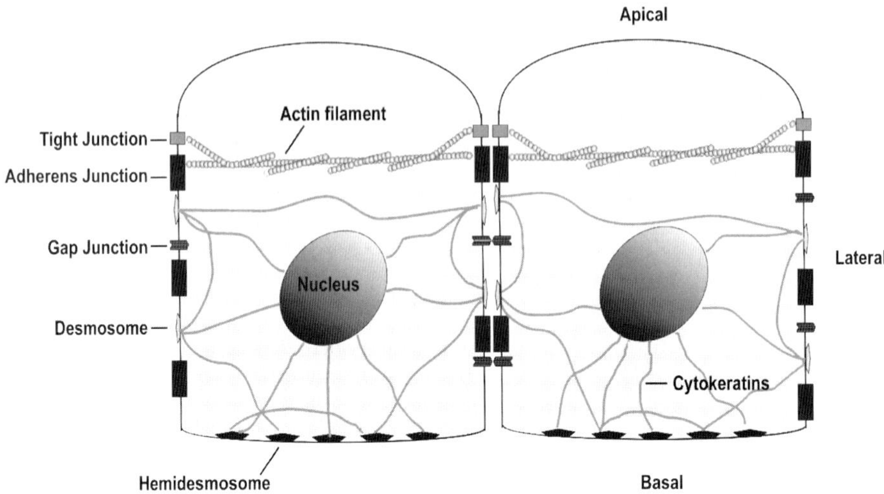

Figure 4. Epithelial intercellular junctions. The figure shows a schematic representation of a polarized epithelial cell. The different types of intercellular junctions as well as hemidesmosomes are shown. It should be noted that the tight junctions and adherens junctions are linked to the actin cytoskeleton, and desmosomes and hemidesmosomes are linked to intermediate filaments.

3.2.1 Cadherin expression and function

Classical cadherins mediate intercellular adhesion in a homotypic and Ca^{2+}-dependent manner through the formation of zipper-like molecular structures on opposing cells (66, 67). Cadherins are stabilized on the plasma membrane through dimerisation (68) and the linkage of the cytoplasmic tails to the F-actin cytoskeleton via the Armadillo-family proteins α- and β-catenin and the actin binding proteins α-actinin and vinculin (69, 70).

Cadherins play a critical role in tissue morphogenesis and homeostasis of tissue architecture and their functional elimination represents a key step in the acquisition of the invasive phenotype for many epithelial tumors. *In vitro*, a strong correlation has been found between the loss of E-cadherin and the loss of the epithelial phenotype, whereas overexpression of E-cadherin in mesenchymal cells induces a mesenchymal-epithelial transition (71, 72). The progression from benign to malignant epithelial tumors and the dedifferentiation of tumor cells *in vivo* is strongly correlated with the loss of function of E-cadherin (73).

The loss of function of E-cadherin in epithelial tumors can have multiple causes (reviewed in 74-76). Since loss-of-function mutations in the E-cadherin gene are sporadic in human tumors and re-expression of E-cadherin in distant metastases is often observed, epigenetic regulation of cadherin expression is probably most relevant in tumor progression. Downregulation at the RNA level can be established through regulation of transcription factors such as SNAIL, Slug, and SIP1. Other mechanisms of gene silencing involve hypermethylation of the E-cadherin promoter. At the post-translational level, cadherins can be regulated by enhanced turnover and degradation, processes that can be triggered by several oncogenic signaling pathways and importantly also by Rho GTPases.

3.2.2 Rho-GTPases regulate cadherin-based intercellular adhesions

Rho GTPases are actively involved in the formation and maintenance of intercellular adhesions. Conceivably, F-actin dynamics mediated by Rho-GTPases influence the assembly, stability and function of the adherens - and the tight junction. However, the effects of Rho GTPases on cell-cell adhesion complexes are not all mediated through rearrangement of the actin cytoskeleton. Although the involvement of Rho proteins in the formation and maintenance of cell-cell contacts are well established, contradictory results make it difficult to form a clear consensus for their roles in these processes (reviewed in 77-79).

Initially, Rac1, Cdc42 and RhoA where found to affect cadherin accumulation in intercellular adhesions of MDCK cells and primary human keratinocytes. Inactivation of RhoA and Rac1 inhibit the formation and break down of already established adherens junctions in MDCK cells (80-82). RhoA is involved in clustering of cadherin receptors (82), whereas Rac mediates actin recruitment to primary contacts, thereby stimulating stabilization and additional binding of cadherins to the cortical actin cytoskeleton (81, 82). In contrast, both Rac1 and RhoA have also been described to disassemble cadherin-based adhesions in human epidermal keratinocytes (83-85). Rac activity stimulates disassembly of the adherens junction in a dose and time dependent manner (86), possibly via enhanced endocytosis of cadherins (84).

In transformed cells, different oncogenic signaling pathways regulate the stability of the adherens junction via Rho GTPases. Again, GTPase signaling can either contribute to disruption or stabilization of the AJ. Tiam1, an activator (GEF) of Rac, induces a mesenchymal-to-epithelial reversion of Ras transformed MDCK cells (34, 87). Over-expression of Tiam1 and thereby activation of Rac blocks cell-cell dissociation (or scattering) by hepatocyte growth factor (HGF/SF) in MDCK cells (87) and it is required for maintenance of the AJ upon overexpression of the viral oncogene E1A (88).

The role of RhoA in support of the transformed phenotype is more unequivocal; Ras-transformation of epithelial cells is often associated with a more contractile fibroblastic phenotype. Enhanced RhoA activity, generating contractile force through the formation of stress fibers, contributes to this phenotype (89, 90). The migratory potential of Ras-transformed cells furthermore is associated with low Rac activity and high Rho activity (89, 91). The activity of RhoA cooperates with HGF and TGFβ in the disruption of cadherin-based adhesions (92, 93). From all these studies it is clear that RhoA and Rac1 may differentially regulate cadherin-based cell-cell adhesions, but the outcome of their activities depends on the cellular context and whether or not they are targets of oncogenic signaling pathways.

Several downstream effectors of Rho GTPases are known to affect cadherins either directly or via the actin cytoskeleton (reviewed in 78). One important target of both Rac1 and Cdc42 that regulates E-cadherin stability is the actin-binding protein IQGAP1 (94, 95). IQGAP1 hampers the stable association of E-cadherin to the actin cytoskeleton. Recent data suggest that IQGAP1 is involved in a positive feedback-loop in which it activates Rac1 downstream of E-cadherin (96). Two downstream effectors of Rho signaling, ROCK and mDia, differentially affect the AJ. Whereas ROCK mediated actin-myosin contraction downstream of RhoC and RhoA disrupts AJ formation, actin polymerization downstream of mDia is required for maintenance and formation of the AJ (91).

3.2.3 Cadherins signal towards Rho-GTPases

Cadherin molecules do not only glue cells together but are also considered as signaling complexes that mediate outside-in signaling. The cytoplasmic domain of E-cadherin binds several molecules involved in intracellular signaling towards differentiation-, growth- and survival pathways. Interestingly, classical cadherins signal in an adhesion dependent fashion directly or indirectly to the Rho-GTPases RhoA, Cdc42 and in particular Rac1.

Early work described a correlation between E-cadherin ligation, the activation of Rac1 (97, 98) and the downregulation of RhoA activity (97, 89) in differentiated epithelial cells. By the use of surfaces coated with recombinant cadherin-specific adhesive ligands, Rho-GTPase signaling has been studied as a direct consequence of cadherin ligation. In cadherin expressing cells that adhered to substrates coated with *Xenopus* C-cadherin (97) or human E-cadherin (99, 100) Rac1 was activated in minutes upon binding to the substrate. Long term adhesion to C-cadherin (97) or adhesion of mouse C2C12 cells to chick N-cadherin (101) led to lower or higher

RhoA activity respectively, which indicates that the outcome of cadherin signaling differs dependent on cell type or cadherin isoforms.

Several molecular mechanisms have been proposed by which cadherins signal to Rho-GTPases. One candidate molecule that is supposed to control Rac1 signaling is PI3 kinase, of which the lipid products are potent activators of several Rac-GEFs. Cadherin dependent adhesion recruits and activates PI3 kinase (100, 102) and the p85 regulatory subunit of PI3 kinase is a direct binding partner for β-catenin (103).

Rac1, Cdc42 and RhoA activity can also be regulated via p120-catenin, the armadillo-family protein that binds the membrane proximal region of the cadherin cytoplasmic tail (104-107). p120 can shuttle between the plasma membrane (where it binds cadherins) and the cytoplasm where it can inactivate RhoA (104), but activates Rac1 (107) and Cdc42 (106), possibly via a direct interaction of p120 with the Rac-GEF Vav2 (105). The cytoplasmic pool of p120 enhances the migratory capacity of epithelial cells in a Rac1/Cdc42 dependent manner, whereas cadherins sequester p120 to the plasma membrane, thereby suppressing its function in cell motility (106). Functional loss of cadherins in epithelial tumors could therefore enhance the migratory potential of these cells via p120 and Rho-GTPase signaling, as recently suggested for invasive breast cancer cells (108).

Loss of expression of E-cadherin is often correlated with the enhanced expression of other types of cadherins, a process called the 'cadherin switch'. This switch is often associated with the loosening of cell-cell contacts and a transition to a migratory phenotype. Since N-cadherin was shown to activate several Rho-GTPases (101, 109) it is likely that GTPase signaling downstream of N-cadherin affects cell migration and epithelial to mesenchymal transition of tumor cells, although no direct evidence for this is currently available. Recently, R-cadherin, a classical cadherin with high homology to N-cadherin, was shown to increase the steady state activity of Rac1 and Cdc42 resulting in increased motility of different R-cadherin expressing cell-lines (110). A potential role of R-cadherin in tumorigenesis has however not been investigated yet.

In summary, Rho GTPases are required for the formation and maintenance of cadherin-based adhesions, that prevents cell invasion and migration. However, Rho-GTPase signaling downstream of cadherins can also positively or negatively affect the migratory capacity of cells by (local) feedback signaling towards the cadherins and the tight junctions. Moreover, various types of cadherins, especially the ones overexpressed in epithelial tumors, might differentially regulate the steady state levels of Rho-GTPases and could thereby enhance cell migration rather then promote stable intercellular adhesion.

3.2.4 Rho-GTPases regulate tight junctions and apicobasal polarity

Loss of cell polarity is another important hallmark of epithelial cancer cells that become invasive. The spatial and functional separation of the apical and baso-lateral plasma membrane of epithelial cells (apicobasal polarity) is established through asymmetric distribution of junctional intercellular adhesion complexes that are linked to the actin cytoskeleton. At the most apical region of the plasma membrane of endothelial and epithelial cells the tight junction creates a barrier for paracellular diffusion of small soluble molecules and it restricts free intramembrane diffusion of integral components of the plasma membrane. Loss of tight junction function occurs at multiple stages in tumor metastasis, between tumor cells themselves but also between cells of the vascular endothelium through which tumor cells can invade the underlying tissue (111). Moreover, leakiness of epithelial tight junctions allows luminal growth factors to reach their receptors at the baso-lateral membrane, causing changes in epithelial cell proliferation and survival. (30, 112).

The integral membrane proteins occludin, JAMs and claudins, that make up the tight junction are connected to the F-actin cytoskeleton (113). Therefore, Rho-mediated cortical actin dynamics influence the assembly, stability and function in regulation of paracellular permeability of the tight junction (114-116). Recent studies indicate that tight junctions can also function as signaling platforms, regulating cell morphology and gene expression (117) and RhoGTPase signaling (118).

Cdc42 plays a direct role in the formation of the tight junction and the establishment of apicobasal polarity in epithelial cells. Activated Cdc42, as a binding partner for the adaptor molecule PAR6, was implicated in the recruitment of the polarity complex (consisting of PAR6, PAR3 and an atypical PKC) to the plasma membrane (119, 120) at regions where initial cell-cell contacts are formed. This signaling complex is of pivotal importance for the sequential formation of the tight junction and full maturation of the initial cell-cell contacts into the cortical adhesion belt (121, 122). Initial cell-cell contacts mediated by cadherins and nectins activate Cdc42 (123) and recruit the polarity complex and stimulate the assembly of the tight junction. Both the formation and maintenance of tight junction and cadherin-based cell-cell adhesions contribute to the strength of the contacts between cells and thereby determine the invasive migratory capacity of epithelial tumor cells.

4. RHO GTPASES IN CELL TRANSFORMATION AND METASTASIS

4.1 Rho proteins in human cancer

The first findings that implied a role for Rho GTPases in cancer-development came from *in vitro* foci formation assays. These assays are used as a model for growth factor- and anchorage-independent tumor cell growth. Foci-formation assays identified many N-terminal-truncated RhoGEFs, such as Dbl, Ect2, Lfc and Vav as proto-oncogenes (reviewed in 124, 125). Furthermore, activating mutants of RhoA, RhoG, Rac1, TC10 and Cdc42 induced foci formation, albeit weakly compared to constitutively active Ras (V12Ras) (126, 127). Consistent with these findings, inhibitory mutants of Rho proteins prevent Ras-induced transformation in soft-agar assays and tumor formation in nude mice (128-132), and V12Rac1 cooperates with constitutively active Raf (RafCAAX) in focus formation (130, 133). These observations suggest that Rho proteins act downstream of Ras in oncogenic transformation of cells.

Unlike Ras, there are no reports of mutated, constitutively active forms of Rho proteins in human tumors. The only Rho protein known to be altered genetically in human tumors is RhoH, which is frequently rearranged in lymphomas (148) (Table 1) or to have mutations in the 5' untranslated region (149), which are predicted to affect RhoH expression. Many reports have shown that other members of the Rho GTPase family are over-expressed in human cancer (see table) and that, in some cases, this correlates with clinical outcome. For instance, in breast cancer and testicular germ-cell cancer RhoA expression levels correlate positively with progression of the disease (134, 140). RhoC is overexpressed in pancreatic ductal adenocarcinoma and inflammatory breast cancer (141-143).

Two reports describe a role for Rho-GEFs in human tumorigenesis (150-152). Tiam1, a Rac-specific GEF was found to have a point mutation in the PH-domain of one of the two alleles, in around 10% (4 of 35) of the renal cell carcinoma examined. The mutation results in an alanine-glycine substitution at amino acid 441, and was suggested to be a dominant gain-of function mutation. However, in three of five renal cell carcinoma cell lines a decrease in the Tiam1 expression was found, including two that contained the A441G mutation (150, 151). The significance of this in light of the gain-of-function activity of this mutation is unclear. Another RhoGEF, leukemia-associated Rho guanine nucleotide exchange factor (LARG) has been isolated as a fusion partner of the mixed-lineage leukemia (MLL) gene in a patient with acute myeloid leukemia. The fusion protein contained the N-terminal part of MLL in frame with the C-terminal part of LARG, which includes the DH/PH domain (important for GEF function), a domain with homology to the Rho GEF Lsc, and a nuclear localization signal (152). It remains to be established whether the MLL-LARG fusion protein is sufficient to induce transformation in myeloid cells and how frequently the locus is affected in human cancer.

Table 1. Aberrant regulation of Rho proteins and their regulators in human cancer.

Gene	GTPase specificity	Type of deregulation	Tumor types
Cdc42		Overexpression	Breast cancer (134)
Rac1		Overexpression	Breast & gastric cancer (135)
Rac1B		Alternative splicing	Colon & breast cancer (136, 137)
Rac2		Overexpression	Head and neck squamous carcinoma (138)
Rac3		Hyperactive (majority of Rac3 in GTP-bound state)	Breast cancer (139)
RhoA		Overexpression	Colon, breast & lung cancer (134) Testicular germ cell cancer (140) Head and neck squamous carcinoma (138) Gastric cancer (135)
RhoC		Overexpression	Inflammatory breast cancer (141,142) Pancreatic ductal adeno-carcinoma (143) Non-small cell lung carcinoma (144) Ovarian cancer (145) Gastric cancer (146) Hepatocellular carcinoma (147)
RhoH		Rearrangement Mutations in 5' UTR	Non-Hodgkin's lymphoma & multiple myeloma (148) Diffuse large B-cell lymphoma (149)
Tiam1	Rac	Point mutation	Renal-cell carcinoma (150, 151)
LARG	Rho	5' End of MLL gene fused to the 3' end of LARG	Acute myeloid leukemia (152)

Several lines of evidence indicate that the cycling of Rho GTPases between GTP- and GDP-bound states might be important for transformation. *In vitro* studies have shown that Rho-GEFs are more potent oncogenes than

GTPases defective Rho proteins, and that a fast GTP-GDP cycling mutant of Cdc42 has a greater transforming capacity than a GTPase defective mutant (132,153). Furthermore, the splice variant of Rac1, Rac1B, which has an increased GTP-GDP cycling rate *in vitro* is highly overexpressed in breast and colon cancer (136). The requirement for cycling between GDP- and GTP-bound states might reflect the cyclic nature of the processes that are regulated by Rho proteins in tumorigenesis, and might provide an explanation for the fact that GTPase defective mutants –analogous to those in oncogenic Ras- have not been identified in tumors.

4.2 Participation of Rho GTPases in different stages of tumorigenesis

4.2.1 Rho proteins in growth and apoptosis

The growth rate of tumors is determined by the difference between cell growth and cell death. Rho proteins have been implicated in the (de)regulation of both processes in normal and tumor cells. Rho proteins regulate cell growth via multiple pathways, including regulation of the expression of cell cycle proteins like cyclin D1 and the CDKIs $p21^{Waf1}$ and $p27^{Kip1}$ (reviewed in 91). In addition, the Rho proteins can modulation of the activity of growth-factor-regulated pathways by regulating the transport and turnover of growth-factor receptors. RhoB and its effector PKC-related kinase 1 (PRK1), affect the kinetics of the epidermal growth-factor receptor (EGFR) internalization after ligand stimulation (154), by coordinately regulating the movement of vesicles along microfilament networks (155). However, many of the studies that led to the conclusions outlined above have been done *in vitro* using fibroblasts, which are cells that rarely become cancerous. Further work is needed to determine via which pathways Rho proteins regulate cell proliferation *in vivo*.

Apoptosis is a counterbalance to mechanisms of cell proliferation and is critically important in regulation of the immune system, development, and normal tissue homeostasis. Cancer cells frequently show enhanced sensitivity to pro-apoptotic stimuli (156), but often become resistant to pro-apoptotic anti cancer therapies over time (157). Rho proteins have been implicated in both pro- and anti-apoptotic signaling, and in the apoptotic process itself (reviewed in 158). More significantly, they are involved in the decision to commit to apoptosis; ectopic expression of active Rac1 can provide a survival signal to protect tumor cell lines or transformed fibroblasts from apoptosis (159-161). Other mechanisms that link Rho proteins to cell survival have been described in non-transformed haematopoetic cells: Rac2 is required for the activation of the pro-survival kinase AKT in mast cells (162), and Rho function prevents p53-dependent apoptosis during T-cell development (163).

In other contexts, Rho proteins might promote pro-apoptotic signaling. RhoB is required for the induction of apoptosis by DNA-damaging agents (164) and farnesyltransferase inhibitors, but not other cytotoxic treatments (165), and Rac is required for FAS-induced apoptosis (166, 167).

4.2.2 Rho proteins in invasion and metastasis

In the tumor progression phase, tumor cells have altered morphological characteristics and, in the case of metastasis, acquire the ability to traverse tissue boundaries. Given the role of Rho proteins in the regulation of cell motility in normal cells, and their aberrant regulation in tumor cells, it is likely that they are involved in the invasive phenotype of tumor cells.

Rho proteins are involved in the loss of epithelial polarity that is evident even in benign tumors, and are also important in the epithelial to mesenchymal transition (EMT) that is observed in more aggressive epithelial tumors (see previous paragraphs on Rho protein signaling in cell-cell contracts). In addition to the disruption of cell polarity and cell-cell junctions, increased motility and the ability to remodel the ECM is required for tumor cells to become locally invasive. RhoA and Rac1 can regulate the function of ezrin, moesin and radixin: these related proteins promote cell motility by linking the actin cytoskeleton to the plasma membrane through the membrane-spanning ECM receptor CD44. RhoA can promote phosphorylation of ezrin by ROCK, leading to its increased association with the cytoskeleton (168), whereas Rac1 promotes the phosphorylation and inhibition of the ezrin antagonist neurofibromatosis 2 (NF2) (169, 170). Several lines of evidence indicate that this is an important event in tumor progression: ezrin and CD44 are frequently overexpressed in metastatic tumor cells, and NF2 is a tumor-suppressor gene (170-174), the deletion of which gives rise to highly metastatic tumors (175). In addition, RhoA and Rac1 can modulate the degradation and remodeling of the ECM either by regulating the levels of matrix metalloproteinases (MMPs) that degrade the ECM or by regulating the levels of their antagonists, tissue inhibitors of metalloproteinases (TIMPs) (151, 176-178).

The ability to enter either the blood or the lymphatic vasculature is required for tumor cells to metastasize to distant sites. RhoA and ROCK are required in both the endothelial and the migrating cells for them to cross the vascular endothelium (179, 180). Overexpression of RhoC leads to increased expression of angiogenic factors in breast epithelial cells (142), which could lead to increased vascularization of the tumor and an increased likelihood of tumor cells entering the blood stream. RhoC overexpression also promotes the ability of melanoma cells to exit the blood and colonize the lungs (181), and interfering mutants — which sequester RhoGEFs away from endogenous Rho — of RhoA and Cdc42 prevent T-cell hybridomas from exiting the blood and colonizing the liver (182).

4.3 Rho protein signaling in mouse models

Since Rho proteins play a major role in regulating cell growth, apoptosis and cell motility in normal cells, it is not surprising that they are involved in the development and progression of tumors. However, it has proven difficult to extrapolate the *in vitro* studies in normal cells to the tumor environment. The literature on Rho protein function in tumor progression is confusing, mainly because different studies have indicated contradictory roles for the Rho proteins. However, upon closer inspection the findings may be explained if Rho proteins have different functions at different stages of tumor development. Recent studies, in which recombinant mice were used, directly implicate Rho proteins in all stages of tumorigenesis, and reveal both tumor-promoting and -suppressing functions.

4.3.1 Rho protein signaling in Ras-induced skin carcinogenesis

To gain insight in the role of small GTPases in tumor formation and progression *in vivo*, two groups examined the consequences of functional deletion of members of Rho GTPase family in the mouse: Tiam1, a GEF for Rac (183), and RhoB (164). The knock out of either gene did not negatively affect mouse development, fertility or wound healing. However, these models did show that deregulated Rho protein signaling could influence various processes involved in tumorigenesis. Both studies applied a two-stage chemical skin carcinogenesis protocol, which induces oncogenic activation of the *c-Ha-Ras* gene in the basal layer of the epidermis followed by the induction of the outgrowth and progression of the initiated keratinocytes.

In the RhoB knockout mice the treatment resulted in the development of increased numbers of skin tumors compared with wild-type mice (164). Moreover, DNA-damaging agents were found to induce less apoptosis in Ras- and E1A-transformed RhoB-deficient primary mouse embryonic fibroblasts. This suggests that RhoB normally suppresses tumorigenesis by promoting apoptosis following cellular stress.

Application of the same protocol to Tiam1-deficient mice had a different result: these mice had much fewer and smaller tumors compared to the wild-type mice (183). Malliri *et al.* showed that Tiam1-deficiency was associated with more apoptotic cells in the tumors during initiation and with impeded proliferation during tumor promotion. These results are consistent with *in vitro* studies showing that Tiam/Rac-signaling is required for Ras-induced tumorigenesis through the stimulation of cell growth and enhancing cell survival subsequent to cellular stress.

Another interesting observation in the Tiam1-knockout mice was that although the mice develop less tumors, a larger portion of these tumors progresses to a more malignant stage than in wild-type mice (183). This implicates that Tiam1-deficiency is favorable in malignant conversion. This hypothesis was confirmed by the observation that the tumors in wild-type mice gradually lost Tiam1-expression progressing from benign papillomas to highly invasive spindle cell carcinomas (183).

4.3.2 Rho protein signaling in lymphoid tumors

RhoH has been found to be frequently translocated or mutated in human lymphoid tumors (148, 149). However, the potential function of RhoH in the formation of lymphoid tumors remains to be established. In mouse models, a role for Rho proteins in the development of lymphomas has been elucidated. In transgenic mice, in which the expression of C3 toxin is driven by the thymocyte-specific Lck promoter, aggressive malignant thymic lymphomas were found (184). Because C3 toxin inactivates RhoA, -B and -C, it is not clear which of the inhibited Rho proteins play a role in promoting tumorigenesis.

In search of genes involved in the process of tumor cell invasion and metastasis, the Tiam-1 gene was identified by retroviral insertional mutagenesis in combination with *in vitro* selection of invasive T-lymphoma variants. Furthermore, it was shown that cell clones, which were invasive *in vitro*, also produced metastases upon injection in nude mice (185). Later, Tiam1 was found to be activated by retroviral insertions in T lymphomas induced by Moloney murine leukaemia virus infection of transgenic Eμ-Pim1 mice, thereby inducing an accelerated onset of T-cell lymphomas. This indicates that Tiam1 can cooperate with Pim1 in *in vivo* lymphomagenesis (186).

4.3.3 Rho proteins in Wnt/APC signaling

Wnt proteins constitute a large family of secreted signaling molecules that play central roles in animal development (187). Well-studied examples of Wnt regulation of embryogenesis involve the canonical β-catenin signaling pathway, which in *Xenopus* induces dorsal axis formation and plays a key role in human carcinogenesis (188). In the Wnt/β-catenin pathway, signals are initiated that result in the transcription of genes that regulate cell growth and differentiation. β-catenin participates in this pathway as a heterodimer with T-cell factor (TCF) transcription factors. In the absence of a Wnt signal, cytosolic β-catenin is degraded through a pathway that is dependent upon adenomatous polyposis coli (APC). However, upon stimulation of the Wnt pathway, cytosolic β-catenin is stabilized. The Wnt pathway is tightly regulated in normal cells, but its regulation is often disturbed during tumorigenesis (reviewed in 189).

Recently, Rho proteins have been implicated in Wnt-signaling pathways. In embryonic development of *Drosophila* and *Xenopus*, Rac and Rho have been implicated in Wnt/Fz signaling in regulating cell polarity and movements: a Wnt-Fz-Dvl-Daam1-Rho-ROCK pathway and a Wnt-Fz-Dvl-Rac-JNK pathway (190-193). A common Wnt-Fz-Dvl signaling cassette branches into these two pathways downstream of Dvl, most likely via distinct Dvl-Rac and Dvl-Rho complexes. The Rho-ROCK pathway regulates the phosphorylation of non-muscle myosin regulatory light chain and thus the assembly of actin filaments, whereas the Rac-JNK pathway regulates cytoskeletal or nuclear events (190).

Furthermore, preliminary results from our laboratory indicate that Tiam1 is a Wnt-responsive gene. Moreover, its deficiency impairs the development of intestinal tumors in APC mutant mice (A. Malliri *et al.*, unpublished data), similarly as shown before for Ras-induced skin tumors. These data suggest that Tiam1 and Rac play a role in the oncogenic Wnt signaling pathway.

Also another recently identified Wnt-transcriptional target is a member of the Rho GTPases family: Wrch-1, a novel Cdc42-like GTPase. Like Wnt-1, Wrch-1 is not normally expressed in the mouse mammary gland, but it could be detected in Wnt-1 induced mouse mammary tumors. This finding is consistent with the involvement of Wrch-1 in Wnt-1-induced transformation and tumor formation, and is further supported by the observation that Wrch-1 can mimic the effect of Wnt-1 in morphological transformation of breast cancer cells (194).

5. CONCLUDING REMARKS

The single most notable feature of Rho GTPases is their participation in so many aspects of cell biology, ranging from fundamental to highly specialized processes. The analysis of individual signal transduction pathways controlled by Rho GTPases still poses many problems, and with so many GEFs, GAPs and effector proteins it has proved difficult to link specific receptors to specific biological responses.

Recently, many efforts have been directed towards the understanding of how spatially localized activation of GTPases is achieved or, when activated, how GTPases are able to 'choose' the correct target(s) or effectors from the large number of possibilities present in the same cell. The proposed models involve positive feedback loops, cooperative signaling pathways, scaffold proteins and intracellular compartmentalization.

Furthermore, it is now clear that Rho GTPases may act in almost every stage of tumorigenesis. The lack of a general consensus as to the relative importance of the various mechanisms that link Rho proteins to tumorigenesis means that the challenge is now to determine the crucial pathways that link Rho proteins to cancer. The use of *in vivo* tumorigenesis models and mouse genetics, together with studies on human tumors, should help to identify the important connections *in vivo*. Such studies can prove the principle that Rho proteins or their effectors are valid anti-tumor targets; the next step will be to develop viable pharmacological or gene-therapy agents to interfere with Rho-protein function. Hopefully, these will be novel useful leads in the development of therapies that can be taken into the clinic.

REFERENCES

1. Schultz J, Milpetz F, Bork P, Ponting CP. SMART, a simple modular architecture research tool: identification of signaling domains. Proc. Natl. Acad. Sci. USA 1998, 95: 5857-5864.
2. Hart MJ, Eva A, Evans T, Aaronson SA, Cerione RA. Catalysis of guanine nucleotide exchange on the CDC42Hs protein by the dbl oncogene product. Nature 1991, 354: 311-314.
3. Kjoller L, Hall A. Signaling to Rho GTPases. Exp. Cell Res. 1999, 253: 166-179.
4. Schmidt A, Hall A. Guanine nucleotide exchange factors for Rho GTPases: turning on the switch. Genes Dev. 2002, 16: 1587-1609.
5. Mertens AE, Roovers RC, Collard JG. Regulation of Tiam1-Rac signalling. FEBS Lett. 2003, 546: 11-16.
6. Moon SY, Zheng Y. Rho GTPase-activating proteins in cell regulation. Trends Cell Biol. 2003, 13: 13-22.
7. Michaelson D, Silletti J, Murphy G, D'Eustachio P, Rush M, Philips MR. Differential localization of Rho GTPases in live cells: regulation by hypervariable regions and RhoGDI binding. J. Cell Biol. 2001, 152: 111-126.
8. Hoffman GR, Nassar N, Cerione RA. Structure of the Rho family GTP-binding protein Cdc42 in complex with the multifunctional regulator RhoGDI. Cell 2000, 100: 345-356.
9. Olofsson B. Rho guanine dissociation inhibitors: pivotal molecules in cellular signalling. Cell Signal. 1999, 11: 545-554.
10. Seabra MC. Membrane association and targeting of prenylated Ras-like GTPases. Cell Signal. 1998, 10: 167-172.
11. Self AJ, Hall A. Measurement of intrinsic nucleotide exchange and GTP hydrolysis rates. Methods Enzymol. 1995, 256: 67-76.

12. Feig LA. Tools of the trade: use of dominant-inhibitory mutants of Ras-family GTPases. Nat. Cell Biol. 1999, 1: E25-E27.
13. Burridge K, Wennerberg K. Rho and Rac take center stage. Cell 2004, 116: 167-179.
14. Barbieri JT, Riese MJ, Aktories K. Bacterial toxins that modify the actin cytoskeleton. Annu. Rev. Cell Dev. Biol. 2002, 18: 315-344.
15. Lauffenburger DA, Horwitz AF. Cell migration: a physically integrated molecular process. Cell 1996, 84: 359-369.
16. Itoh RE, Kurokawa K, Ohba Y, Yoshizaki H, Mochizuki N, Matsuda M. Activation of rac and cdc42 video imaged by fluorescent resonance energy transfer-based single-molecule probes in the membrane of living cells. Mol. Cell Biol. 2002, 22: 6582-6591.
17. Etienne-Manneville S, Hall A. Rho GTPases in cell biology. Nature 2002, 420: 629-635.
18. Srinivasan S, Wang F, Glavas S, Ott A, Hofmann F, Aktories K, Kalman D, Bourne HR. Rac and Cdc42 play distinct roles in regulating PI(3,4,5)P3 and polarity during neutrophil chemotaxis. J. Cell Biol. 2003, 160: 375-385.
19. Rodriguez OC, Schaefer AW, Mandato CA, Forscher P, Bement WM, Waterman-Storer CM. Conserved microtubule-actin interactions in cell movement and morphogenesis. Nat. Cell Biol. 2003, 5: 599-609.
20. Kim SK. Cell polarity: new PARtners for Cdc42 and Rac. Nat. Cell Biol. 2000, 2: E143-E145.
21. Etienne-Manneville S, Hall A. Cell polarity: Par6, aPKC and cytoskeletal crosstalk. Curr. Opin. Cell Biol. 2003, 15: 67-72.
22. Allen WE, Zicha D, Ridley AJ, Jones GE. A role for Cdc42 in macrophage chemotaxis. J. Cell Biol. 1998, 141: 1147-1157.
23. Nobes CD, Hall A. Rho GTPases control polarity, protrusion, and adhesion during cell movement. J. Cell Biol. 1999, 144: 1235-1244.
24. Daub H, Gevaert K, Vandekerckhove J, Sobel A, Hall A. Rac/Cdc42 and p65PAK regulate the microtubule-destabilizing protein stathmin through phosphorylation at serine 16. J. Biol. Chem. 2001, 276: 1677-1680.
25. Waterman-Storer CM, Worthylake RA, Liu BP, Burridge K, Salmon ED. Microtubule growth activates Rac1 to promote lamellipodial protrusion in fibroblasts. Nat. Cell Biol. 1999, 1: 45-50.
26. Liu BP, Chrzanowska-Wodnicka M, Burridge K. Microtubule depolymerization induces stress fibers, focal adhesions, and DNA synthesis via the GTP-binding protein Rho. Cell Adhes. Commun. 1998, 5: 249-255.
27. Ishizaki T, Morishima Y, Okamoto M, Furuyashiki T, Kato T, Narumiya S. Coordination of microtubules and the actin cytoskeleton by the Rho effector mDia1. Nat. Cell Biol. 2001, 3: 8-14.
28. Glaven JA, Whitehead I, Bagrodia S, Kay R, Cerione RA. The Dbl-related protein, Lfc, localizes to microtubules and mediates the activation of Rac signaling pathways in cells. J. Biol. Chem. 1999, 274: 2279-2285.
29. van Horck FP, Ahmadian MR, Haeusler LC, Moolenaar WH, Kranenburg O. Characterization of p190RhoGEF, a RhoA-specific guanine nucleotide exchange factor that interacts with microtubules. J. Biol. Chem. 2001, 276: 4948-4956.
30. Pollard TD, Blanchoin L, Mullins RD. Molecular mechanisms controlling actin filament dynamics in nonmuscle cells. Annu. Rev. Biophys. Biomol. Struct. 2000, 29: 545-576.
31. Welch MD, Mullins RD. Cellular control of actin nucleation. Annu. Rev. Cell Dev. Biol. 2002, 18: 247-288.
32. Knight B, Laukaitis C, Akhtar N, Hotchin NA, Edlund M, Horwitz AR. Visualizing muscle cell migration in situ. Curr. Biol. 2000, 10: 576-585.

33. Kraynov VS, Chamberlain C, Bokoch GM, Schwartz MA, Slabaugh S, Hahn KM. Localized Rac activation dynamics visualized in living cells. Science 2000, 290: 333-337.
34. Sander EE, van Delft S, ten Klooster JP, Reid T, Van der Kammen RA, Michiels F, Collard JG. Matrix-dependent Tiam1/Rac signaling in epithelial cells promotes either cell-cell adhesion or cell migration and is regulated by phosphatidylinositol 3-kinase. J. Cell Biol. 1998, 143: 1385-1398.
35. Rickert P, Weiner OD, Wang F, Bourne HR, Servant G. Leukocytes navigate by compass: roles of PI3Kgamma and its lipid products. Trends Cell Biol. 2000, 10: 466-473.
36. Royal I, Lamarche-Vane N, Lamorte L, Kaibuchi K, Park M. Activation of cdc42, rac, PAK, and rho-kinase in response to hepatocyte growth factor differentially regulates epithelial cell colony spreading and dissociation. Mol. Biol. Cell 2000, 11: 1709-1725.
37. Meili R, Ellsworth C, Lee S, Reddy TB, Ma H, Firtel RA. Chemoattractant-mediated transient activation and membrane localization of Akt/PKB is required for efficient chemotaxis to cAMP in Dictyostelium. EMBO J. 1999, 18: 2092-2105.
38. Haugh JM, Codazzi F, Teruel M, Meyer T. Spatial sensing in fibroblasts mediated by 3' phosphoinositides. J. Cell Biol. 2000, 151: 1269-1280.
39. Servant G, Weiner OD, Herzmark P, Balla T, Sedat JW, Bourne HR. Polarization of chemoattractant receptor signaling during neutrophil chemotaxis. Science 2000, 287: 1037-1040.
40. Das B, Shu X, Day GJ, Han J, Krishna UM, Falck JR, Broek D. Control of intramolecular interactions between the pleckstrin homology and Dbl homology domains of Vav and Sos1 regulates Rac binding. J. Biol. Chem. 2000, 275: 15074-15081.
41. Fleming IN, Gray A, Downes CP. Regulation of the Rac1-specific exchange factor Tiam1 involves both phosphoinositide 3-kinase-dependent and -independent components. Biochem J. 2000, 351: 173-182.
42. Sells MA, Knaus UG, Bagrodia S, Ambrose DM, Bokoch GM, Chernoff J. Human p21-activated kinase (Pak1) regulates actin organization in mammalian cells. Curr. Biol. 1997, 7: 202-210.
43. Manser E, Huang HY, Loo TH, Chen XQ, Dong JM, Leung T, Lim L. Expression of constitutively active alpha-PAK reveals effects of the kinase on actin and focal complexes. Mol. Cell Biol. 1997, 17: 1129-1143.
44. Edwards DC, Sanders LC, Bokoch GM, Gill GN. Activation of LIM-kinase by Pak1 couples Rac/Cdc42 GTPase signalling to actin cytoskeletal dynamics. Nat. Cell Biol. 1999, 1: 253-259.
45. Stanyon CA, Bernard O. LIM-kinase1. Int. J. Biochem. Cell Biol. 1999, 31: 389-394.
46. Mermall V, Post PL, Mooseker MS. Unconventional myosins in cell movement, membrane traffic, and signal transduction. Science 1998, 279: 527-533.
47. van Leeuwen FN, van Delft S, Kain HE, Van der Kammen RA, Collard JG. Rac regulates phosphorylation of the myosin-II heavy chain, actinomyosin disassembly and cell spreading. Nat. Cell Biol. 1999, 1: 242-248.
48. Sander EE, ten Klooster JP, van Delft S, Van der Kammen RA, Collard JG. Rac downregulates Rho activity: reciprocal balance between both GTPases determines cellular morphology and migratory behavior. J. Cell Biol. 1999, 147: 1009-1022.
49. Daniels RH, Bokoch GM. p21-activated protein kinase: a crucial component of morphological signaling? Trends Biochem. Sci. 1999, 24: 350-355.
50. Kiosses WB, Daniels RH, Otey C, Bokoch GM, Schwartz MA. A role for p21-activated kinase in endothelial cell migration. J. Cell Biol. 1999, 147: 831-844.

51. Nobes CD, Hall A. Rho, rac, and cdc42 GTPases regulate the assembly of multimolecular focal complexes associated with actin stress fibers, lamellipodia, and filopodia. Cell 1995, 81: 53-62.
52. Allen WE, Jones GE, Pollard JW, Ridley AJ. Rho, Rac and Cdc42 regulate actin organization and cell adhesion in macrophages. J. Cell Sci. 1997, 110 (Pt 6): 707-720.
53. Rottner K, Hall A, Small JV. Interplay between Rac and Rho in the control of substrate contact dynamics. Curr. Biol. 1999, 9: 640-648.
54. Chrzanowska-Wodnicka M, Burridge K. Rho-stimulated contractility drives the formation of stress fibers and focal adhesions. J. Cell Biol. 1996, 133: 1403-1415.
55. Adams JC, Schwartz MA. Stimulation of fascin spikes by thrombospondin-1 is mediated by the GTPases Rac and Cdc42. J. Cell Biol, 2000, 150: 807-822.
56. Ridley A. Rho GTPases. Integrating integrin signaling. J. Cell Biol. 2000, 150: F107-F109.
57. Wenk MB, Midwood KS, Schwarzbauer JE. Tenascin-C suppresses Rho activation. J. Cell Biol. 2000, 150: 913-920.
58. Schwartz MA, Shattil SJ. Signaling networks linking integrins and rho family GTPases. Trends Biochem. Sci. 2000, 25: 388-391.
59. Cox EA, Sastry SK, Huttenlocher A. Integrin-mediated adhesion regulates cell polarity and membrane protrusion through the Rho family of GTPases. Mol. Biol. Cell 2001, 12: 265-277.
60. Zhao JH, Guan JL. Role of focal adhesion kinase in signaling by the extracellular matrix. Prog. Mol. Subcell Biol. 2000, 25: 37-55.
61. Mitchison TJ, Cramer LP. Actin-based cell motility and cell locomotion. Cell 1996, 84: 371-379.
62. Kaibuchi K. Regulation of cytoskeleton and cell adhesion by Rho targets. Prog. Mol. Subcell Biol. 1999, 22: 23-38.
63. Amano M, Fukata Y, Kaibuchi K. Regulation and functions of Rho-associated kinase. Exp. Cell Res. 2000, 261: 44-51.
64. Hansen SH, Zegers MM, Woodrow M, Rodriguez-Viciana P, Chardin P, Mostov KE, McMahon M. Induced expression of Rnd3 is associated with transformation of polarized epithelial cells by the Raf-MEK-extracellular signal-regulated kinase pathway. Mol. Cell Biol. 2000, 20: 9364-9375.
65. Totsukawa G, Yamakita Y, Yamashiro S, Hartshorne DJ, Sasaki Y, Matsumura F. Distinct roles of ROCK (Rho-kinase) and MLCK in spatial regulation of MLC phosphorylation for assembly of stress fibers and focal adhesions in 3T3 fibroblasts. J. Cell Biol. 2000, 150: 797-806.
66. Adams CL, Nelson WJ. Cytomechanics of cadherin-mediated cell-cell adhesion. Curr. Opin. Cell Biol. 1998, 10: 572-577.
67. Gumbiner BM. Regulation of cadherin adhesive activity. J. Cell Biol. 2000, 148: 399-404.
68. Takeda H, Shimoyama Y, Nagafuchi A, Hirohashi S. E-cadherin functions as a cis-dimer at the cell-cell adhesive interface in vivo. Nat. Struct. Biol. 1999, 6: 310-312.
69. Tsukita S, Tsukita S, Nagafuchi A, Yonemura S. Molecular linkage between cadherins and actin filaments in cell-cell adherens junctions. Curr. Opin. Cell Biol. 1992, 4: 834-839.
70. Kemler R. From cadherins to catenins: cytoplasmic protein interactions and regulation of cell adhesion. Trends Genet. 1993, 9: 317-321.
71. Behrens J, Mareel MM, Van Roy FM, Birchmeier W. Dissecting tumor cell invasion: epithelial cells acquire invasive properties after the loss of uvomorulin-mediated cell-cell adhesion. J. Cell Biol. 1989, 108: 2435-2447.

72. Vleminckx K, Vakaet L, Jr., Mareel M, Fiers W, van Roy F. Genetic manipulation of E-cadherin expression by epithelial tumor cells reveals an invasion suppressor role. Cell 1991, 66: 107-119.
73. Hajra KM, Fearon ER. Cadherin and catenin alterations in human cancer. Genes Chromosomes Cancer 2002, 34: 255-268.
74. Cavallaro U, Christofori G. Cell adhesion and signalling by cadherins and Ig-CAMs in cancer. Nat. Rev. Cancer 2004, 4: 118-132.
75. Beavon IR. The E-cadherin-catenin complex in tumour metastasis: structure, function and regulation. Eur. J. Cancer 2000, 36: 1607-1620.
76. Mareel M, Leroy A. Clinical, cellular, and molecular aspects of cancer invasion. Physiol. Rev. 2003, 83: 337-376.
77. Fukata M, Kaibuchi K. Rho-family GTPases in cadherin-mediated cell-cell adhesion. Nat. Rev. Mol. Cell Biol. 2001, 2: 887-897.
78. Lozano E, Betson M, Braga VM. Tumor progression: Small GTPases and loss of cell-cell adhesion. Bioessays 2003, 25: 452-463.
79. van Aelst L, Symons M. Role of Rho family GTPases in epithelial morphogenesis. Genes Dev. 2002, 16: 1032-1054.
80. Kuroda S, Fukata M, Fujii K, Nakamura T, Izawa I, Kaibuchi K. Regulation of cell-cell adhesion of MDCK cells by Cdc42 and Rac1 small GTPases. Biochem. Biophys. Res. Commun. 1997, 240: 430-435.
81. Takaishi K, Sasaki T, Kotani H, Nishioka H, Takai Y. Regulation of cell-cell adhesion by rac and rho small G proteins in MDCK cells. J. Cell Biol. 1997, 139: 1047-1059.
82. Braga VM, Machesky LM, Hall A, Hotchin NA. The small GTPases Rho and Rac are required for the establishment of cadherin-dependent cell-cell contacts. J. Cell Biol. 1997, 137: 1421-1431.
83. Akhtar N, Hudson KR, Hotchin NA. Co-localization of Rac1 and E-cadherin in human epidermal keratinocytes. Cell Adhes. Commun. 2000, 7: 465-476.
84. Akhtar N, Hotchin NA. RAC1 regulates adherens junctions through endocytosis of E-cadherin. Mol. Biol. Cell 2001, 12: 847-862.
85. Braga VM, Del Maschio A, Machesky L, Dejana E. Regulation of cadherin function by Rho and Rac: modulation by junction maturation and cellular context. Mol. Biol. Cell 1999, 10: 9-22.
86. Braga VM, Betson M, Li X, Lamarche-Vane N. Activation of the small GTPase Rac is sufficient to disrupt cadherin-dependent cell-cell adhesion in normal human keratinocytes. Mol. Biol. Cell 2000, 11: 3703-3721.
87. Hordijk PL, ten Klooster JP, Van der Kammen RA, Michiels F, Oomen LC, Collard JG. Inhibition of invasion of epithelial cells by Tiam1-Rac signaling. Science 1997, 278: 1464-1466.
88. Malliri A, van Es S, Huveneers S, Collard JG. The Rac exchange factor Tiam1 is required for the establishment and maintenance of cadherin-based adhesions. J. Biol. Chem. 2004, 279: 30092-30098.
89. Zondag GC, Evers EE, ten Klooster JP, Janssen L, van Der K, Collard JG. Oncogenic Ras Downregulates Rac Activity, which Leads to Increased Rho Activity and Epithelial-Mesenchymal Transition. J. Cell Biol. 2000, 149: 775-782.
90. Zhong C, Kinch MS, Burridge K. Rho-stimulated contractility contributes to the fibroblastic phenotype of Ras-transformed epithelial cells. Mol. Biol. Cell 1997, 8: 2329-2344.
91. Sahai E, Olson MF, Marshall CJ. Cross-talk between Ras and Rho signalling pathways in transformation favours proliferation and increased motility. EMBO J. 2001, 20: 755-766.

92. Takaishi K, Sasaki T, Kato M, Yamochi W, Kuroda S, Nakamura T, Takeichi M, Takai Y. Involvement of Rho p21 small GTP-binding protein and its regulator in the HGF-induced cell motility. Oncogene, 1994, 9: 273-279.
93. Bhowmick NA, Ghiassi M, Bakin A, Aakre M, Lundquist CA, Engel ME, Arteaga CL, Moses HL. Transforming growth factor-beta1 mediates epithelial to mesenchymal transdifferentiation through a RhoA-dependent mechanism. Mol. Biol. Cell, 2001, 12: 27-36.
94. Fukata M, Kuroda S, Nakagawa M, Kawajiri A, Itoh N, Shoji I, Matsuura Y, Yonehara S, Fujisawa H, Kikuchi A, Kaibuchi K. Cdc42 and Rac1 regulate the interaction of IQGAP1 with beta-catenin. J. Biol. Chem. 1999, 274: 26044-26050.
95. Kuroda S, Fukata M, Nakagawa M, Kaibuchi K. Cdc42, Rac1, and their effector IQGAP1 as molecular switches for cadherin-mediated cell-cell adhesion. Biochem. Biophys. Res. Commun. 1999, 262: 1-6.
96. Noritake J, Fukata M, Sato K, Nakagawa M, Watanabe T, Izumi N, Wang S, Fukata Y, Kaibuchi K. Positive role of IQGAP1, an effector of Rac1, in actin-meshwork formation at sites of cell-cell contact. Mol. Biol. Cell 2004, 15: 1065-1076.
97. Noren NK, Niessen CM, Gumbiner BM, Burridge K. Cadherin engagement regulates Rho family GTPases. J. Biol. Chem. 2001, 276: 33305-33308.
98. Nakagawa M, Fukata M, Yamaga M, Itoh N, Kaibuchi K. Recruitment and activation of Rac1 by the formation of E-cadherin-mediated cell-cell adhesion sites. J. Cell Sci. 2001, 114: 1829-1838.
99. Kovacs EM, Goodwin M, Ali RG, Paterson AD, Yap AS. Cadherin-directed actin assembly: E-cadherin physically associates with the Arp2/3 complex to direct actin assembly in nascent adhesive contacts. Curr. Biol. 2002, 12: 379-382.
100. Kovacs EM, Ali RG, McCormack AJ, Yap AS. E-cadherin homophilic ligation directly signals through Rac and phosphatidylinositol 3-kinase to regulate adhesive contacts. J. Biol. Chem. 2002, 277: 6708-6718.
101. Charrasse S, Meriane M, Comunale F, Blangy A, Gauthier-Rouviere C. N-cadherin-dependent cell-cell contact regulates Rho GTPases and beta-catenin localization in mouse C2C12 myoblasts. J. Cell Biol. 2002, 158: 953-965.
102. Pece S, Chiariello M, Murga C, Gutkind JS. Activation of the protein kinase Akt/PKB by the formation of E-cadherin-mediated cell-cell junctions. Evidence for the association of phosphatidylinositol 3-kinase with the E-cadherin adhesion complex. J. Biol. Chem. 1999, 274: 19347-19351.
103. Woodfield RJ, Hodgkin MN, Akhtar N, Morse MA, Fuller KJ, Saqib K, Thompson NT, Wakelam MJ. The p85 subunit of phosphoinositide 3-kinase is associated with beta-catenin in the cadherin-based adhesion complex. Biochem. J. 2001, 360: 335-344.
104. Anastasiadis PZ, Moon SY, Thoreson MA, Mariner DJ, Crawford HC, Zheng Y, Reynolds AB. Inhibition of RhoA by p120 catenin. Nat. Cell Biol. 2000, 2: 637-644.
105. Noren NK, Liu BP, Burridge K, Kreft B. p120 catenin regulates the actin cytoskeleton via Rho family GTPases. J. Cell Biol. 2000, 150: 567-580.
106. Grosheva I, Shtutman M, Elbaum M, Bershadsky AD. p120 catenin affects cell motility via modulation of activity of Rho-family GTPases: a link between cell-cell contact formation and regulation of cell locomotion. J. Cell Sci. 2001, 114: 695-707.
107. Goodwin M, Kovacs EM, Thoreson MA, Reynolds AB, Yap AS. Minimal mutation of the cytoplasmic tail inhibits the ability of E-cadherin to activate Rac but not phosphatidylinositol 3-kinase: direct evidence of a role for cadherin-activated Rac signaling in adhesion and contact formation. J. Biol. Chem. 2003, 278: 20533-20539.

108. Shibata T, Kokubu A, Sekine S, Kanai Y, Hirohashi S. Cytoplasmic p120ctn regulates the invasive phenotypes of E-cadherin-deficient breast cancer. Am. J. Pathol. 2004, 164: 2269-2278.
109. Gavard J, Lambert M, Grosheva I, Marthiens V, Irinopoulou T, Riou JF, Bershadsky A, Mege RM. Lamellipodium extension and cadherin adhesion: two cell responses to cadherin activation relying on distinct signalling pathways. J. Cell Sci. 2004, 117: 257-270.
110. Johnson E, Theisen CS, Johnson KR, Wheelock MJ. R-cadherin Influences Cell Motility via Rho Family GTPases. J. Biol. Chem. 2004, 279: 31041-31049.
111. Martin TA, Jiang WG. Tight junctions and their role in cancer metastasis. Histol Histopathol. 2001, 16: 1183-1195.
112. Mullin JM. Epithelial barriers, compartmentation, and cancer. Sci. STKE. 2004; 2004: e2.
113. Schneeberger EE, Lynch RD. The tight junction: a multifunctional complex. Am. J. Physiol. Cell Physiol. 2004, 286: C1213-C1228.
114. Jou TS, Schneeberger EE, Nelson WJ. Structural and functional regulation of tight junctions by RhoA and Rac1 small GTPases. J. Cell Biol. 1998, 142: 101-115.
115. Bruewer M, Hopkins AM, Hobert ME, Nusrat A, Madara JL. RhoA, Rac1, and Cdc42 exert distinct effects on epithelial barrier via selective structural and biochemical modulation of junctional proteins and F-actin. Am. J. Physiol. Cell Physiol. 2004, 287: C327-C335.
116. Braga VM. Cell-cell adhesion and signalling. Curr. Opin. Cell Biol. 2002, 14: 546-556.
117. Matter K, Balda MS. Signalling to and from tight junctions. Nat. Rev. Mol. Cell Biol. 2003, 4: 225-236.
118. Benais-Pont G, Punn A, Flores-Maldonado C, Eckert J, Raposo G, Fleming TP, Cereijido M, Balda MS, Matter K. Identification of a tight junction-associated guanine nucleotide exchange factor that activates Rho and regulates paracellular permeability. J. Cell Biol. 2003, 160: 729-740.
119. Lin D, Edwards AS, Fawcett JP, Mbamalu G, Scott JD, Pawson T. A mammalian PAR-3-PAR-6 complex implicated in Cdc42/Rac1 and aPKC signalling and cell polarity. Nat. Cell Biol. 2000, 2: 540-547.
120. Joberty G, Petersen C, Gao L, Macara IG. The cell-polarity protein Par6 links Par3 and atypical protein kinase C to Cdc42. Nat. Cell Biol. 2000, 2: 531-539.
121. Suzuki A, Ishiyama C, Hashiba K, Shimizu M, Ebnet K, Ohno S. aPKC kinase activity is required for the asymmetric differentiation of the premature junctional complex during epithelial cell polarization. J. Cell Sci. 2002, 115: 3565-3573.
122. Suzuki A, Yamanaka T, Hirose T, Manabe N, Mizuno K, Shimizu M, Akimoto K, Izumi Y, Ohnishi T, Ohno S. Atypical protein kinase C is involved in the evolutionarily conserved par protein complex and plays a critical role in establishing epithelia-specific junctional structures. J. Cell Biol. 2001, 152: 1183-1196.
123. Fukuhara A, Shimizu K, Kawakatsu T, Fukuhara T, Takai Y. Involvement of nectin-activated Cdc42 small G protein in organization of adherens and tight junctions in Madin-Darby canine kidney cells. J. Biol. Chem. 2003, 278: 51885-51893.
124. Whitehead IP, Campbell S, Rossman KL, Der CJ. Dbl family proteins. Biochim. Biophys. Acta. 1997, 1332: F1-23.
125. Stam JC, Collard JG. The DH protein family, exchange factors for Rho-like GTPases. Prog. Mol. Subcell. Biol. 1999, 22: 51-83.
126. Perona R, Esteve P, Jimenez B, Ballestero RP, Cajal SR. Tumorigenic Activity of rho Genes from Aplysia-Californica. Oncogene 1993, 8: 1285-1292.

127. van Leeuwen FN, Van der Kammen RA, Habets GG, Collard JG. Oncogenic activity of Tiam1 and Rac1 in NIH3T3 cells. Oncogene 1995, 11: 2215-2221.
128. Khosravi-Far R, Solski PA, Clark GJ, Kinch MS, Der CJ. Activation of Rac1, RhoA, and mitogen-activated protein kinases is required for Ras transformation. Mol. Cell Biol. 1995, 15: 6443-6453.
129. Murphy GA, Solski PA, Jillian SA, Perez de la Ossa P, D'Eustachio P, Der CJ, Rush MG. Cellular functions of TC10, a Rho family GTPase: regulation of morphology, signal transduction and cell growth. Oncogene 1999, 18: 3831-3845.
130. Qiu RG, Chen J, Kirn D, McCormick F, Symons M. An essential role for Rac in Ras transformation. Nature 1995, 374: 457-459.
131. Qiu RG, Abo A, McCormick F, Symons M. Cdc42 regulates anchorage-independent growth and is necessary for Ras transformation. Mol. Cell Biol. 1997, 17: 3449-3458.
132. Roux P, Gauthier-Rouviere C, Doucet-Brutin S, Fort P. The small GTPases Cdc42Hs, Rac1 and RhoG delineate Raf-independent pathways that cooperate to transform NIH3T3 cells. Curr. Biol. 1997, 7: 629-637.
133. Khosravi-Far R, Solski PA, Clark GJ, Kinch MS, Der CJ. Activation of rac and Rho, and mitogen activated protein kinases, are required for Ras transformation. Mol. Cellular Biol. 1995, 15: 6443-6453.
134. Fritz G, Just I, Kaina B. Rho GTPases are over-expressed in human tumors. Int. J. Cancer 1999, 81: 682-687.
135. Pan Y, Bi F, Liu N, Xue Y, Yao X, Zheng Y, Fan D. Expression of seven main Rho family members in gastric carcinoma. Biochem. Biophys. Res. Commun. 2004, 315: 686-691.
136. Schnelzer A, Prechtel D, Knaus U, Dehne K, Gerhard M, Graeff H, Harbeck N, Schmitt M, Lengyel E. Rac1 in human breast cancer: overexpression, mutation analysis, and characterization of a new isoform, Rac1b. Oncogene 2000.Jun.15.;19.(26.):3013.-20. 2000; 19: 3013-3020.
137. Jordan P, Brazao R, Boavida MG, Gespach C, Chastre E. Cloning of a novel human Rac1b splice variant with increased expression in colorectal tumors. Oncogene 1999, 18: 6835-6839.
138. Abraham MT, Kuriakose MA, Sacks PG, Yee H, Chiriboga L, Bearer EL, Delacure MD. Motility-related proteins as markers for head and neck squamous cell cancer. Laryngoscope 2001, 111: 1285-1289.
139. Mira JP, Benard V, Groffen J, Sanders LC, Knaus UG. Endogenous, hyperactive Rac3 controls proliferation of breast cancer cells by a p21-activated kinase-dependent pathway. Proc. Natl. Acad. Sci. USA 2000, 97: 185-189.
140. Kamai T, Arai K, Tsujii T, Honda M, Yoshida K. Overexpression of RhoA mRNA is associated with advanced stage in testicular germ cell tumour. BJU Int. 2001, 87: 227-231.
141. van Golen KL, Bao LW, Pan Q, Miller FR, Wu ZF, Merajver SD. Mitogen activated protein kinase pathway is involved in RhoC GTPase induced motility, invasion and angiogenesis in inflammatory breast cancer. Clin. Exp. Metastasis 2002, 19: 301-311.
142. van Golen KL, Wu ZF, Qiao XT, Bao LW, Merajver SD. RhoC GTPase, a novel transforming oncogene for human mammary epithelial cells that partially recapitulates the inflammatory breast cancer phenotype. Cancer Res. 2000, 60: 5832-5838.
143. Suwa H, Ohshio G, Imamura T, Watanabe G, Arii S, Imamura M, Narumiya S, Hiai H, Fukumoto M. Overexpression of the rhoC gene correlates with progression of ductal adenocarcinoma of the pancreas. Br. J. Cancer 1998, 77: 147-152.

144. Shikada Y, Yoshino I, Okamoto T, Fukuyama S, Kameyama T, Maehara Y. Higher expression of RhoC is related to invasiveness in non-small cell lung carcinoma. Clin. Cancer Res. 2003, 9: 5282-5286.
145. Horiuchi A, Imai T, Wang C, Ohira S, Feng Y, Nikaido T, Konishi I. Up-regulation of small GTPases, RhoA and RhoC, is associated with tumor progression in ovarian carcinoma. Lab. Invest. 2003, 83: 861-870.
146. Kondo T, Sentani K, Oue N, Yoshida K, Nakayama H, Yasui W. Expression of RHOC is associated with metastasis of gastric carcinomas. Pathobiology 2004, 71: 19-25.
147. Wang W, Yang LY, Yang ZL, Huang GW, Lu WQ. Expression and significance of RhoC gene in hepatocellular carcinoma. World J. Gastroenterol 2003, 9: 1950-1953.
148. Preudhomme C, Roumier C, Hildebrand MP, Dallery-Prudhomme E, Lantoine D, Lai JL, Daudignon A, Adenis C, Bauters F, Fenaux P, Kerckaert JP, Galiegue-Zouitina S. Nonrandom 4p13 rearrangements of the RhoH/TTF gene, encoding a GTP-binding protein, in non-Hodgkin's lymphoma and multiple myeloma. Oncogene 2000, 19: 2023-2032.
149. Pasqualucci L, Neumeister P, Goossens T, Nanjangud G, Chaganti RS, Kuppers R, Dalla-Favera R. Hypermutation of multiple proto-oncogenes in B-cell diffuse large-cell lymphomas. Nature 2001, 412: 341-346.
150. Engers R, Zwaka TP, Gohr L, Weber A, Gerharz CD, Gabbert HE. Tiam1 mutations in human renal-cell carcinomas. Int. J. Cancer 2000, 88: 369-376.
151. Engers R, Springer E, Michiels F, Collard JG, Gabbert HE. Rac Affects Invasion of Human Renal Cell Carcinomas by Up-regulating Tissue Inhibitor of Metalloproteinases (TIMP)-1 and TIMP-2 Expression. J. Biol. Chem. 2001, 276: 41889-41897.
152. Kourlas PJ, Strout MP, Becknell B, Veronese ML, Croce CM, Theil KS, Krahe R, Ruutu T, Knuutila S, Bloomfield CD, Caligiuri MA. Identification of a gene at 11q23 encoding a guanine nucleotide exchange factor: evidence for its fusion with MLL in acute myeloid leukemia. Proc. Natl. Acad. Sci. USA 2000, 97: 2145-2150.
153. Lin R, Bagrodia S, Cerione R, Manor D. Novel Cdc42Hs mutant induces cellular transformation. Curr. Biol. 1997, 7: 794-797.
154. Gampel A, Parker PJ, Mellor H. Regulation of epidermal growth factor receptor traffic by the small GTPase rhoB. Curr. Biol. 1999, 9: 955-958.
155. Vignal E, De Toledo M, Comunale F, Landopoulou A, Gauthier-Rouviere C, Blangy A, Fort P. Characterization of TCL, a new GTPase of the Rho family related to TC10 and Cdc42. J. Biol. Chem. 2000, 17, 275: 36457-64.
156. Evan G, Littlewood T. A matter of life and cell death. Science 1998, 281: 1317-1322.
157. Green DR, Evan GI. A matter of life and death. Cancer Cell 2002, 1: 19-30.
158. Coleman ML, Olson MF. Rho GTPase signalling pathways in the morphological changes associated with apoptosis. Cell Death.Differ. 2002, 9: 493-504.
159. Coniglio SJ, Jou TS, Symons M. Rac1 protects epithelial cells against anoikis. J. Biol. Chem. 2001, 276: 28113-28120.
160. Joneson T, Bar-Sagi D. Suppression of Ras-induced apoptosis by the Rac GTPase. Mol. Cell Biol. 1999, 19: 5892-5901.
161. Pervaiz S, Cao J, Chao OS, Chin YY, Clement MV. Activation of the RacGTPase inhibits apoptosis in human tumor cells. Oncogene 2001, 20: 6263-6268.
162. Yang FC, Kapur R, King AJ, Tao W, Kim C, Borneo J, Breese R, Marshall M, Dinauer MC, Williams DA. Rac2 stimulates Akt activation affecting BAD/Bcl-XL expression while mediating survival and actin function in primary mast cells. Immunity 2000, 12: 557-568.

163. Costello PS, Cleverley SC, Galandrini R, Henning SW, Cantrell DA. The GTPase rho controls a p53-dependent survival checkpoint during thymopoiesis. J. Exp. Med. 2000, 192: 77-85.
164. Liu AX, Rane N, Liu JP, Prendergast GC. RhoB is dispensable for mouse development, but it modifies susceptibility to tumor formation as well as cell adhesion and growth factor signaling in transformed cells. Mol. Cell Biol. 2001, 21: 6906-6912.
165. Prendergast GC. Actin' up: RhoB in cancer and apoptosis. Nat. Rev. Cancer 2001, 1: 162-168.
166. Subauste MC, Von Herrath M, Benard V, Chamberlain CE, Chuang TH, Chu K, Bokoch GM, Hahn KM. Rho family proteins modulate rapid apoptosis induced by cytotoxic T lymphocytes and Fas. J. Biol. Chem. 2000, 275: 9725-9733.
167. Brenner B, Koppenhoefer U, Weinstock C, Linderkamp O, Lang F, Gulbins E. Fas- or ceramide-induced apoptosis is mediated by a Rac1-regulated activation of Jun N-terminal kinase/p38 kinases and GADD153. J. Biol. Chem. 1997, 272: 22173-22181.
168. Matsui T, Maeda M, Doi Y, Yonemura S, Amano M, Kaibuchi K, Tsukita S, Tsukita S. Rho-kinase phosphorylates COOH-terminal threonines of ezrin/radixin/moesin (ERM) proteins and regulates their head-to-tail association. J. Cell Biol. 1998, 140: 647-657.
169. Shaw RJ, Paez JG, Curto M, Yaktine A, Pruitt WM, Saotome I, O'Bryan JP, Gupta V, Ratner N, Der CJ, Jacks T, McClatchey AI. The Nf2 tumor suppressor, merlin, functions in Rac-dependent signaling. Dev. Cell 2001, 1: 63-72.
170. Xiao GH, Beeser A, Chernoff J, Testa JR. p21-activated kinase links Rac/Cdc42 signaling to merlin. J. Biol. Chem. 2002, 277: 883-886.
171. Akisawa N, Nishimori I, Iwamura T, Onishi S, Hollingsworth MA. High levels of ezrin expressed by human pancreatic adenocarcinoma cell lines with high metastatic potential. Biochem. Biophys. Res. Commun. 1999, 258: 395-400.
172. Clarke G, Ryan E, O'Keane JC, Crowe J, Mathuna PM. Mortality association of enhanced CD44v6 expression is not mediated through occult lymphatic spread in stage II colorectal cancer. J. Gastroenterol Hepatol. 2000, 15: 1028-1031.
173. Harada N, Mizoi T, Kinouchi M, Hoshi K, Ishii S, Shiiba K, Sasaki I, Matsuno S. Introduction of antisense CD44S CDNA down-regulates expression of overall CD44 isoforms and inhibits tumor growth and metastasis in highly metastatic colon carcinoma cells. Int. J. Cancer 2001, 91: 67-75.
174. Khanna C, Khan J, Nguyen P, Prehn J, Caylor J, Yeung C, Trepel J, Meltzer P, Helman L. Metastasis-associated differences in gene expression in a murine model of osteosarcoma. Cancer Res. 2001, 61: 3750-3759.
175. McClatchey AI, Saotome I, Mercer K, Crowley D, Gusella JF, Bronson RT, Jacks T. Mice heterozygous for a mutation at the Nf2 tumor suppressor locus develop a range of highly metastatic tumors. Genes. Dev. 1998, 12: 1121-1133.
176. Kheradmand F, Werner E, Tremble P, Symons M, Werb Z. Role of Rac1 and oxygen radicals in collagenase-1 expression induced by cell shape change. Science 1998, 280: 898-902.
177. Matsumoto Y, Tanaka K, Harimaya K, Nakatani F, Matsuda S, Iwamoto Y. Small GTP-binding protein, Rho, both increased and decreased cellular motility, activation of matrix metalloproteinase 2 and invasion of human osteosarcoma cells. Jpn. J. Cancer Res. 2001, 92: 429-438.
178. Zhuge Y, Xu J. Rac1 mediates type I collagen-dependent MMP-2 activation. role in cell invasion across collagen barrier. J. Biol. Chem. 2001, 276: 16248-16256.

179. Adamson P, Etienne S, Couraud PO, Calder V, Greenwood J. Lymphocyte migration through brain endothelial cell monolayers involves signaling through endothelial ICAM-1 via a rho-dependent pathway. J. Immunol. 1999, 162: 2964-2973.
180. Worthylake RA, Lemoine S, Watson JM, Burridge K. RhoA is required for monocyte tail retraction during transendothelial migration. J. Cell Biol. 2001, 154: 147-160.
181. Clark EA, Golub TR, Lander ES, Hynes RO. Genomic analysis of metastasis reveals an essential role for RhoC. Nature 2000, 406: 532-535.
182. Soede RD, Zeelenberg IS, Wijnands YM, Kamp M, Roos E. Stromal cell-derived factor-1-induced LFA-1 activation during in vivo migration of T cell hybridoma cells requires Gq/11, RhoA, and myosin, as well as Gi and Cdc42. J. Immunol. 2001, 166: 4293-4301.
183. Malliri A, Van der Kammen RA, Clark K, Van Der Valk M, Michiels F, Collard JG. Mice deficient in the Rac activator Tiam1 are resistant to Ras-induced skin tumours. Nature 2002, 417: 867-871.
184. Cleverley SC, Costello PS, Henning SW, Cantrell DA. Loss of Rho function in the thymus is accompanied by the development of thymic lymphoma. Oncogene 2000, 19: 13-20.
185. Habets GG, Scholtes EH, Zuydgeest D, Van der Kammen RA, Stam JC, Berns A, Collard JG. Identification of an invasion-inducing gene, Tiam-1, that encodes a protein with homology to GDP-GTP exchangers for Rho-like proteins. Cell 1994, 77: 537-549.
186. Jonkers J, Korswagen HC, Acton D, Breuer M, Berns A. Activation of a novel proto-oncogene, Frat1, contributes to progression of mouse T-cell lymphomas. EMBO J. 1997, 16: 441-450.
187. Wodarz A, Nusse R. Mechanisms of Wnt signaling in development. Annu. Rev. Cell Dev. Biol. 1998, 14: 59-88.
188. Miller JR, Rowning BA, Larabell CA, Yang-Snyder JA, Bates RL, Moon RT. Establishment of the dorsal-ventral axis in Xenopus embryos coincides with the dorsal enrichment of dishevelled that is dependent on cortical rotation. J. Cell Biol. 1999, 146: 427-437.
189. van Noort M, Clevers H. TCF transcription factors, mediators of Wnt-signaling in development and cancer. Dev. Biol. 2002, 244: 1-8.
190. Habas R, Dawid IB, He X. Coactivation of Rac and Rho by Wnt/Frizzled signaling is required for vertebrate gastrulation. Genes Dev. 2003, 17: 295-309.
191. Habas R, Kato Y, He X. Wnt/Frizzled activation of Rho regulates vertebrate gastrulation and requires a novel Formin homology protein Daam1. Cell 2001, 107: 843-854.
192. Fanto M, Weber U, Strutt DI, Mlodzik M. Nuclear signaling by Rac and Rho GTPases is required in the establishment of epithelial planar polarity in the Drosophila eye. Curr. Biol. 2000, 10: 979-988.
193. Fanto M, McNeill H. Planar polarity from flies to vertebrates. J. Cell Sci. 2004, 117: 527-533.
194. Tao W, Pennica D, Xu L, Kalejta RF, Levine AJ. Wrch-1, a novel member of the Rho gene family that is regulated by Wnt-1. Genes Dev. 2001, 15: 1796-1807.

Chapter 10

GLIOBLASTOMA CELL MOTILITY: THE ROLE OF FAK AND CELLULAR SRC

Michelle R. Stettner*, Meera Natarajan* and Candece L. Gladson
*Division of Neuropathology, Department of Pathology, The University of Alabama at Birmingham, Birmingham, AL, USA (*co-first authors)*

Abstract: Understanding the signaling events that promote glioblastoma tumor cell motility is crucial, as invasion of these tumors into the normal brain contributes significantly to the morbidity and mortality observed in patients with these tumors. Also, the molecular mechanisms that promote glioblastoma cell motility/migration likely also promote invasion. Here we discuss the role of two non-receptor tyrosine kinases, focal adhesion kinase (FAK) and cellular Src, in promoting glioblastoma cell motility/migration. We discuss the activation of FAK, the elevated levels of FAK protein found in these tumors, the domains in FAK that are necessary for it to promote cell motility/migration, and studies dissecting FAK function in glioblastoma tumors. We also discuss data indicating Src family members have specific functions and the role of Lyn in promoting glioblastoma cell migration. FAK and cellular Src could be important new targets for glioblastoma therapy.

Key words: FAK, Src, Lyn, glioblastoma, astrocytoma, glioma, integrins

1. INTRODUCTION: CLASSIFICATION AND PROGNOSIS OF ASTROCYTIC TUMORS

Gliomas are tumors derived from differentiated glial cells (astrocytes, oligodendrocytes and ependymal cells), or from early glial cells prior to differentiation (1). The most common type of glioma is the astrocytoma, which is composed of cells expressing glial fibrillary acidic protein (GFAP). The World Health Organization (WHO) classifies astrocytomas into four histologic grades based on the degree of nuclear atypia, the cellularity, the

A. Wells (ed.), Cell Motility in Cancer Invasion and Metastasis, 221-243.

mitotic index, the presence of angiogenesis and the presence of necrosis (2). Grade I astrocytic tumors (pilocytic astrocytomas) often have an excellent prognosis and can frequently be cured through surgical resection; Grade II astrocytic tumors typically progress in grade and have a 5- to 7-year median survival; Grade III astrocytic tumors (anaplastic astrocytomas) are aggressive tumors, and have a 3-year median survival; and Grade IV astrocytic tumors (glioblastomas) are highly aggressive and have a one-year median survival. Grade IV tumors (glioblastomas) can be subdivided into those that arise *de novo* (≈ 60% of the cases), and are referred to as primary glioblastomas, and those tumors that arise from lower grade astrocytic tumors (≈ 40%) and are referred to as secondary glioblastomas (3). The current standard treatment for malignant astrocytomas (Grades III and IV) is surgical resection and radiotherapy; these treatments have little effect on patient survival.

2. COMMON GENETIC ABNORMALITIES IN ASTROCYTIC TUMORS

The etiologic events causing astrocytoma formation have not been clearly defined, but they are thought to involve genetic alterations resulting in the disruption of cell cycle arrest pathways and aberrant kinase signaling (4). The loss or mutation of the p53 tumor suppressor gene, located on chromosome 17p13.1, has been detected in all grades of astrocytomas and represents an early genetic event in many of these tumors (5-8). The p53 gene encodes a transcription factor that upon stress, e.g., DNA damage, is activated leading to cell cycle arrest, or apoptosis. The cell growth arrest activity of p53 allows the activation of the DNA repair system in the cell. In approximately 10% of malignant astrocytomas (Grades III and IV) that lack a p53 mutation, there is the overexpression of MDM2, which can bind and degrade wild-type p53 (9-11). In a significant number of *de novo* glioblastomas (≈ 60%), a mutated or absent INK4a-ARF locus is found (12-15). The INK4a-ARF locus encodes the tumor suppressor genes, $p14^{ARF}$ ($p19^{ARF}$ in mice) and $p16^{INK4a}$, both of which are important in cell cycle arrest at the G_1 to S phase (16, 17). Overexpression of growth factors and growth factor receptors also occurs in malignant astrocytomas (Grades III and IV), and can in some instances promote glioma formation (18). For example, targeting the PDGF-B gene to nestin-expressing early glial cells promotes malignant glioma formation (19). The PDGF receptor (PDGFr) and its ligands PDGF-A and -B are upregulated in low grade and some malignant astrocytoma tumors, suggesting an autocrine loop that promotes PDGFr signaling (20, 21). Overexpression of the PDGFr has been associated with loss of heterozygosity of chromosome 17p and p53 mutations in secondary glioblastomas (5). The epidermal growth factor receptor (EGFR), located on human chromosome 7, locus 7p12-p14, is amplified in 40-60% of *de novo* or primary glioblastomas, and in some instances it is mutated such that a constitutively active form of the EGFR is expressed (6, 21-23). EGFR gene amplification correlates with a poorer prognosis in patients with anaplastic astrocytoma, or glioblastoma (24).

3. INTEGRIN AND GROWTH FACTOR RECEPTORS PROMOTE MALIGNANT ASTROCYTOMA/GLIOBLASTOMA CELL MIGRATION

3.1 Integrins

Integrins are a widely dispersed family of heterodimeric adhesion receptors that respond to molecular, as well as mechanical stimuli, by activating intracellular signaling pathways and mediating cell adhesion and migration (25-27). These receptors also participate in the regulation of cell growth and survival. The integrin α and β subunits typically have short cytoplasmic tails which are necessary for the signaling that occurs with the engagement of extracellular matrix ligands, or when cooperating with growth factor receptors (28). Integrin receptor engagement typically results in the organization of the actin cytoskeleton and the assembly of a signaling complex at the submembranous region of the cell termed a focal contact or focal adhesion. Multiple integrins have been reported to be expressed on glioblastoma tumor cells in tissue biopsies, as well as on malignant astrocytoma/glioblastoma cell lines *in vitro*, including αvβ3, αvβ5, α5β1, α2β1, α3β1, α6β1, and αvβ1 (29-37). We and others have shown previously that the expression of integrins αvβ3 and αvβ5 is elevated on malignant astrocytoma cells in Grades III and IV astrocytoma tumor biopsies, as compared to Grade II astrocytoma tumor biopsies and normal brain (30, 31). Furthermore, integrin αvβ3 on malignant astrocytoma/glioblastoma cells propagated *in vitro* mediates attachment and migration toward vitronectin and osteopontin, and integrin αvβ5 expressed on some malignant astrocytoma/glioblastoma cell lines can participate in this attachment and migration (30, 31, 38).

Functional studies with blocking antibodies directed toward the β1 integrin subunit have shown an inhibition of adhesion, motility and invasion of cultured glioma cells plated on multiple extracellular matrix substrates (laminin, collagen type IV, fibronectin and vitronectin), suggesting a role for one or more β1 integrin(s) in neoplastic glial cell migration into the brain (39). The inhibition observed with a neutralizing anti-β1 antibody was greatest in cells adherent to collagen type IV (39). Another study utilizing neutralizing antibodies directed toward the αv or the β1 integrin subunits reported complete inhibition of glioblastoma cell migration on most substrates, suggesting the αv and β1 integrins play a crucial role in glioblastoma tumor cell infiltration into the normal brain (32). As the αv and β1 integrin subunit each pair with multiple beta or alpha subunits, respectively (e.g., αv can pair with β3, β5, β1, β6, and β8 subunits, and the β1 subunit can pair with α5, αv, α1, α2, α3, α4, α6, and α8 subunits), the inhibition of cell migration or invasion with blocking anti-αv or anti-β1 antibodies is typically strong. In addition, other cell surface molecules can cooperate with or modulate integrin-mediated cell attachment, migration, and invasion, such as CD44, the urokinase receptor and CD151/PETA or SFA-1 (40-42).

3.2 Growth Factor Receptor Promotion of Malignant Astrocytoma/Glioblastoma Cell Migration

Studies examining chemotactic migration of malignant astrocytoma/ glioblastoma cells propagated *in vitro* have shown that multiple growth factors can promote cell motility, including EGF and PDGF-BB (43, 44). In some instances these growth factors and their receptors are expressed by the tumor cells in biopsy samples, suggesting an autocrine mechanism whereby the tumor can promote its own migration (6, 20-23). Comparisons between reported studies to determine the most potent chemoattractant for glioblastoma cell migration is difficult, as the tumor cell lines used for these *in vitro* studies are multiple, and the integrin and growth factor receptor repertoires expressed are frequently different. In addition, the type of migration assay utilized impacts the results. For example, some studies have examined migration through filters in two-well Boyden-type chambers where cells sense a gradient of a chemoattractant in the bottom chamber, other studies have utilized a wounded monolayer assay plus/minus the addition of growth factor, while yet other studies have utilized a radial migration assay plus/minus the addition of growth factor. The extracellular matrix protein the tumor cells are adherent to, or the integrin receptor engaged, at the time of growth factor addition can also impact the results of chemotactic migration assays (44).

3.3 Integrin Cooperation with Growth Factor Receptors

Current thinking in the integrin field suggests integrin receptors cooperate with growth factor receptors to maximally mediate cell migration (45). Several examples of a specific cooperation between an integrin receptor and a growth factor receptor have been described. For example, the αvβ3 integrin has been shown to specifically cooperate with the PDGFrβ on fibroblasts, endothelial cells, oligodendrocytes and malignant astrocytoma/glioblastoma cells to mediate migration (46-48). In addition, integrin αvβ3 has been shown to specifically cooperate with the VEGFr-2 on endothelial cells, and the IGF-1 on porcine aortic smooth muscle cells to promote angiogenesis and cell migration, respectively (49, 50). Also, integrin α5β1 has been shown to specifically cooperate with the EGFR on interstitial epithelial cells (51).

As malignant astrocytoma/glioblastoma cells express integrin αvβ3 and the PDGFr, we investigated recently the specific cooperation of these two receptors on these cells. Using the U-87MG human malignant astrocytoma/glioblastoma cells, we found that integrin αvβ3 and the PDGFrβ specifically cooperate in promoting cell migration into a wounded (scratched) monolayer (44). In the assay utilized, integrin αvβ3 mediated 70% of the PDGF-stimulated migration while integrin αvβ5 mediated approximately 30% of this migration. Furthermore, based on co-immunoprecipitation studies, integrin αvβ3 and the PDGFrβ were found to specifically associate on these tumor cells (44). When the U-87MG cells were adherent to other integrin ligands, such as collagen or laminin, and scratched in a two-dimensional wound healing assay, PDGF-BB stimulation failed to significantly promote cell migration, as compared to cells adherent to vitronectin (engagement of integrin αvβ3), followed by scratching and PDGF BB stimulation. The molecular basis for this specific cooperation of integrin αvβ3 and the PDGFrβ is due, at least in part, to the direct association of the extracellular domain of the β3 subunit with the extracellular domain of the PDGFrβ (52). This association is constitutive and independent of PDGFrβ activation (44, 52).

4. FAK PROMOTES MALIGNANT ASTROCYTOMA/GLIOBLASTOMA CELL MIGRATION

4.1 FAK Domain Structure and Activation

Focal adhesion kinase (FAK) is a non-receptor cytoplasmic tyrosine kinase that has been shown to promote cell migration in multiple cell types (53-55). The domain structure of the FAK molecule is shown in the schematic, Figure 1. FAK is composed of an amino-terminal FERM domain, followed by a proline-rich region, the kinase domain, two additional proline-rich regions (denoted as PR1 and PR2), and a focal adhesion targeting (FAT) domain. The FERM domain mediates interactions of FAK with membrane proteins, such as integrins and the activated EGFR and PDGFr, as well as modulates the kinase activity of FAK by interacting with the kinase domain (56-58).

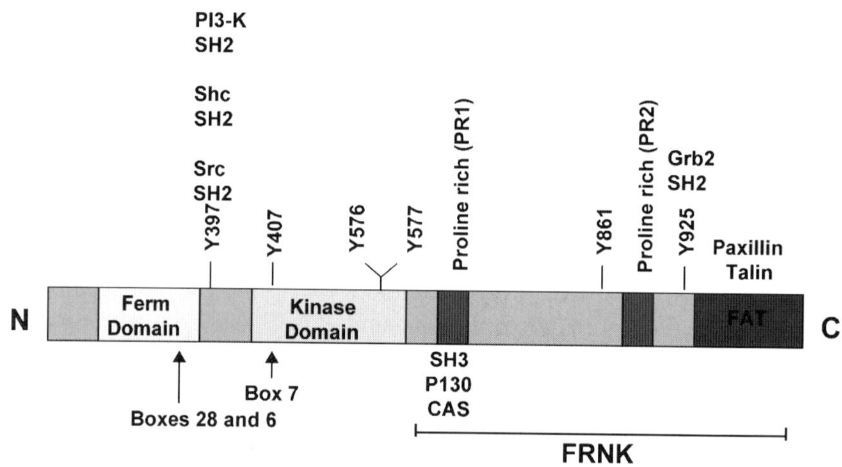

Figure 1. The domain structure of the FAK molecule. FERM domain, homologous to the band 4.1 protein. PR1 and PR2, denotes proline-rich regions 1 and 2 in the carboxyl-terminus. FAT, denotes the focal adhesion targeting domain. FRNK, denotes the FAK-related non-kinase construc. (The amino-terminal proline-rich domain is not shown). Boxes 28, 6 and 7 represent sites of altermative splicing where peptides of 28, 6 and 7 residues are inserted in the normal adult brain.

The amino-terminal proline-rich region mediates interactions with SH3-domain containing molecules, such as cellular Src; and the PR1 and PR2 in the carboxyl-terminal domain also mediates interactions with SH3-domain containing molecules, such as CAS family members (59, 60). The FAT domain is necessary for the recruitment of FAK to focal contacts (immature adhesion complexes) or focal adhesions (mature adhesion complexes), and this is due, at least in part, to an interaction of the FAT domain with paxillin and talin (61-63). The carboxyl-terminal region of the FAK molecule including PR1, PR2 and the FAT domain, referred to as FAK-related non-kinase (FRNK), is expressed in chicken fibroblasts as an alternative splice product of the FAK gene (64). Expression of FAT or FRNK in cells results in their competing with endogenous FAK for localization to focal contacts/focal adhesions, and as FAK signaling occurs at focal contacts/focal adhesions, expression of FAT or FRNK inhibits the activation and signaling of endogenous FAK (63, 64). FAK is activated through autophosphorylation on Y397 (65). FAK activation occurs in a temporally-related manner to the clustering of cell surface integrin receptors, which is triggered by integrin binding to its extracellular matrix ligand (66, 67). Activation of various growth factor receptors and G protein coupled receptors can also activate FAK, and in addi-tion FAK can be activated by Src phosphorylation of Y397 (68). Studies using the c-Src/c-Yes/Fyn (SYF)-null mouse embryo fibroblasts (MEF) indicate FAK activity is reduced with integin engagement, supporting a role for cellular Src in FAK activation (68). In non-malignant or non-transformed cells, FAK activation is adhesion-(integrin-engagement)-dependent (reviewed in 68). However, in malignant cells, such as malignant astrocytoma/glioblastoma cells, FAK activation is not strictly adhesion-dependent (69), suggesting FAK is regulated differently in malignant cells. Activation of FAK is necessary for its promotion of cell migration (reviewed in 55). Activation of FAK, or phosphorylation of Y397, leads to the SH2-domain mediated binding of cellular Src to this site, followed by cellular Src phosphorylation of Y576/577 in the kinase domain activation loop promoting maximal activation of FAK (70). Cellular Src can also phosphorylate FAK on Y407, Y861, and Y925 creating SH2-domain binding sites for interacting proteins. The function of these phosphorylated residues (pY407, pY861 and pY925) in FAK is not clear. Recently, in Ras-transformed NIH3T3 cells, phosphorylation of Y861 was shown to be necessary for cell migration (71). In fibroblasts, cellular Src phosphorylation of Y925 in FAK leads to Erk activation (72).

4.2 FAK Expression in Malignant Astrocytic Tumors

This lab first reported elevated levels of FAK protein in five of five anaplastic astrocytoma (Grade III) biopsy samples and in six of six glioblastoma (Grade IV) biopsy samples, as compared to six normal brains, as detected by immunohistochemical analysis of formalin-fixed and paraffin-embedded biopsy samples (73). Jones et al. (74) also reported increased FAK expression in glioblastoma tumor biopsies by immunostaining, as well as an increased expression of FAK in glioblastoma tumor biopsies following radiotherapy. We subsequently confirmed our immunohistochemical finding of elevated FAK expression in malignant astrocytic tumors by immunoblot analysis. We demonstrated elevated levels of normalized FAK protein in anaplastic astrocytoma (Grade III) tumors (an estimated 2.5-fold increase), as compared to 10 normal brain samples (69). We also demonstrated elevated FAK activity in these tumors as compared to the normal brain by assessing phosphorylation of Y397. Other investigators have also reported elevated levels of FAK protein in these tumors by western blot analysis, Rutka et al. (75) reported an elevated level of FAK protein in one glioblastoma biopsy sample as compared to one normal brain sample, and Zagzag et al. (76) reported enhanced expression of FAK protein in a subset of glioblastoma biopsy samples as compared to the normal brain. In addition, the latter group reported that the expression of FAK protein was elevated at the invasive tumor edge in high-grade gliomas. The finding of elevated FAK protein and activity in anaplastic astrocytoma (Grade III) and glioblastoma (Grade IV) tumor biopsy samples, suggests that FAK signaling could be important in the biology of these tumors.

4.3 FAK Promotes Migration of Malignant Astrocytoma/Glioblastoma Cells

FAK is thought to play an integrative role in coordinating the signals generated from integrin receptors and growth factor receptors (77). This concept is supported by studies in which the overexpression of FAK has been shown to promote haptotactic and chemotactic migration (73, 78-80). This is also supported by studies using MEF isolated from the FAK knockout mouse, in which significantly reduced cell migration was found that could be rescued by the re-expression of FAK (70, 81, 82).

FAK promotion of cell migration in fibroblasts is thought to be due to its promotion of focal contact/focal adhesion turnover and the promotion of membrane protrusion through Rac activation. In fibroblasts FAK promotion of focal contact/focal adhesion turnover is due, at least in part, to its ability to promote the localization of calpain to focal contacts/focal adhesions (83, 84). Calpain can cleave talin and FAK (85, 86). The interactions of FAK, cellular Src, and the docking molecules p130CAS and paxillin promote focal contact/focal adhesion turnover at the leading edge of the cell in fibroblasts (87). FAK recruits CAS family members to its PR1 and PR2 proline-rich domains and cellular Src phosphorylation of the substrate domain in CAS family members occurs most efficiently when they are bound to FAK (88). The latter event results in the recruitment of an adaptor molecule to the phosphorylated CAS substrate domain, such as Crk, the activation of a guanine exchange factor (such as Dock 180) and the activation of Rac promoting membrane protrusion (89).

We have shown that the overexpression of FAK in the U-251MG malignant astrocytoma/glioblastoma cells promotes haptotactic migration (73), and downregulation of FAK signaling by expression of the FAT domain has been shown by others to inhibit migration of LN-401 glioblastoma cells (43). FAK also promotes chemotactic migration of malignant astrocytoma/glioblastoma cells. Jones et al. (43) reported that EGF (2 ng/ml) increased the migration of the LN-401 glioblastoma cells by 10-fold, and transfection with the FAT domain of FAK reduced this EGF-induced increase in migration to 2-fold. Also, proteolytic cleavage of FAK by activated caspase-3 inhibits glioblastoma cell migration, supporting a role for intact FAK in promoting the migration of these cells (90).

In the U-87MG malignant astrocytoma/glioblastoma cells, FAK has been shown to promote focal adhesion turnover (disassembly and remodeling), and this requires inside to out Ca^{2+} signaling (91). In these studies, Ca^{2+} spikes were temporally associated with the disassembly and movement of focal adhesions, suggesting they trigger this event (91). Focal adhesion disassembly occurred simultaneously with cell-edge membrane movement and retraction of associated stress fibers. FRNK transfection did not affect the generation of Ca^{2+} spikes, although the cell morphology was altered and focal adhesions were observed to be fewer in number and larger in size. More recently, using the U-87MG cells, this group showed that disassembly of individual focal adhesions is associated with local increases in Ca^{+2}, and that transient increases in Ca^{+2} triggers focal adhesion disassembly (92). Furthermore, there is a loss of FAK from disassembling focal adhesions, and the dynamics of FAK movement between cytosolic and focal adhesion compartments is rapid ($\simeq$ 17 seconds) as compared to the time necessary for focal adhesion formation and disassembly ($\simeq$ 5 minutes) (92). Calpain is expressed in glioblastoma tumors (93), and its potential role in cleavage of focal adhesion proteins, as well as the role of CAS family members in promoting focal adhesion turnover needs to be evaluated in malignant astrocytoma/glioblastoma cell migration.

Molecules that inhibit FAK phosphorylation would be expected to inhibit migration of cells. Indeed, Zagzag et al. (94) reported recently that geldanamycin inhibits FAK phosphorylation and migration of three different malignant astrocytoma/glioblastoma cell lines (U-87MG, U-251MG, and LN229). The exact mechanism of the inhibition of FAK phosphorylation is not clear, but it is likely related to the effect of geldanamycin on the heat shock protein (Hsp 90). These data suggest FAK could be an important therapeutic target for new anti-glioma therapy.

Other FAK signaling mechanisms have been linked to cell migration. Ling et al. (95) demonstrated that the complex formation of PI3-kinase and FAK observed in U-251MG malignant astrocytoma/glioblastoma cells when adherent to specific matrix proteins correlated with the ability of PI3-kinase inhibitors to block haptotactic cell migration, suggesting that the association of FAK and PI3-kinase is important for the promotion of cell migration. Other investigators have also reported that PI3-kinase activity promotes cell migration (96).

4.4 The Role of a FAK Family Member Pyk2 in Malignant Astrocytoma/ Glioblastoma Cell Migration

Lipinski et al. (97) reported recently that Pyk2 activity was elevated in two malignant astrocytoma/glioblastoma cell lines (SF767 and T98G) as compared to FAK, and that the overexpression of Pyk2 in these cells promoted migration. This suggests additional studies evaluating the role of FAK as compared to Pyk2 need to be performed in glioblastoma cells.

4.5 FAK Promotes Migration of Brain Microvessel Endothelial Cells (Promotion of Angiogenesis)

Angiogenesis is necessary for the invasion and metastasis of malignant tumors, and it is defined as an increase in microvessel density (98). Angiogenesis in tumors requires the sprouting, migration and proliferation of host endothelial cells (98). Angiogenesis is a histologiiccharacteristic of malignant astrocytoma/glioblastoma tumors (2). To determine the role of FAK in modulating angiogenesis of brain microvascular endothelial cells, we examined the effect of FRNK transfection. We found that FRNK expression inhibited tube formation and branching of brain microvessel endothelial cells propagated in collagen gels (Fig. 2) (99). We then investigated the mechanism for this inhibition and found that FRNK expression inhibited haptotactic migration of the brain microvessel endothelial cells towards fibronectin and collagen (Fig. 2) (99). Other investigators have shown recently that FAK is necessary for blood vessel morphogenesis and that FAK promotes tube formation and branching of non-brain endothelial cells by promoting fibronectin matrix assembly (100, 101). Thus, likely FAK promotion of angiogenesis occurs through more than one mechanism.

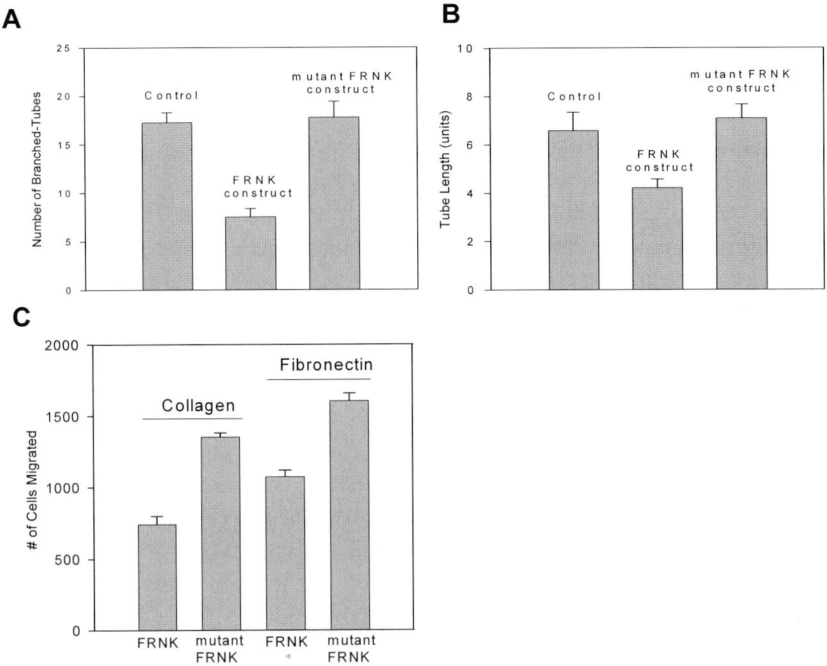

Figure 2. FAK promotes tube formation and branching, as well as cell migration, of mouse brain microvascular endothelial (IBE) cells. A & B, IBE mouse brain microvessel endothelial cells transfected with the FRNK or mutant FRNK construct were allowed to recover for 20 hours, followed by harvest, resuspension in serum-free media, and aliquoting onto collagen gels along with 5 ng/ml FGF2 in serum-free media. Two hours later, the media was removed and a second collagen gel was poured over the cells, followed by serum-free media and incubation for 48 hours (33°C, 5% CO2). Subsequently, tube formation and branching were photographed using a Nikon video microscope and analyzed using the Universal Metamorph software. Conditions were assayed in replicas of four and tube formation and branching analyzed in 5 fields at 10X magnification. Data is presented as the mean ± S.E.M. C, Alternatively, after FRNK or mutant FRNK transfection of IBE cells and recovery for 20 hrs, IBE cells were harvested, and plated in serum-free media onto 8 micron pore filters that had been coated previously on the bottom surface with either collagen or fibronectin and the cells allowed to migrate for 3 hours (33°C, 5% COs). Subsequently, migrated cells on the bottom surface of the filter were stained and counted. The conditions were assayed in replicas of four, and the data analyzed and presented as the mean ± the S.E.M. (Reprinted with permission from Haskell et al., 2003 Clin. Cancer Res. 9:2157-2165).

5. CELLULAR SRC PROMOTES CELL MIGRATION

5.1 Src Domain Structure

Src family members are non-receptor tyrosine kinases that are involved in signal transduction pathways that modulate a variety of cellular processes, such as adhesion, migration, proliferation, and differentiation. This is thought to be due to their association with multiple cell surface receptors (including growth factor receptors and integrins), as well as the actin cytoskeleton (reviewed in 102). There are 9 known Src family members: c-Src, Fyn, Lyn, c-Yes, Hck, Fgr, Blk, Lck, and Yrk, as well as a related molecule Rak (Frk) (102). Src family members share a similar structure consisting of six distinct functional domains (Fig. 3): the Src homology domain 4 (SH4) signals for lipid modification to provide hydrophobicity and membrane association, followed by a highly divergent unique domain that is Src family member specific, an SH3 domain, an SH2 domain, a catalytic domain that contains an autophosphorylation site important for Src kinase activity, and a carboxyl-terminal negative regulatory domain (reviewed in 103). These shared structural elements lead to a common mechanism of activation and regulation.

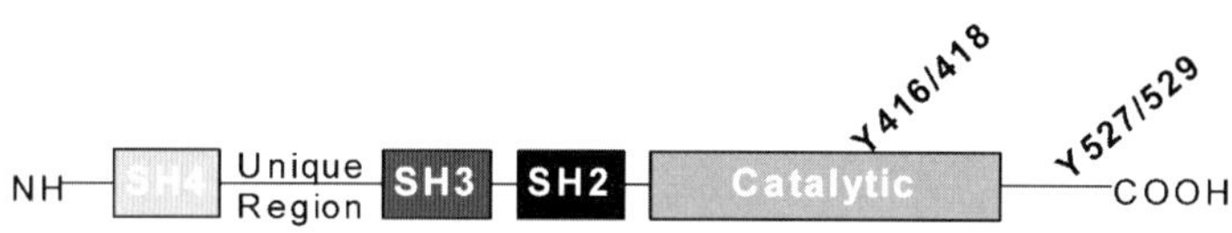

Domain	Function
SH4	Membrane Association
Unique	Region Highly Divergent
SH3	Mediate protein:protein interactions (P-X-X-P)
SH2	Mediate protein:protein interactions (P-Tyr)
Catalytic	Tyrosine Kinase Activity
C-terminus	Negative Regulation

Figure 3. The domain structure of cellular Src. The amino-terminal SH4 domain mediates membrane localization, the unique region is where the greatest diversity among Src family members lies, the SH3 domain mediates protein-protein interactions with proline-rich regions, the SH2 domain recognizes specific tyrosine phosphorylated peptides, followed by the kinase domain and the carboxyl-terminal negative regulatory peptide. Tyrosine residue 416/418 in the kinase domain is the autoactivation site in mouse/human c-Src and tyrosine residue 527/529 in the carboxyl-terminal negative regulatory carboxyl-terminal negatively regulatory site in mouse/human c-Src.

5.2 Examples of Src Family Member Specific Functions

Accumulating evidence suggests there are Src family member specific cellular functions. For example, Fyn promotes oligodendrocyte differentiation of progenitor cells (104), expression of constitutively active Fyn promotes differentiation of neuroblastoma tumor cells (105), and in the A431 epidermoid carcinoma cells Fyn activity is necessary for hemidesmosome disassembly (106). In addition, Src family members can associate specifically with certain cell surface receptors (107-109).

5.3 Migration and the Role of Cellular Src

The role of Src family members in cell migration has been implied by its role in phosphorylating FAK and the adaptor molecules that bind FAK (such as CAS family members), as well as the role of FAK in promoting integrin-mediated and growth factor-stimulated migration. The important role of Src and the Src/FAK complex in promoting cell migration is supported by the observation that a FAK mutant lacking the Src binding site Y397 (mutated to phenylalanine) is unable to restore the migratory phenotype to MEF isolated from the FAK knockout mouse (70). This suggests that Src binding to phosphorylated Y397 of FAK is important in the signals that promote cell migration. This concept is also supported by studies utilizing mouse fibroblasts isolated from the c-Src knockout animal, these fibroblasts are less motile than the wild-type fibroblasts and re-expression of c-Src compensates this abnormality (110, 111). Furthermore, MEF isolated from the c-Src/c-Yes/Fyn triple knockout mouse demonstrate significantly reduced cell motility (112).

5.4 The Src Family Member Lyn Promotes Glioblastoma Cell Migration

We have reported recently that PDGF-BB stimulation of U-87MG human malignant astrocytoma/glioblastoma cells promotes $\alpha v\beta 3$-mediated cell migration of cells adherent to vitronectin through the specific cooperation of the PDGFrβ and integrin $\alpha v\beta 3$ in a Src kinase-dependent manner (Fig. 4) (44). We characterized the Src family member repertoire on the U-87MG and U-251MG human malignant astrocytoma/glioblastoma cells and demonstrated expression of Fyn, Lyn and small amounts of c-Src. To determine whether Lyn or Fyn was necessary for the PDGF-stimulated integrin $\alpha v\beta 3$-mediated migration of these cells, we downregulated Lyn versus Fyn with siRNA (44). Downregulation of Lyn mRNA and protein but not Fyn significantly inhibited this migration, indicating that Lyn is necessary for the cooperative signaling of the PDGFrβ and integrin αvβ3 that promotes glioblastoma cell migration (Fig. 5). Lyn has also been implicated in the migration of B cells (113), and leukocytes (114). In support of a role for Lyn in the malidnant phenotype of glioblastoma tumors, we have recently shown that lyn kinase activity is significantly elevated in glioblastoma tumors as compared to the normal brain and as compared to anaplastic astrocytoma tumors (Grade III) (115).

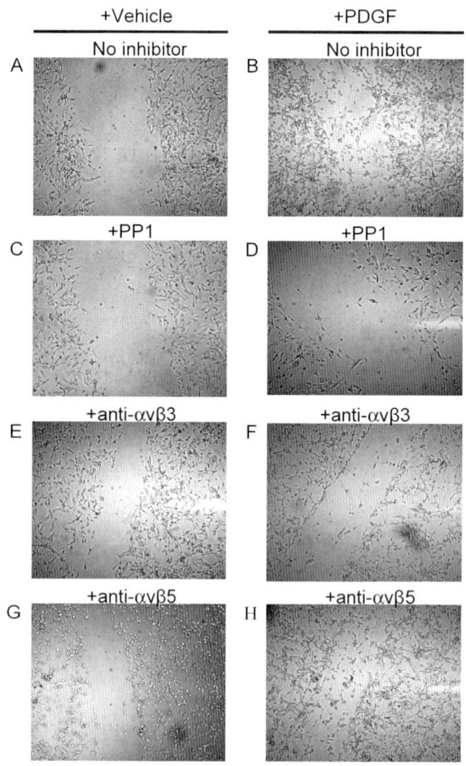

Figure 4. PDGF stimulation promotes integrin αvβ3-mediated U-87MG cell migration into a scratched area in a Src kinase-dependent manner. A confluent monolayer of serum-starved U-87MG cells adherent to vitronectin in serum-free media was scratched, treated with the Src kinase inhibitor, PP1 (5 nM, the IC50; for 30 min) or vehicle, followed by incubation with 20 μg/ml anti-integrin specific mAb or mouse IgG (20 min, 37° C), and immediately treated with PDGF BB (83 pM, 10 min, 37° C) or vehicle, followed by washing and incubation (37° C, 5%, CO2, 24 h). A, C, E, and G, + Vehicle; and B, D, F, and H, + PDGF treatment. At 24 h post-scratching, the scratched area was photographed. Conditions were assayed in replicas of six, and the experiment was repeated two times (Reprinted with permission from Ding et al., 2003 J. Biol. Chem. 278:39882-39891).

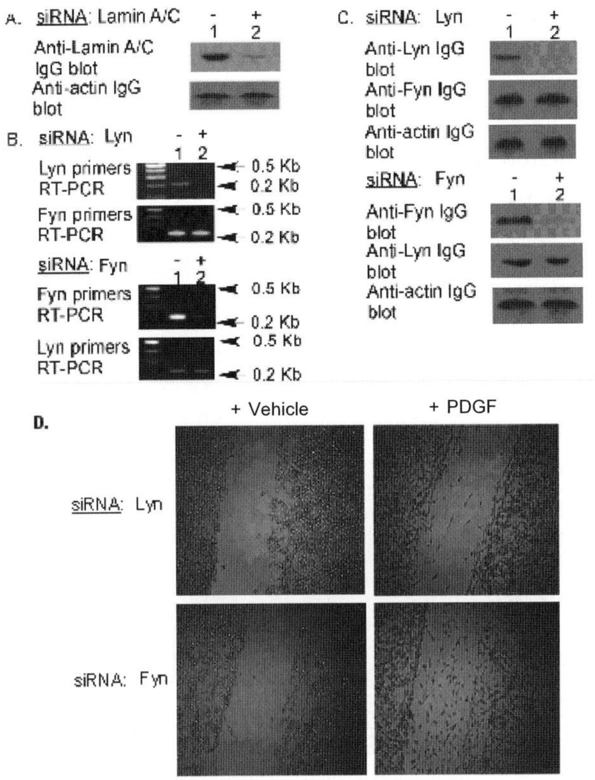

Figure 5. Downregulation of Lyn mRNA and protein by siRNA inhibits PDGF-stimula-ted integrin $\alpha v\beta 3$-mediated cell migration. U-87MG cells were transfected with 160 nM Lamin A/C siRNA (A), 320 nM Lyn siRNA (B-D), 320 nM Fyn siRNA (B-D), or with 320 nM non-specific control siRNA (A-C). U-87MG cells 48 h post-transfection were detergent lysed, subjected to disulfide-reduced 10% SDS-PAGE and blotted with mAb anti-Lamin A/C (A), stripped, and reprobed with mAb anti-actin (A), or blotted with rabbit anti-Lyn or anti-Fyn-specific IgG (C), stripped, and reprobed with mAb anti-actin (C). The experiment was repeated four times. B, total RNA from U-87MG cells 48 h post-transfection was extracted and subjected to RT-PCR with Lyn- or Fyn-specific primers, followed by electrophoresis of identical PCR product volumes on a 2% agarose gel. The experiment was repeated four times. D, A confluent monolayer of serum-starved U-87MG cells adherent to vitronectin in serum-free media was scratched 20 h after transfection with 320 nM Lyn or Fyn siRNA, stimulated with 83 pM PDGF (10 min, 37° C) or vehicle, followed by washing and incubation (37° C, 5% CO_2). At 24 h post-scratching, the scratched area was photographed. Conditions were assayed in replicas of six and the experiment was repeated two times (Reprinted with permission from Ding et al., 2003 J. Biol. Chem. 278:39882-39891).

6. SUMMARY

A number of molecules contribute to the migratory phenotype. We have summarized here what is known regarding the role of FAK and cellular Src in promoting the migration of malignant astrocytoma/glioblastoma cells. Future studies will hopefully define how FAK and cellular Src (specifically Lyn) are regulated in these tumors, and identify molecules that can specifically inhibit their function and be used in new therapeutic approaches.

ACKNOWLEDGEMENT

This work was supported by grants CA97247, project 5 and CA97110 from the National Institutes of Health, The National Cancer Institute, to CLG.

REFERENCES

1. Louis DN, Holland EC, Cairncross JF. Glioma Classification: A Molecular Reappraisal. Am. J. Pathol. 2001, 159: 779-786.
2. Kleihues P, Sobin LH. World Health Organization classification of tumors. Cancer 2000, 88: 2887.
3. Kleihues P, P Ohgaki H, H. Primary and secondary glioblastomas: from concept to clinical diagnosis. Neuro-oncol. 1999, 1: 44-51.
4. Holland EC. Gliomagenesis: genetic alterations and mouse models. Nat. Rev. Genet. 2001, 2: 120-129.
5. Hermanson M, Funa K, Koopmann J, Maintz D, Waha A, Westermark B, Heldin CH, Wiestler OD, Louis DN, von Deimling A, Nister M. Association of loss of heterozygosity on chromosome 17p with high platelet-derived growth factor alpha receptor expression in human malignant gliomas. Cancer Res. 1996, 56: 164-171.
6. Watanabe K, Tachibana O, Sata K, Yonekawa Y, Kleihues P, Ohgaki H. Overexpression of the EGF receptor and p53 mutations are mutually exclusive in the evolution of primary and secondary glioblastomas. Brain Pathol. 1996, 6: 217-223, discussion 23-4.
7. Bogler HJ, Huang P, Kleihues, Cavenee WK. The p53 gene and its role in human brain tumors. Glia. 1995, 15: 308-327.
8. Chattopadhyay P, Rathore A, Mathur M, Sarkar C, Mahapatra AK, Sinha S. Loss of heterozygosity of a locus on 17p13.3, independent of p53, is associated with higher grades of astrocytic tumours. Oncogene 1997, 15: 871-874.
9. Biernat W, Kleihues P, Yonekawa Y, Ohgaki H. Amplification and overexpression of MDM2 in primary (de novo) glioblastomas. J. Neuropathol. Exp. Neurol. 1997, 56: 180-185.
10. Reifenberger G, Liu L, Ichimura K, Schmidt EE, Collins VP. Amplification and overexpression of the MDM2 gene in a subset of human malignant gliomas without p53 mutations. Cancer Res. 1993, 53: 2736-2739.

11. Zhang Y, Xiong Y, Yarbrough WG. ARF promotes MDM2 degradation and stabilizes p53: ARF-INK4a locus deletion impairs both the Rb and p53 tumor suppression pathways. Cell 1998, 92: 725-734.
12. Louis DN, von Deimling A, Chung RY, Rubio MP, Whaley JM, Eibl RH, Ohgaki H, Wiestler OD, Thor AD, Seizinger BR. Comparative study of p53 gene and protein alterations in human astrocytic tumors. J. Neuropathol. Exp. Neurol. 1993, 52: 31-38.
13. He J, Allen JR, Collins VP, Allalunis-Turner MJ, Godbout R, Day, 3rd RS, James CD. CDK4 amplification is an alternative mechanism to p16 gene homozygous deletion in glioma cell lines. Cancer Res. 1994, 54: 5804-5807.
14. Hayashi Y, Ueki K, Waha A, Wiestler OD, Louis DN, von Deimling A. Association of EGFR gene amplification and CDKN2 (p16/MTS1) gene deletion in glioblastoma multiforme. Brain Pathol. 1997, 7: 871-875.
15. Ono Y, Tamiya T, Ichikawa T, Kunishio K, Matsumoto K, Furuta T, Ohmoto T, Ueki K, Louis DN. Malignant astrocytomas with homozygous CDKN2/p16 gene deletions have higher Ki-67 proliferation indices. J. Neuropathol. Exp. Neurol. 1996, 55: 1026-1031.
16. Ivanchuk SM, Mondal S, Dirks PB, Rutka JT. The INK4A/ARF locus: role in cell cycle control and apoptosis and implications for glioma growth. J. Neurooncol. 2001, 51: 219-229.
17. Rutka JT, Akiyama Y, Lee SP, Ivanchuk S, Tsugu A, Hamel PA. Alterations of the p53 and pRB pathways in human astrocytoma. Brain Tumor Pathol. 2000, 17: 65-70.
18. Maher EA, Furnari FB, Bachoo RM, Rowitch DH, Louis DN, Cavenee WK, DePinho RA. Malignant glioma: genetics and biology of a grave matter. Genes Dev. 2001, 15: 1311-1333.
19. Uhrbom L, Hesselager G, Nister M, Westermark B. Induction of brain tumors in mice using a recombinant platelet-derived growth factor B-chain retrovirus. Cancer Res. 1998, 58: 5275-5279.
20. Hermansson M, Nister M, Betsholtz C, Heldin CH, Westermark B, Funa K. Endothelial cell hyperplasia in human glioblastoma: coexpression of mRNA for platelet-derived growth factor (PDGF) B chain and PDGF receptor suggests autocrine growth stimulation. Proc. Natl. Acad. Sci. USA 1988, 85: 7748-7752.
21. Guha A, Dashner K, Black PM, Wagner JA, Stiles CD. Expression of PDGF and PDGF receptors in human astrocytoma operation specimens supports the existence of an autocrine loop. Int. J. Cancer 1995, 60: 168-73.
22. Bigner AH, Humphrey PA, Wong AJ, Vogelstein B, Mark J, Friedman HS, Bigner DD. Characterization of the epidermal growth factor receptor in human glioma cell lines and xenografts. Cancer Res. 1990, 50: 8017-8022.
23. Libermann TA, Nusbaum HR, Razon N, Kris R, Lax I, Soreq H, Whittle N, Waterfield MD, Ullrich A, Schlessinger J. Amplification, enhanced expression and possible rearrangement of EGF receptor gene in primary human brain tumours of glial origin. Nature 1985, 313: 144-147.
24. Liu KJ, Chen CT, Hu WS, Hung YM, Hsu CY, Chuang BF, Juang SH. Expression of cytoplasmic-domain substituted epidermal growth factor receptor inhibits tumorigenicity of EGFR-overexpressed human glioblastoma multiforme. Int. J. Oncol. 2004, 24: 581-590.
25. Giancotti FG, Ruoslahti E. Integrin signaling. Science 1999, 285: 1028-1032. Review.
26. Clark EA, Brugge JS. Integrins and signal transduction pathways: the road taken. Science 1995, 268: 233- 239.
27. Schwartz MA, Schaller MD, Ginsberg MH. Integrins: emerging paradigms of signal transduction. Ann. Rev. Cell. Dev. Biol. 1995, 11: 549-599.

28. Hynes RO. Integrins: bidirectional, allosteric signaling machines. Cell 2002, 110: 673-687.
29. Uhm JH, Gladson CL, Rao JS. The role of integrins in the malignant phenotype of gliomas. Front Biosci. 1999, 4: D188-199.
30. Gladson CL, Cheresh DA. Glioblastoma expression of vitronectin and the alpha v beta 3 integrin. Adhesion mechanism for transformed glial cells. J. Clin. Invest. 1991, 88: 1924-1932.
31. Gladson CL, Wilcox JN, Sanders L, Gillespie GY, Cheresh DA. Cerebral microenvironment influences expression of the vitronectin gene in astrocytic tumors. J. Cell Sci. 1995, 108: 947-956.
32. Friedlander DR, Zagzag D, Shiff B, Cohen H, Allen JC, Kelly PJ, Grumet M. Migration of brain tumor cells on extracellular matrix proteins in vitro correlates with tumor type and grade and involves alphaV and beta1 integrins. Cancer Res. 1996, 56: 1939-1947.
33. Gingras MC, Roussel E, Bruner JM, Branch CD, Moser RP. Comparison of cell adhesion molecule expression between glioblastoma multiforme and autologous normal brain tissue. J. Neuroimmunol. 1995, 57: 143-153.
34. Kim MO, Yun SJ, Kim IS, Sohn S, Lee EH. Transforming growth factor-beta-inducible gene-h3 (beta(ig)-h3) promotes cell adhesion of human astrocytoma cells in vitro: implication of alpha6beta4 integrin. Neurosci. Lett. 2003, 336: 93-96.
35. Giese A, Loo MA, Norman SA, Treasurywala S, Berens ME. Contrasting migratory response of astrocytoma cells to tenascin mediated by different integrins. J. Cell Sci. 1996, 109: 2161-2168.
36. Paulus W, Baur I, Schuppan D, Roggendorf W. Characterization of integrin receptors in normal and neoplastic human brain. Am. J. Pathol. 1993, 143: 154-163.
37. Deryugina EI, Bourdon MA. Tenascin mediates human glioma cell migration and modulates cell migration on fibronectin. J. Cell Sci. 1996, 109: 643-652.
38. Ding Q, Stewart J Jr, Prince CW, Chang PL, Trikha M, Han X, Grammer JR, Gladson CL. Promotion of malignant astrocytoma cell migration by osteopontin expressed in the normal brain: differences in integrin signaling during cell adhesion to osteopontin versus vitronectin. Cancer Res. 2002, 62: 5336-5343.
39. Rooprai HK, Vanmeter T, Panou C, Schnull S, Trillo-Pazos G, Davies D, Pilkington GJ. The role of integrin receptors in aspects of glioma invasion in vitro. Int. J. Dev. Neurosci. 1999, 17: 613-623.
40. Merzak A, Koocheckpour S, Pilkington GJ. CD44 mediates human glioma cell adhesion and invasion in vitro. Cancer Res. 1994, 54: 3988-3992.
41. Simon DI, Wei Y, Zhang L, Rao NK, Xu H, Chen Z, Liu Q, Rosenberg S, Chapman HA. Identification of a urokinase receptor-integrin interaction site. Promiscuous regulator of integrin function. J. Biol. Chem. 2000, 275: 10228-10234.
42. Kohno M, Hasegawa H, Miyake M, Yamamoto T, Fujita S. CD151 enhances cell motility and metastasis of cancer cells in the presence of focal adhesion kinase. Int. J. Cancer 2002, 97: 336-43.
43. Jones G, Machado J. Jr, Merlo A. Loss of focal adhesion kinase (FAK) inhibits epidermal growth factor receptor-dependent migration and induces aggregation of NH_2-terminal FAK in the nuclei of apoptotic glioblastoma cells. Cancer Res. 2001, 61: 4978-4981.
44. Ding Q, Stewart J. Jr, Olman MA, Klobe MR, Gladson CL. The pattern of enhancement of Src kinase activity on platelet-derived growth factor stimulation of

glioblastoma cells is affected by the integrin engaged. J. Biol. Chem. 2003, 278: 39882-39891.
45. Sieg DJ, Hauck CR, Ilic D, Klingbeil CL, Schaefer E, Damsky CH, Schlaepfer DD. FAK integrates growth-factor and integrin signals to promote cell migration. Nat. Cell Biol. 2000, 2: 249-56.
46. Baron W, Shattil SJ, ffrench-Constant C. The oligodendrocyte precursor mitogen PDGF stimulates proliferation by activation of alpha(v)beta3 integrins. Embo. J. 2002, 21: 1957-1966.
47. Schneller M, Vuori K, Ruoslahti E. Alpha(v)beta3 integrin associates with activated insulin and PDGFbeta receptors and potentiates the biological activity of PDGF. Embo. J. 1997, 16: 5600-5607.
48. Woodward AS, Garcia-Cardena G, Leong M, Madri JA, Sessa WC, Languino LR. The synergistic activity of alphavbeta3 integrin and PDGF receptor increases cell migration. J. Cell Sci. 1998, 111: 469-478.
49. Soldi R, Mitola S, Strasly M, Defilippi P, Tarone G, Bussolino F. Role of alphavbeta3 integrin in the activation of vascular endothelial growth factor receptor-2. Embo. J. 1999, 18: 882-892.
50. Maile LA, Clemmons DR. The alphaVbeta3 integrin regulates insulin-like growth factor I (IGF-I) receptor phosphorylation by altering the rate of recruitment of the Src-homology 2-containing phosphotyrosine phosphatase-2 to the activated IGF-I receptor. Endocrinology 2002, 143: 4259-4264.
51. Lee JW, Juliano RL. The alpha5beta1 integrin selectively enhances epidermal growth factor signaling to the phosphatidylinositol-3-kinase/Akt pathway in intestinal epithelial cells. Biochim. Biophys. Acta. 2002, 1542: 23-31.
52. Borges E, Jan Y, Ruoslahti E. Platelet-derived growth factor receptor beta and vascular endothelial growth factor receptor 2 bind to the beta 3 integrin through its extracellular domain. J. Biol. Chem. 2000, 275: 39867-39873.
53. Cary LA, Chang JF, Guan JL. Stimulation of cell migration by overexpression of focal adhesion kinase and its association with Src and Fyn. J. Cell Sci. 1996, 108: 1787-1794.
54. Gilmore AP, Romer LH. Inhibition of focal adhesion kinase (FAK) signaling in focal adhesions decreases cell motility and proliferation. Mol. Biol. Cell 1996, 7: 1209-1224.
55. Schlaepfer DD, Mitra SK, Ilic D. Control of motile and invasive cell phenotypes by focal adhesion kinase. Biochim. Biophys. Acta. 2004, Jul 5;1692: 77-102.
56. Hauck CR, Hsia DA, Schlaepfer DD. FAK facilitates PDGF-BB stimulated ERK2 activation required for chemotaxis migration of vascular smooth muscle cells. J. Biol. Chem. 2000, 275: 41092-41099.
57. Cooper LA, Shen TL, Guan JL. Regulation of focal adhesion kinase by its amino-terminal domain through an autoinhibitory interaction. Mol. Cell Biol. 2003, 23: 8030-8041.
58. Dunty JM, Gabarra-Niecko V, King ML, Ceccarelli DF, Eck MJ, Schaller MD. FERM domain interaction promotes FAK signaling. Mol. Cell Biol. 2004, 24: 5353-5368.
59. Polte TR, Hanks SK. Interaction between focal adhesion kinase and Crk-associated tyrosine kinase substrate p130Cas. Proc. Natl. Acad. Sci. USA 1995, 92: 10678-82.
60. O'Neill GM, Fashena SJ, Golemis EA. Integrin signalling: a new Cas(t) of characters enters the stage. Trends Cell Biol. 2000, 10: 111-119.
61. Hildebrand JD, Schaller MD, Parsons JT. Paxillin, a tyrosine phosphorylated focal adhesion-associated protein binds to the carboxyl terminal domain of focal adhesion kinase. Mol. Biol. Cell 1995, 6: 637-647.

62. Chen HC, Appeddu PA, Parsons JT, Hildebrand JD, Schaller MD, Guan JL. Interaction of focal adhesion kinase with the cytoskeletal protein talin. J. Biol. Chem. 1995, 270: 16995-16999.
63. Shen Y, Schaller MD. Focal adhesion targeting: the critical determinant of FAK regulation and substrate phosphorylation. Mol. Biol. Cell 1999, 10: 2507-2518.
64. Schaller MD, Borgman CA, Parsons JT. Autonomous expression of a noncatalytic domain of the focal adhesion-associated protein tyrosine kinase pp125FAK. Mol. Cell Biol. 1993, 13: 785-791.
65. Guan JL, Shalloway D. Regulation of focal adhesion-associate protein tyrosine kinase by both cellular adhesion and oncogenic transformation. Nature 1992, 358: 690-692.
66. Guan JL, Trevithick JE, Hynes RO. Fibronectin/integrin interaction induces tyrosine phosphorylation of a 120 kDa protein. Cell Regulation 1991, 2: 951-964.
67. Kornberg LJ, Earp HS, Parsons JT, Schaller M, Juliano RL. Cell adhesion or integrin clustering increases phosphorylation of a focal adhesion-associated tyrosine kinase. J. Biol. Chem. 1992, 267: 23439-23442.
68. Hanks SK, Ryzhova L, Shin NY, Brabek J. Focal adhesion kinase signaling activities and their implications in the control of cell survival and motility. Front Biosci. 2003, 8: d982-96.
69. Hecker TP, Grammer JR, Gillespie GY, Stewart J, Jr, Gladson CL. Focal adhesion kinase enhances signaling through the Shc/extracellular signal-regulated kinase pathway in anaplastic astrocytoma tumor biopsy samples. Cancer Res. 2002, 62: 2699-2707.
70. Owen JD, Ruest PJ, Fry DW, Hanks SK. Induced focal adhesion kinase (FAK) expression in FAK-null cells enhances cell spreading and migration requiring both auto- and activation loop phosphorylation sites and inhibits adhesion-dependent tyrosine phosphorylation of Pyk2. Mol. Cell Biol. 1999, 19: 4806-4818.
71. Lim Y, Han I, Jeon J, Park H, Bahk YY, Oh ES. Phosphorylation of focal adhesion kinase at tyrosine 861 is crucial for ras transformation of fibroblasts. J. Biol. Chem. 2004, 279: 29060-29065. Epub 2004 May 03.
72. Schlaepfer DD, Hunter T. Focal adhesion kinase overexpression enhances ras-dependent integrin signaling to ERK2/mitogen-activated protein kinase through interactions with and activation of c-Src. J. Biol. Chem. 1997, 272: 13189-13195.
73. Wang D, Grammer R, Cobbs CS, Stewart JE, Jr, Liu Z, Rhoden R, Hecker TP, Ding Q, Gladson CL. p125 focal adhesion kinase promotes malignant astrocytoma cell proliferation in vivo. J. Cell Sci. 2000, 113: 4221-4230.
74. Jones G, Machado J, Jr, Tolnay M, Merlo A. PTEN-independent induction of caspase-mediated cell death and reduced invasion by the focal adhesion targeting domain (FAT) in human astrocytic brain tumors which highly express focal adhesion kinase (FAK). Cancer Res. 2001, 61: 5688-5691.
75. Rutka JT, Muller M, Hubbard SL, Forsdike J, Dirks PB, Jung S, Tsugu A, Ivanchuk S, Costello P, Mondal S, Ackerley C, Becker LE. Astrocytoma adhesion to extracellular matrix: functional significance of integrin and focal adhesion kinase expression. J. Neuropathol. Exp. Neurol. 1999, 58: 198-209.
76. Zagzag D, Friedlander DR, Margolis B, Grumet M, Semenza GL, Zhong H, Simons JW, Holash J, Wiegand SJ, Yancopoulos GD. Molecular events implicated in brain tumor angiogenesis and invasion. Pediatr. Neurosurg. 2000, 33: 49-55.
77. Sieg DJ, Hauck CR, Ilic D, Klingbeil CK, Schaefer E, Damsky CH, Schlaepfer DD. FAK integrates growth-factor and integrin signals to promote cell migration. Nat. Cell Biol. 2000, 2: 249-256.

78. Cary LA, Chang JF, Guan JL. Stimulation of cell migration by overexpression of focal adhesion kinase and its association with Src and Fyn. J. Cell Sci. 1996, 109: 1787-1794.
79. Schlaepfer DD, Hauck CR, Sieg DJ. Signaling through focal adhesion kinase. Biophy. Mol. Bio. 1999, 71: 435-478.
80. Hauck CR, Sieg DJ, Hsia DA, Loftus JC, Gaarde WA, Monia BP, Schlaepfer DD. Inhibition of focal adhesion kinase expression or activity disrupts epidermal growth factor-stimulated signaling promoting the migration of invasive human carcinoma cells. Cancer Res. 2001, 61: 7079-7090.
81. Ilic D, Furuta Y, Kanazawa S, Takeda N, Sobue K, Nakatsuji N, Nomura S, Fujimoto J, Okada M, Yamamoto T. Reduced cell motility and enhanced focal adhesion contact formation in cells from FAK-deficient mice. Nature 1995, 377: 539-544.
82. Ilic D, Kanazawa S, Furuta Y, Yamamoto T, Aizawa S. Impairment of mobility in endodermal cells by FAK deficiency. Exp. Cell Res. 1996, 222: 298-303.
83. Dourdin N, Bhatt AK, Dutt P, Greer PA, Arthur JS, Elce JS, Huttenlocher A. Reduced cell migration and disruption of the actin cytoskeleton in calpain-deficient embryonic fibroblasts. J. Biol. Chem. 2001, 276: 48382-48388. Epub 2001 Oct 15.
84. Carragher NO, Westhoff MA, Fincham VJ, Schaller MD, Frame MC. A novel role for FAK as a protease-targeting adaptor protein: regulation by p42 ERK and Src. Curr. Biol. 2003, 13: 1442-1450.
85. O'Halloran T, Beckerle MC, Burridge K. Identification of talin as a major cytoplasmic protein implicated in platelet activation. Nature 1985, 317: 449-451.
86. Carragher NO, Fonseca BD, Frame MC. Calpain activity is generally elevated during transformation but has oncogene-specific biological functions. Neoplasia. 2004, 6: 53-73.
87. Webb DJ, Donais K, Whitmore LA, Thomas SM, Turner CE, Parsons JT, Horwitz AF. FAK-Src signalling through paxillin, ERK and MLCK regulates adhesion disassembly. Nat. Cell Biol. 2004, 6: 154-161. Epub 2004 Jan 25.
88. Cary LA, Han DC, Polte TR, Hanks SK, Guan JL. Identification of p130Cas as a mediator of focal adhesion kinase-promoted cell migration. J. Cell Biol. 1998, 140: 211-221.
89. Kiyaokawa E, Hashimoto Y, Kobayashi S, Sugimura H, Kurata T, Matsuda M. Activation of Rac1 by a Crk SH3-binding protein, DOCK180. Genes Devel. 1998, 12: 3331-3336.
90. Wick W, Wick A, Schulz JB, Dichgans J, Rodemann HP, Weller M. Prevention of irradiation-induced glioma cell invasion by temozolomide involves caspase 3 activity and cleavage of focal adhesion kinase. Cancer Res. 2002, 62: 1915-1919.
91. Giannone G, Ronde P, Gaire M, Haiech J, Takeda K. Calcium oscillatins trigger focal adhesion disassembly in human U-87 astrocytoma cells. J. Biol. Chem. 2002, 277: 26364-26371.
92. Giannone G, Ronde P, Gaire M, Beaudouin J, Haiech J, Ellenberg J, Takeda K. Calcium rises locally trigger focal adhesion disassembly and enhance residency of focal adhesion kinase at focal adhesions. J. Biol. Chem. 2004, 279: 28715-28723. Epub 2004 Apr 21.
93. Ray SK, Patel SJ, Welsh CT, Wilford GG, Hogan EL, Banik NL. Molecular evidence of apoptic death in malignant brain tumors including glioblastoma multiforme: upregulation of calpain and caspase-3. J. Neurosci. Res. 2002, 69: 197-206.
94. Zagzag D, Nomura M, Friedlander DR, Blanco CY, Gagner JP, Nomura N, Newcomb EW. Geldanamycin inhibits migration of glioma cells in vitro: a potential role for hypoxia-inducible factor (HIF-1alpha) in glioma cell invasion. J. Cell Physiol. 2003, 196: 394-402.

95. Ling J, Liu Z, Wang D, Gladson CL. Malignant astrocytoma cell attachment and migration to various matrix proteins is differentially sensitive to phosphoinositide 3-OH kinase inhibitors. J. Cell Biochem. 1999, 73: 533-44.
96. Brader S, Eccles SA. Phosphoinositide 3-kinase signalling pathways in tumor progression, invasion and angiogenesis. Tumori. 2004, 90: 2-8.
97. Lipinski CA, Tran NL, Bay C, Kloss J, McDonough WS, Beaudry C, Berens ME, Loftus JC. Differential role of proline-rich tyrosine kinase 2 and focal adhesion kinase in determining glioblastoma migration and proliferation. Mol. Cancer Res. 2003, 1: 323-332.
98. Sarcevic B. Angiogenesis in malignant tumors. Acta. Med. Croatica 2001, 55: 187-190.
99. Haskell H, Natarajan M, Hecker TP, Ding Q, Stewart J, Jr, Grammer R, Gladson CL. Focal adhesion kinase is expressed in the angiogenic blood vessels of malignant astrocytic tumors in vivo and promotes capillary tube formation of brain microvascular endothelial cells. Clin. Cancer Res. 2003, 9: 2157-2165.
100. Ilic D, Kovacic B, Johkura K, Schlaepfer DD, Tomasevic N, Han Q, Kim JB, Howerton K, Baumbusch C, Ogiwara N, Streblow DN, Nelson JA, Dazin P, Shino Y, Sasaki K, Damsky CH. FAK promotes organization of fibronectin matrix and fibrillar adhesions. J. Cell Sci. 2004, 117: 177-187.
101. Ilic D, Kovacic B, McDonagh S, Jin F, Baumbusch C, Gardner DG, Damsky CH. Focal adhesion kinase is required for blood vessel morphogenesis. Circ. Res. 2003, 92: 300-307.
102. Thomas SM, Brugge JS. Cellular functions regulated by Src family kinases. Annu. Rev. Cell Dev. Biol. 1997, 13: 513-609.
103. Brown MT, Cooper JA. Regulation, substrates and functions of src. Biochim. Biophys. Acta. 1996, 1287: 121-49.
104. Wolf RM, Wilkes JJ, Chao MV, Resh MD. Tyrosine phosphorylation of p190 RhoGAP by Fyn regulates oligodendrocyte differentiation. J. Neurobiol. 2001, 49: 62-78.
105. Berwanger B, Hartmann O, Bergmann E, Bernard S, Nielsen D, Krause M, Kartal A, Flynn D, Wiedemeyer R, Schwab M, Schafer H, Christiansen H, Eilers M. Loss of a FYN-regulated differentiation and growth arrest pathway in advanced stage neuroblastoma. Cancer Cell 2002, 2: 377-86.
106. Mariotti A, Kedeshian PA, Dans M, Curatola AM, Gagnoux-Palacios L, Giancotti FG. EGF-R signaling through Fyn kinase disrupts the function of integrin alpha6beta4 at hemidesmosomes: role in epithelial cell migration and carcinoma invasion. J. Cell Biol. 2001, 155: 447–458.
107. Kralisz U, Cierniewski CS. Association of pp60c-src with alpha IIb beta 3 in resting platelets. Biochem. Mol. Biol. Int. 1998, 45: 735-743.
108. Briddon SJ, Watson SP. Evidence for the involvement of p59fyn and p53/56lyn in collagen receptor signaling in human platelets. Biochem. J. 1999, 338: 203-209.
109. Vines CM, Potter JW, Xu Y, Geahlen RL, Costello PS, Tybulewicz VL, Lowell CA, Chang PW, Gresham HD, Willman CL. Inhibition of beta 2 integrin receptor and Syk kinase signaling in monocytes by the Src family kinase Fgr. Immunity 2001, 15: 507-519.
110. Klinghoffer RA, Sachsenmaier C, Cooper JA, Soriano P. Src family kinases are required for integrin but not PDGFR signal transduction. EMBO J. 1999, 18: 2459-2471.
111. Hall CL, Lange LA, Prober DA, Zhang S, Turley EA. pp60(c-src) is required for cell locomotion regulated by the hyaluronanreceptor RHAMM. Oncogene 1996, 13: 2213-2224.
112. Lowell CA, Soriano P. Knockouts of Src-family kinases: stiff bones, wimpy T cells, and bad memories. Genes. Dev. 1996, 10: 1845-1857.

113. Seo S, Buckler J, Erikson J. Novel roles for Lyn in B cell migration and lipopolysaccharide responsiveness revealed using anti-double-stranded DNA Ig transgenic mice. J. Immunol. 2001, 166: 3710-3716.
114. Suzuki T, Shoji S, Yamamoto K, Nada S, Okada M, Yamamoto T, Honda Z. Essential roles of Lyn in fibronectin-mediated filamentous actin assembly and cell motility in mast cells. J. Immunol. 1998, 161: 3694-3701.
115. Stettner MR, Wang W, Nabors LB, Bharara S, Flynn DC, Grammer JR, Gillespie GY Gladson CL. Lyn kinase activity is the predominant cellular Src kinase activity in glioblastoma tumor cells. Cancer Res. (in press, July). 2005.

Chapter 11

MOTILITY IN HEAD AND NECK CARCINOMA

Sufi Mary Thomas[1] and Jennifer Rubin Grandis[1, 2]
Departments of Otolaryngology[1] and Pharmacology[2], University of Pittsburgh, University of Pittsburgh Cancer Institute and University of Pittsburgh Medical Center, Pittsburgh, PA, USA

Abstract: Head and neck squamous cell carcinoma (HNSCC) arise in areas where locoregional spread of tumors interferes with functions vital for survival. These tumors are highly aggressive with poor five-year survival rates. Understanding the mechanisms whereby HSNCC invade into the surrounding tissues may help identify novel therapeutic targets for management and prevention of tumor dissemination. Several molecules including growth factor receptors, cytokines and matrix degrading enzymes have been examined for their role in HNSCC cell invasion. *In vivo* animal models are being developed to test novel therapeutic strategies and investigate the mechanisms of HNSCC invasion. In this chapter an overview of the pathways implicated in HNSCC invasion, the animal models commonly used and some of the preclinical therapeutic approaches targeting HNSCC invasion are presented.

Key words: autocrine signaling, paracrine signaling, EGF receptor, MMP, cytokines, src

1. INTRODUCTION

Head and neck squamous cell carcinomas (HNSCC) are malignancies that arise in the upper aero-digestive mucosa including the larynx, hypopharynx, oropharynx and oral cavity. These tumors are aggressive and difficult to treat with conventional treatment modalities. Tumors arising in the head and neck region frequently invade surrounding tissues resulting in complications including airway obstruction and rupture of the great vessels. HNSCC cells degrade the surrounding basement membrane via activation of specific enzymes. This process is essential for HNSCC cell motility and lymph node involvement. Understanding the molecular mechanisms that

A. Wells (ed.), Cell Motility in Cancer Invasion and Metastasis, 245-264.

mediate HNSCC motility and invasion may enable the development of novel therapeutic agents that target key regulators of this process. Abrogation of invasion could potentially reduce lymph node metastasis and increase the efficacy of conventional treatments. Several molecules have been implicated in HNSCC invasion. These can be broadly divided into three categories including membrane bound receptors, motility associated proteins and basement membrane degrading enzymes (1). In this chapter, we will discuss each class of motility mediators and present the *in vivo* animal models that are being used to investigate the mechanisms of motility and invasion in HNSCC. In addition, we will review the preclinical studies investigating novel strategies targeting invasion of HNSCC.

2. BIOMARKERS IMPORTANT FOR HNSCC MOTILITY AND INVASION

There is increasing evidence to suggest that molecular alterations contribute to increased invasiveness of HNSCC tumors. Over the last decade a number of molecules have been implicated in mediating HNSCC invasion (Table 1). HNSCC tumors are heterogenous and different regions of the same tumor may have distinct protein expression profiles depending upon several factors including nutrient availability. In addition there is some degree of variability in RNA and protein expression profiles of HNSCC tumors from different patients. This could be due to various factors including genetic predisposition and differences in exposure to environmental carcinogens. HNSCC invasion is a complex process involving multiple proteins that are either membrane bound, intracellular or secreted. Membrane bound proteins are usually receptors that respond to extracellular ligand stimulation triggering downstream signaling events leading to invasion. Mobilization of cells requires restructuring of the cytoskeletal framework and de-adhesion from the extracellular matrix. Proteins involved in maintaining the cytoskeleton framework of the cells have also been implicated in HNSCC invasion. On loss of cellular rigidity, cells adopt a rounded morphology and either secrete proteases that degrade the extracellular matrix or move through the extracellular matrix in an amoeboid manner (2). Although the exact mechanism whereby cells invade surrounding stroma is not completely understood, emerging evidence suggests that membrane receptors, intracellular proteins and secreted enzymes all work in a coordinated fashion to enable HNSCC invasion.

Table 1. Molecules implicated in HNSCC motility and invasion.

Protein	Description	Downstream targets	Expression in Metastasis	Modulation associated with invasion
EGFR	Growth factor receptor	MAPK, PLCγ-1	Increased	Yes
Integrin	Membrane receptor	FAK, MAPK	Increased	Yes
TβR-II	Growth factor receptor	MMP	Increased	Yes
uPAR	Growth factor receptor	MAPK, MMP	Increased	Yes
Cadherins	Membrane receptor	Catenins	Decreased	Yes
c-Met	Growth factor receptor	p21 rho	Increased	Yes
FAK	Intracellular protein	Paxillin, MAPK	Increased	Yes
PP2A	Intracellular protein	Serine-threonine residues	Increased	Yes
E2F	Transcription factor	DNA	Increased	Yes
Laminin	Extracellular glycoprotein	Integrin receptor	Increased	Yes
Collagen XVII	Transmembrane protein	Unknown	Decreased	Unknown
Moesin	Intracellular protein	Unknown	Decreased	Unknown
S100A4	Intracellular protein	Unknown	Increased	Unknown

EGFR epidermal growth factor receptor; MAPK mitogen activated protein kinase, PLCγ-1 phospholipase C gamma-1, FAK focal adhesion kinase, TβR-II TGFβ receptor II, MMP matrix metallotoprotease, uPAR urokinase plasminogen activator receptor, c-Met hepatocyte growth factor receptor, PP2A protein phosphatase 2 A.

2.1 Membrane Bound Receptors and Proteins

Receptors bind ligands in the extracellular milieu triggering signal transduction cascades. Autocrine and paracrine stimulation may result in simultaneous activation of several receptors on the cell surface. Cross talk among receptors may lead to additive or synergistic activation of downstream signaling pathways. Several membrane bound receptors including growth factor receptors and integrins are involved in HNSCC invasion. Members of the ErbB receptor family including epidermal growth factor receptor (EGFR) and ErbB2 (HER2) have been extensively studied for their role in proliferation and invasion of a number of different tumor types including carcinomas of the breast, prostate, colon and HNSCC.

2.1.1 Epidermal growth factor receptor (EGFR)

EGFR is overexpressed in more than 90% of HNSCC tumors (3). Several ligands have been shown to bind to EGFR including the epidermal growth factor (EGF), transforming growth factor-α (TGF-α), heparin binding-EGF (HB-EGF), amphiregulin, betacellulin and epiregulin. EGFR ligands stimulate the receptor in an autocrine as well as paracrine manner. Physiologic levels of EGF (0.9 - 3 ng/ml) were reported to stimulate cell migration in HNSCC cells *in vitro* (4). In addition EGF (5) and betacellulin (6) have been reported to stimulate HNSCC invasion.

Ligand binding to the receptor triggers receptor dimerization. Since members of the ErbB family form homo or heterodimers on ligand binding, it is possible that signaling triggered by various receptor combinations significantly impacts invasion of HNSCC cells. There is some evidence to suggest that co-localization and overexpression of EGFR and ErbB2 (HER2) in HNSCC cells increases invasion and nodal metastasis (7, 8). Receptor activation triggers signaling cascades downstream of the receptor.

Autocrine stimulation of EGFR increases HNSCC proliferation and anti-apoptotic effects via downstream effectors including MAPK, signal transducers and activators of transcription (STAT) and Akt (9). In addition, EGFR activation also increases HNSCC cell invasion (Figure 1).

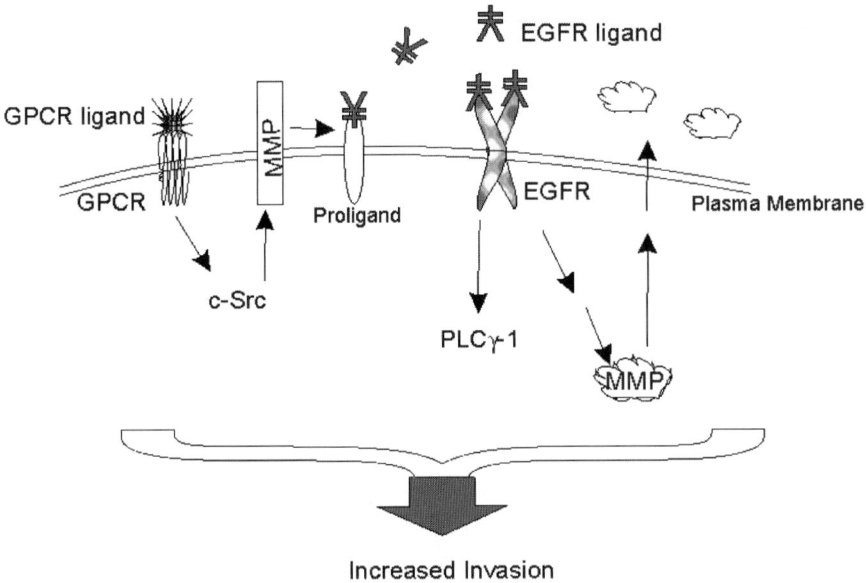

Figure 1. EGFR signaling contributes to invasion of HNSCC tumor cells. GPCR ligand stimulation activates c-Src and matrix metalloproteases (MMP) in HNSCC cells. MMPs cleave membrane bound EGFR proligand including TGF-α and Amphiregulin. Ligand binding activates EGFR thereby triggering downstream signaling cascades including PLCγ-1 and MMPs that promote HNSCC invasion.

The downstream molecules that mediate EGFR ligand-induced motility are not well defined. Phospholipase C gamma-1 (PLCγ-1) is a phosphoinositide specific phospholipase that on activation hydrolyses phosphatidylinositol (4, 5) biphosphate (PIP2) into inositol-triphosphate (IP3) and diacylglycerol (DAG). IP3 mediates the release of Ca^{+2} ions from intracellular stores and DAG activates PKC.These events are followed by actin remodeling within the cell enabling HNSCC cell motility. We have recently demonstrated that PLCγ-1 is downstream of EGFR and involved in HNSCC invasion on EGFR ligand (EGF) stimulation (5). Abrogation of PLCγ-1 activity with inhibitors abrogated EGFR-mediated HNSCC cell invasion through Matrigel. Further, HNSCC tumor tissue was demonstrated to express higher levels of both phosphorylated as well as total PLCγ-1 levels compared to normal adjacent mucosa from the same patient. PLCγ-1 may be an important downstream effector of EGFR-mediated invasion in HNSCC cells. EGFR also modulates HNSCC motility via upregulation of matrix metalloproteases (MMPs) specifically MMP 9 (10, 11). Several G-protein-coupled receptors including the gastrin-releasing peptide receptor (GRPR) activate EGFR via release of EGFR ligand from the membrane surface (12). In addition to stimulating cell proliferation via EGFR, GRPR also increased HNSCC cell invasion *in vitro* (13 2004 2004). We have demonstrated that inhibition of c-Src in HNSCC cells abrogates GRP stimulated invasion in HNSCC cells indicating that c-Src plays a role in HNSCC invasion. HNSCC cells respond to ligand binding to EGFR by activating signaling molecules that drive the cell into proliferation and invasion. EGFR is therefore an attractive target for prevention of metastasis.

2.1.2 Integrins

Integrins are cell surface receptors that bind ligands in the extracellular matrix (ECM). They are transmembrane proteins composed of non-covalently associated α and β subunits involved in cell adhesion and signal transduction (14). Integrin ligands are ECM components and ligand-binding links the cell cytoskeleton with the ECM. Activation of the integrin receptor triggers integrin clustering and phosphorylation of intracellular protein tyrosine kinases. In addition to migratory functions integrins also affect cell growth, apoptosis, differentiation and angiogenesis (1). Integrin expression has been demonstrated to correlate with HNSCC tumor invasion and metastasis (15). HNSCC cells use multiple integrins to mediate cell migration (16).

Altered expression of several integrins has been reported in HNSCC tumors compared to normal oral epithelium (1). Specifically, decreased expression of integrins α2β1, α4, α5 and α6 have been reported in HNSCC tumors compared to normal oral epithelium or to tissue from involved lymph nodes (17, 18). In the majority of reports α-subunit downmodulation increases HNSCC cell detachment from the matrix and facilitates invasion. For example, inhibition of the αv, α4, α5 and α6 with antisense oligonucleotides or with neutralizing antibodies has been reported to abrogate HNSCC motility and invasion (17). Overexpression of the β6 integrin subunit in HNSCC cells that were poorly invasive increased cell adhesion to fibronectin and increased migration and invasion *in vitro* and *in vivo* (19, 20). Integrin receptors have been reported to interact with growth factor signaling pathways.

Integrin αvβ4 found in hemidesmosomes increases HNSCC cell attachment to the ECM. Stimulation of HNSCC cells with EGF has been reported to reduce the number of hemidesmosomes and increase cell detachment from the substratum (21). Further, it was observed that EGFR mediated phosphorylation of the β4 subunit of integrin α6β4 via the src-family kinase protein Fyn resulting in destabilization of hemidesmosomes increasing HNSCC cell motility. Additional investigations are required to elucidate the pathways whereby growth factors modulate integrin receptors in HNSCC cells.

2.1.3 Transforming growth factor-β receptor II (TβR-II)

Transforming growth factor receptor-β (I and II) are transmembrane proteins belonging to the serine-threonine kinase family. The gene encoding TβR is a putative tumor suppressor that can be mutated in several tumor types including colorectal cancer and HNSCC (22, 23). High mRNA levels of TβR ligand-TGF-β in HNSCC tumors have been reported to correlate with non-metastatic tumors (24). Ligand binding triggers TβR-II heteromerization with TβR-I and receptor activation by transphosphorylation. The activated receptor complex phosphorylates intracellular proteins Smad 2/3 that then oligomerize with Smad 4. Smad protein complexes translocate to the nucleus and regulate gene transcription resulting in inhibition of cellular growth. HNSCC tumor cells overcome the inhibitory effects of TβR-II via several mechanisms by downmodulating the expression of TGF-β (25) and TβR-II (26) or by mutation of genes encoding TβR, Smad 2 and 4 (23, 26-28).

Recently reports indicate that TGF-β may regulate invasion of HNSCC cells. TGF-β has been demonstrated to induce transdifferentiation of stromal fibroblasts into myofibroblasts (29). Myofibroblasts have been detected in the stroma at the invasive edge of HNSCC tumors. These cells secrete chemokines, cytokines and several growth factors including hepatocyte growth factor (HGF). HGF was demonstrated to bind to c-Met receptors on HNSCC tumor cells inducing secretion of proteases and invasion. HNSCC cells interact with the stromal fibroblasts inducing changes that facilitate tumor invasion into the ECM. TGF-β may directly influence HNSCC migration by increasing MMP activation in HNSCC cells (30). Further investigations are warranted to fully understand the role of TGF-β in HNSCC invasion.

2.1.4 Urokinase-type plasminogen activator receptor (uPAR)

uPAR is composed of 3 extracellular domains attached to the plasma membrane via a glycosyl phosphatidyl inositol anchor. Urokinase plasminogen activator (uPA) binds uPAR facilitating multiple cellular processes including cell proliferation and adhesion. Expression of uPAR has been reported to be higher in human HNSCC tumors by Northern hybridization and immunohistochemistry (31, 32). Further analysis revealed no correlation of uPAR and uPA expression with several histopathologic parameters including metastasis in the patient population studied (31).

However, in a separate report uPA and uPAR expressing HNSCC tumors were highly invasive and demonstrated lymph node involvement (32). The role of uPA and its receptor in modulating HNSCC invasion has also been studied *in vitro*.

HNSCC cell lines expressing high levels of uPA and uPAR were reported to demonstrate increased invasion and had constitutively activated Erk (33) Ligand stimulation of uPAR increases association of the receptor with α5β1 integrin which in turn activates Erk via downstream intracellular signaling cascades (34). In addition to activating Erk, uPAR also activates matrix metalloprotease (MMP)-3 and –9 in HNSCC cell lines that express β6 integrin (30). Although the mechanisms whereby uPAR and uPA modulate HNSCC invasion are not completely defined, they appear to associate with integrins and activate proteases that enable HNSCC cells to degrade the extracellular matrix.

2.1.5 Cadherins

Membrane bound proteins involved in invasion are often present in adherens junctions. These multiprotein complexes mediate several roles including adhesion between cells (35). Dis-assembly of adherens junctions on cells is essential for cell detachment from the substratum once motility is initiated. Cadherins are a family of transmembrane proteins that mediate calcium-dependent cell-cell adhesion found in the adherens junction of cells (36). The extracellular domain mediates homotypic adhesions with cadherins expressed on neighboring cells. The intracellular domain links the adherens junction to the cytoskeleton via association with intracellular proteins including α, -β and γ-catenins. In addition to adhesion functions catenins also mediate intracellular signaling. Three members of the cadherin family include E-cadherin, N-cadherin and P-cadherin expressed in epithelial, neuronal and placental cells respectively. E-cadherin is often modulated in invading HNSCC cells (37). Biologically aggressive tumors demonstrating increased invasion and lymph node involvement in tumors that demonstrate have reduced E-cadherin expression and/or deranged E-cadherin-β-catenin complex formation (38).

E-cadherin modulates HNSCC tumor cell motility as demonstrated by down-regulation of membrane type–I matrix metalloproteases (MT1-MMP) (39). Reduction in levels of E-cadherin, α, -β and -γ catenin have been reported in metastatic versus non-metastatic HNSCC tumors (40-43). Reduced γ-catenin levels have been reported to have predictive value for nodal metastasis (43). Reduction in levels of E-cadherins was recently reported to be due to methylation of 5'CpG islands in the promoter region of E-cadherin (44).

2.1.6 Hepatocyte growth factor receptor (c-Met)

HGF overexpression has been observed at the mRNA and protein levels in HNSCC tumor tissues compared to normal adjacent mucosa or oral cavity tissue from healthy patients (45, 46). In addition, higher serum levels of HGF have been reported in HNSCC patients compared to healthy subjects (45). Increasing evidence indicates that c-Met is a mediator of HNSCC invasion. HNSCC cell lines stimulated with HGF demonstrate an increase in scattering as well as invasion through basement membrane and collagen matrix *in vitro* (45, 46). Kitaju H et al, reported that activation of the c-Met receptor triggers the activation and translocation of p21 rho (Rho) from the nucleus to the membrane increasing HNSCC invasion (47). Signaling pathways downstream of HGF and its interactions with other tyrosine kinase and G-protein-coupled receptors in HNSCC cells are incompletely understood and further studies are required.

2.2 Motility-Associated Intracellular and Secreted Proteins

In order to move, HNSCC cells lose their epithelial morphology and become rounded. This is a complex process involving modifications in a number of intracellular proteins. Some of these proteins are signal transducers that carry signals from the cell membrane to the nucleus, affecting invasion of HNSCC. Other proteins are involved in regulating the cytoskeletal framework of the cell and respond to signals from the cytoplasm or the nucleus, undergoing changes that involve rearrangement of cell cytoskeletal framework or the extracellular matrix. These proteins may be intracellular or secreted into the extracellular matrix.

2.3 Intracellular proteins involved in HNSCC invasion

Several intracellular proteins that are involved in signal transduction are implicated in HNSCC motility and invasion. Many of these signal transducers are preferentially upregulated in metastatic cells. The repertoire of signaling molecules that play a role in HNSCC cells is not well defined. A few studies have elucidated some of the molecules important for invasion of HNSCC and these will be discussed here.

2.3.1 Focal adhesion kinase (FAK)

FAK is an intracellular non-receptor tyrosine kinase that localizes to cellular structures called focal adhesions and plays a role in invasion of HNSCC cells. FAK is found in association with integrins and is in the phosphorylated state when cells adhere to the substratum. De-phosphorylation of FAK results in cell de-adhesion and motility (reviewed in 48). Oral carcinoma cells express elevated FAK levels (49). Overexpression of FAK in poorly invasive HNSCC increased their invasive rate by 4.5-fold. In addition, immunohistochemical analysis of HNSCC tumors revealed an increase in FAK expression compared to normal adjacent mucosa. EGFR-mediated signaling has been reported to decrease focal adhesion kinase (FAK) phosphorylation increasing invasion of KB oral squamous carcinoma cells (50). However in a separate study, FAK and focal adhesion protein-paxillin-mediated migration of HNSCC was dependant on MAPK activity (51). HGF has also been reported to increases FAK phosphorylation and invasion in HNSCC cells (47). FAK upregulation and activation may increase HNSCC invasion.

2.3.2 Protein phosphatase-2A (PP-2A)

Signal transduction in most cases involves phosphorylation and dephosphorylation of proteins. Enzymes that dephosphorylate signaling proteins help abort the signal transduction process. Protein phosphatase-2A (PP-2A) is an enzyme responsible for de-phosphorylation of serine-threonine residues on proteins (52). PP-2A has been reported to co-localize with microtubules and negatively regulate invasion of HNSCC cells (53) Selective inhibition of PP-2A was demonstrated to increase HNSCC invasion through laminin and vitronectin (53). Forced elevation of PP-2A levels was reported to increase HNSCC cell adhesion on laminin and vitronectin indicating modulation of integrin activity in HNSCC cells (53).

2.3.3 E2F-1

E2F-1 is a transcription factor that binds DNA facilitating DNA synthesis and cell proliferation (54). In resting cells E2F-1 is associated with the product of the retinoblastoma gene (pRb) (55). When cells enter the cell cycle, cyclin D binds and activates cyclin dependant kinases (Cdks) that in turn phosphorylate pRB in the pRB/E2F-2 complex. Activation of pRb results in release of E2F-1 from the complex. The role of E2F-1 has been tested in HNSCC cells (56). Ten HNSCC cell lines expressing E2F-1 were analyzed for their invasive potential. Highly invasive tumors expressed higher levels of E2F-1. The invasive potential of poorly invasive HNSCC cells with low E2F-1 levels was increased significantly when the cells overexpressed E2F-1 suggesting that E2F-1 contributes to HNSCC cell invasion.

2.4 Cytoskeletal and Secreted Proteins

Cancer cells are capable of modifying the extracellular matrix to facilitate growth and motility. In addition to secreting matrix-degrading enzymes cells also secrete components of the extracellular matrix. The newly secreted matrix components provide an anchor for the cells to bind to during the process of motility. Some of the components further stimulate HNSCC tumor cell motility by binding to and stimulating integrins on the membrane surface. Some of these proteins implicated in HNSCC invasion are discussed here.

2.4.1 Laminin

Laminin is an extracellular glycoprotein that is one of the constituents of the basement membrane. It is structurally composed of 3 non-identical α, β and γ polypeptide chains (57). Laminin 5 is synthesized and deposited by keratinocytes as a component of the basement membrane (58). It serves as an anchor for epithelial cells on the basement membrane and is a substrate for integrin $\alpha3\beta1$. Increased expression of laminin 5 $\gamma2$ has been reported to correlate with a high invasive potential in HNSCC tumors (59). Further studies have demonstrated a correlation between gene amplification of EGFR and laminin 5 $\gamma2$ in HNSCC cell lines (60). Stimulation of HNSCC cells with EGFR ligand TGF-α has been reported to increase laminin 5 $\gamma2$ expression levels and EGFR inhibition abrogates TGF-α-mediated laminin 5 $\gamma2$ expression indicating that EGFR induces laminin 5 $\gamma2$ secetion and deposition in the extracellular matrix of HNSCC cells.

2.4.2 Other proteins implicated in HNSCC invasion

There are a few isolated reports on other proteins that play a role in HNSCC invasion. These include transmembrane protein collagen XVII and intracellular proteins moesin and S100A4. Collagen XVII is a transmembrane protein involved in formation of hemidesmosomes (61). It interacts with proteins present in hemidesmosomes including integrin $\alpha6\beta4$. Collagen XVII has been reported to anchor the cell onto the basement membrane. An upregulation of collagen XVII expression has been demonstrated in HNSCC tumors (62). Collagen XVII expression was expressed in the entire proliferative zone of the primary tumor while loss of expression was observed in metastatic tumor (63).

Actin filaments form a part of the cell cytoskeleton closely associated with the plasma membrane. A family of proteins called ERM (ezrin/radixin/moesin) proteins are involved in linking actin filaments to the plasma membrane (64). The amino-terminal domain of the protein binds actin and a carboxyl-terminal domain binds to integral membrane proteins including CD44, CD43 and ICAM-2 (65). The expression pattern of moesin, one of the members of the ERM protein family, has been investigated in HNSCC cell lines, tumors and metastatic lymph nodes (66). Moesin was localized to the membrane in primary HNSCC tumors and non-metastatic cells while in metastatic lymph nodes and cells moesin was diffusely distributed in the cytoplasm. The mechanism behind the altered pattern of expression in metastatic versus non-metastatic tumors has not been defined. Moesin appears to be modulated in HNSCC tumors and is associated with a metastatic phenotype.

S100A4 is an intracellular calcium binding protein that has been implicated in invasion of a number of tumor types including carcinomas of the breast and colon. Its role in tumor cell invasion results from its ability to activate cytoskeletal non-muscle myosin. Although very little known about the mechanism whereby S100A4 mediates motility, forced expression of the protein in non-metastatic cells changes the phenotype into that of a highly metastatic cell. Expression of S100A4 has been investigated in HNSCC tumors correlated with clinoco-pathologic parameters including lymph node involvement (67). Although only 26% of HNSCC tumors expressed S100A4, the expression strongly correlated with lymph node involvement.

2.4.3 Basement Membrane Degrading Enzymes

Tumor cells secrete proteases that degrade the extracellular matrix components facilitating their movement through the interstitial spaces between tissues. Matrix metalloproteases (MMPs) require the presence of Zn^{+2} ions for enzyme activity. There are 4 sub-classes of MMPs collagenases, gelatinases, stromelysins and membrane type I MMPs (MT-1MMP). MMPs degrade basement membrane components including collagens. MMPs are secreted as inactive pro-enzymes. Pro-MMPs are proteolytically cleaved into active proteases. MT1-MMP, MMP2 and MMP9 have been predominantly reported to be expressed in HNSCC tumors (68, 69). Several growth factors have been demonstrated to induce expression of MMPs including EGFR (11) uPAR (30) and granulocyte-colony stimulating factor (70).

3. CURRENT TOOLS USED TO STUDY HNSCC INVASION

One of the major hurdles in testing novel therapeutic strategies for HNSCC is the dearth of *in vivo* models that closely resemble human disease. Subcutaneously transplanted HNSCC tumors in mice rarely invade the peritoneal wall and have not been reported to metastasize (71). The models described here have been used to study HNSCC invasion *in vivo*. In general, these models are limited in that human HNSCC tumors are implanted in immune compromised animals. Interactions between the complete repertoire of immune cells and their effects on HNSCC tumor invasion cannot be adequately investigated in these animals. There is still an urgent need for an animal model that would closely resemble human HNSCC disease.

3.1 Orthotopic floor of the mouth model

Among the currently used models, orthotopically implanted tumors come closest to resembling invasion in HNSCC patients. Mice are typically injected with 2×10^5 to 5×10^6 HNSCC cells in 25 – 100 μl in either growth medium or buffered saline. Several different strategies have been described for implanting tumors in the head and neck region of mice. Transcutaneous injections of HNSCC cells as depicted in Figure 2 into the myohyoid muscle near the floor of the mouth or the submental space have also been demonstrated to give rise to tumors that metastasize to the cervical lymph nodes (19, 72). The advantage of this technique is that it involves minimal nodesdiscomfort to the animals and can be administered with relative ease in the absence of sedation. Intraoral injections into the tongue or the anterior region of the floor of the mouth and myohyloid muscle have been reported to generate HNSCC tumors that metastasize to the cervical lymph nodes and lungs (73). HNSCC tumors injected into the floor of the mouth have been reported to metastasize to the regional cervical lymph nodes (>80% of the animals) and to the lung in 18% of the animals (66).

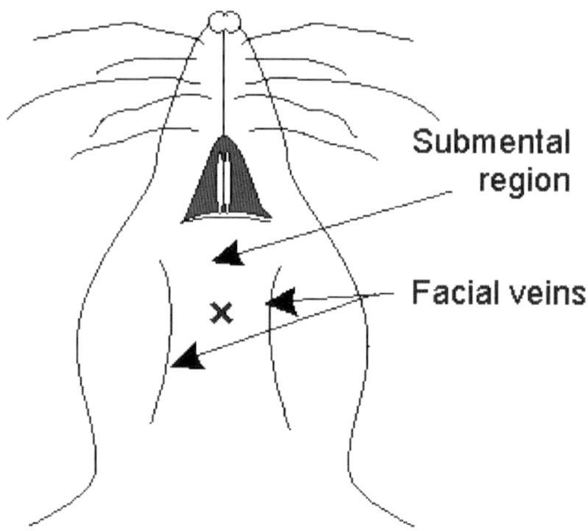

Figure 2. HNSCC orthotopic nude mouse model. HNSCC tumor cells (2 x 105 to 5 x 106) are inoculated into the submental region (X denotes site of injection). Transdermal injections of HNSCC cells between the two facial veins running parallel to each other generates orthotopic tumors in the head and neck region that metastasize to regional lymph.

3.1.1 Rat tracheal implant model

De-epithelialized rat tracheas transplanted into athymic nude mice have also been used to study HNSCC invasion (56). Rat tracheal tubes from Fisher 344 rats were sealed at one end and HNSCC cells (5 x 105 cells) introduced into the lumen of the tracheal tube. The open end was sealed and the rat trachea implanted subcutaneously into athymic nude mice. After a period of 8-12 weeks the animals were sacrificed and the implants retrieved, fixed and cross-sectioned. The degree of penetration of the invasive HNSCC cells into the tracheal wall was estimated on histological examination.

3.2 Gene expression profiling in metastatic and non-metastatic tumors and cell lines

Although a number of molecules are implicated in HNSCC invasion, the precise biomarkers that predict lymph node metastasis remain unknown. Molecular profiling studies have identified patterns of gene expression predominant in metastatic lymph nodes compared to non-metastatic tumors. Gene expression signatures from HNSCC metastatic tumors include the EGFR signaling pathway, adhesion and extracellular matrix proteins and gene expression patterns induced by cigarette smoke (74, 75). These gene expression signatures could be further developed to provide important clues to predict a metastatic disease in HNSCC patients with early stage tumors (76).

4. NOVEL THERAPEUTIC STRATEGIES FOR INVASIVE HNSCC

Therapeutic approaches that target HNSCC invasion may improve tumor management via surgical excision. In addition to reducing invasion in animal models of HNSCC, reduction in tumor growth has also been reported. Various inhibitors have been used to target signaling pathways involved in invasion of HNSCC cells. The broad-spectrum MMP inhibitor Marimastat (BB-2516) has been tested in clinical trials of patients with carcinomas of the ovary, prostate, and coloreactal region (77). Marimastat has been shown to inhibit the growth of HNSCC tumors *in vitro* (78). Further, Marimastat decreased lymph node metastasis and activation of MMP2 increasing survival *in vivo* (79). Calcium-influx is one of the early events triggered on EGF and other growth factor stimulation of HNSCC cells. Targeting calcium influx into HNSCC cells with carboxyamido-trizole (CAI) reduced MMP levels and lymph node metastasis of HNSCC tumors *in vivo* (80).

Effector cells of the immune system have been investigated for their ability to reduce HNSCC invasion in addition to their anti-tumor effects (81). Human cytotoxic T-lymphocytes and IL-2 induced natural killer cells were injected intratumorally into HNSCC tumors *in vivo*. Lymph node metastasis and tumor volumes were significantly reduced with immunotherapy (81). Thus there is strong evidence to suggest that it is feasible to target cell invasion in HNSCC patients.

5. SUMMARY

HNSCC cell invasion is a complex process involving structural proteins, antigens, receptors, signaling molecules and stromal cells surrounding the tumor (Table 1). Most of these molecules and cells interact with each other enabling HNSCC cells to acquire features necessary for cell motility. The precise mechanisms whereby HNSCC motility is orchestrated are not completely understood. Despite this dearth of knowledge, targeting molecules important for HNSCC motility reduces lymph node involvement and increases survival in preclinical animal models. Improved understanding of mechanism and the development of strategies that target molecules that regulate metastasis may improve HNSCC survival.

REFERENCES

1. Ziober BL, Silverman SS, Jr., Kramer RH. Adhesive mechanisms regulating invasion and metastasis in oral cancer. Crit. Rev. Oral. Biol. Med. 2001, 12: 499-510.
2. Friedl P, Borgmann S, Brocker EB. Amoeboid leukocyte crawling through extracellular matrix: lessons from the Dictyostelium paradigm of cell movement. J. Leukoc. Biol. 2001, 70: 491-509.
3. Grandis JR, Tweardy DJ. Elevated levels of transforming growth factor alpha and epidermal growth factor receptor messenger RNA are early markers of carcinogenesis in head and neck cancer. Cancer Res. 1993, 53: 3579-84.
4. Ohshima M, et al. Physiologic levels of epidermal growth factor in saliva stimulate cell migration of an oral epithelial cell line, HO-1-N-1. Eur. J. Oral Sci. 2002, 110: 130-6.
5. Thomas SM, et al. Epidermal growth factor receptor-stimulated activation of phospholipase Cg-1 promotes invasion of head and neck squamous cell carcinoma. Cancer Res. 2003, 51: 921-929.
6. O-charoenrat P, et al. Signaling pathways required for matrix metalloproteinase-9 induction by betacellulin in head-and-neck squamous carcinoma cells. Int. J. Cancer 2004, 111: 174-83.
7. Bei R, et al. Co-localization of multiple ErbB receptors in stratified epithelium of oral squamous cell carcinoma. J. Pathol. 2001, 195: 343-8.
8. O-charoenrat P, et al. C-erbB receptors in squamous cell carcinomas of the head and neck: clinical significance and correlation with matrix metalloproteinases and vascular endothelial growth factors. Oral Oncol. 2002, 38: 73-80.
9. Grandis JR, Sok JC. Signaling through the epidermal growth factor receptor during the development of malignancy. Pharmacol. Ther. 2004, 102: 37-46.
10. O-charoenrat P, et al. Epidermal growth factor-like ligands differentially up-regulate matrix metalloproteinase 9 in head and neck squamous carcinoma cells. Cancer Res. 2000, 60: 1121-8.
11. O-charoenrat P, et al. Differential modulation of proliferation, matrix metalloproteinase expression and invasion of human head and neck squamous carcinoma cells by c-erbB ligands. Clin. Exp. Metastasis 1999, 17: 631-9.
12. Lui VW, et al. Mitogenic effects of gastrin-releasing peptide in head and neck squamous cancer cells are mediated by activation of the epidermal growth factor receptor. Oncogene, 2003, 22: 6183-93.
13. Zhang Q, et al. Src family kinases mediate EGFR ligand cleavage, proliferation and invasion of cancer cells. Cancer Res. In Press, 2004.
14. Keely PJ, et al. R-Ras signals through specific integrin alpha cytoplasmic domains to promote migration and invasion of breast epithelial cells. J. Cell Biol. 1999, 145: 1077-88.
15. Shinohara M, et al. Expression of integrins in squamous cell carcinoma of the oral cavity. Correlations with tumor invasion and metastasis. Am. J. Clin. Pathol. 1999, 111: 75-88.

16. Koivisto L, et al. Integrins alpha5beta1, alphavbeta1, and alphavbeta6 collaborate in squamous carcinoma cell spreading and migration on fibronectin. Exp. Cell Res. 2000, 255: 10-7.
17. Zhang Y, et al. Functional differences between integrin alpha4 and integrins alpha5/alphav in modulating the motility of human oral squamous carcinoma cells in response to the V region and heparin-binding domain of fibronectin. Exp. Cell Res. 2004, 295: 48-58.
18. Maragou P, et al. Alteration of integrin expression in oral squamous cell carcinomas. Oral Dis. 1999, 5: 20-6.
19. Ramos DM, et al. Expression of integrin beta 6 enhances invasive behavior in oral squamous cell carcinoma. Matrix. Biol. 2002, 21: 297-307.
20. Thomas GJ, Jones J, Speight PM. Integrins and oral cancer. Oral Oncol. 1997, 33: 381-8.
21. Mariotti A, et al. EGF-R signaling through Fyn kinase disrupts the function of integrin alpha6beta4 at hemidesmosomes: role in epithelial cell migration and carcinoma invasion. J. Cell Biol. 2001, 155: 447-58.
22. Tannergard P, et al. Tumorigenesis in colorectal tumors from patients with hereditary non-polyposis colorectal cancer. Hum. Genet. 1997, 101: 51-5.
23. Chen T, et al. Novel inactivating mutations of transforming growth factor-beta type I receptor gene in head-and-neck cancer metastases. Int. J. Cancer 2001, 93: 653-61.
24. Pasini FS, et al. Transforming growth factor beta1, urokinase-type plasminogen activator and plasminogen activator inhibitor-1 mRNA expression in head and neck squamous carcinoma and normal adjacent mucosa. Head Neck 2001, 23: 725-32.
25. Paterson IC, et al. Decreased expression of TGF-beta cell surface receptors during progression of human oral squamous cell carcinoma. J. Pathol. 2001, 193: 458-67.
26. Garrigue-Antar L, et al. Missense mutations of the transforming growth factor beta type II receptor in human head and neck squamous carcinoma cells. Cancer Res. 1995, 55: 3982-7.
27. Muro-Cacho CA, et al. Defective transforming growth factor beta signaling pathway in head and neck squamous cell carcinoma as evidenced by the lack of expression of activated Smad2. Clin. Cancer Res. 2001, 7: 1618-26.
28. Kim SK, et al. DPC4, a candidate tumor suppressor gene, is altered infrequently in head and neck squamous cell carcinoma. Cancer Res. 1996, 56: 2519-21.
29. Lewis MP, et al. Tumour-derived TGF-beta1 modulates myofibroblast differentiation and promotes HGF/SF-dependent invasion of squamous carcinoma cells. Br. J. Cancer 2004, 90: 822-32.
30. Dang D, et al. Matrix metalloproteinases and TGFbeta1 modulate oral tumor cell matrix. Biochem. Biophys. Res. Commun. 2004, 316: 937-42.
31. Schmidt M, et al. Urokinase receptor up-regulation in head and neck squamous cell carcinoma. Head Neck 2000, 22: 498-504.
32. Nozaki S, et al. Immunohistochemical localization of a urokinase-type plasminogen activator system in squamous cell carcinoma of the oral cavity: association with mode of invasion and lymph node metastasis. Oral Oncol. 1998, 34: 58-62.
33. Ghosh S, et al. Loss of adhesion-regulated proteinase production is correlated with invasive activity in oral squamous cell carcinoma. Cancer 2002, 95: 2524-33.

34. Aguirre-Ghiso J.A, et al. Urokinase receptor and fibronectin regulate the ERK(MAPK) to p38(MAPK) activity ratios that determine carcinoma cell proliferation or dormancy in vivo. Mol. Biol. Cell 2001, 12: 863-79.
35. Peifer M. Cancer, catenins, and cuticle pattern: a complex connection. Science 1993, 262: 1667-8.
36. Takeichi M. Cadherins: a molecular family essential for selective cell-cell adhesion and animal morphogenesis. Trends in Genetics 1987, 3:8: 213-217.
37. Andrews NA, et al. Expression of the E-cadherin-catenin cell adhesion complex in primary squamous cell carcinomas of the head and neck and their nodal metastases. Br. J. Cancer 1997, 75: 1474-80.
38. Berx G, Nollet F, van Roy F. Dysregulation of the E-cadherin/catenin complex by irreversible mutations in human carcinomas. Cell Adhes. Commun. 6: 171-84.
39. Ara T, et al. Membrane type 1-matrix metalloproteinase expression is regulated by E-cadherin through the suppression of mitogen-activated protein kinase cascade. Cancer Lett. 2000, 157: 115-21.
40. Bankfalvi A, et al. Gains and losses of adhesion molecules (CD44, E-cadherin, and beta-catenin) during oral carcinogenesis and tumour progression. J. Pathol. 2002, 198: 343-51.
41. Bankfalvi A, et al. Deranged expression of the E-cadherin/beta-catenin complex and the epidermal growth factor receptor in the clinical evolution and progression of oral squamous cell carcinomas. J. Oral Pathol. Med. 2002, 31: 450-7.
42. Tanaka N, et al. Expression of E-cadherin, alpha-catenin, and beta-catenin in the process of lymph node metastasis in oral squamous cell carcinoma. Br. J. Cancer 2003, 89: 557-63.
43. Chow V, et al. A comparative study of the clinicopathological significance of E-cadherin and catenins (alpha, beta, gamma) expression in the surgical management of oral tongue carcinoma. J. Cancer Res. Clin. Oncol. 2001, 127: 59-63.
44. Chen Q, et al. Promoter methylation regulates cadherin switching in squamous cell carcinoma. Biochem. Biophys. Res. Commun. 2004, 315: 850-6.
45. Uchida D, et al. Role of HGF/c-met system in invasion and metastasis of oral squamous cell carcinoma cells in vitro and its clinical significance. Int. J. Cancer 2001, 93: 489-96.
46. Morello S, et al. MET receptor is overexpressed but not mutated in oral squamous cell carcinomas. J. Cell Physiol. 2001, 189: 285-90.
47. Kitajo H, et al. Rho regulates the hepatocyte growth factor/scatter factor-stimulated cell motility of human oral squamous cell carcinoma cells. Oncol. Rep. 2003, 10: 1351-6.
48. Kornberg LJ. Focal adhesion kinase and its potential involvement in tumor invasion and metastasis. Head Neck 1998, 20: 745-52.
49. Schneider GB, et al. Elevated focal adhesion kinase expression facilitates oral tumor cell invasion. Cancer 2002, 95: 2508-15.
50. Lu Z, et al. Epidermal growth factor-induced tumor cell invasion and metastasis initiated by dephosphorylation and downregulation of focal adhesion kinase. Mol. Cell Biol. 2001, 21: 4016-31.
51. Crowe DL, Ohannessian A. Recruitment of focal adhesion kinase and paxillin to beta1 integrin promotes cancer cell migration via mitogen activated protein kinase activation. BMC Cancer 2004, 4: 18.

52. Cohen P, Cohen PT. Protein phosphatases come of age. J. Biol. Chem. 1989, 264: 21435-8.
53. Meisinger J, et al. Protein phosphatase-2A association with microtubules and its role in restricting the invasiveness of human head and neck squamous cell carcinoma cells. Cancer Lett. 1997, 111: 87-95.
54. Weinberg RA. The retinoblastoma protein and cell cycle control. Cell 1995, 81: 323-30.
55. Nevins JR. E2F: a link between the Rb tumor suppressor protein and viral oncoproteins. Science 1992, 258: 424-9.
56. Zhang SY, et al. E2F-1 gene transfer enhances invasiveness of human head and neck carcinoma cell lines. Cancer Res. 2000, 60: 5972-6.
57. Beck K, Hunter I, Engel J. Structure and function of laminin: anatomy of a multidomain glycoprotein. Faseb. J. 1990, 4: 148-60.
58. Zhang K, Kramer RH. Laminin 5 deposition promotes keratinocyte motility. Exp. Cell Res. 1996, 227: 309-22.
59. Koshikawa N, et al. Overexpression of laminin gamma2 chain monomer in invading gastric carcinoma cells. Cancer Res. 1999, 59: 5596-601.
60. Ono Y, et al. Epidermal growth factor receptor gene amplification is correlated with laminin-5 gamma2 chain expression in oral squamous cell carcinoma cell lines. Cancer Lett. 2002, 175: 197-204.
61. Borradori L, Sonnenberg A. Structure and function of hemidesmosomes: more than simple adhesion complexes. J. Invest. Dermatol. 1999, 112: 411-8.
62. Parikka M, et al. Alterations of collagen XVII expression during transformation of oral epithelium to dysplasia and carcinoma. J. Histochem. Cytochem. 2003, 51: 921-9.
63. Herold-Mende C, et al. Metastatic growth of squamous cell carcinomas is correlated with upregulation and redistribution of hemidesmosomal components. Cell Tissue Res. 2001, 306: 399-408.
64. Tsukita S, Yonemura S. ERM (ezrin/radixin/moesin) family: from cytoskeleton to signal transduction. Curr. Opin. Cell Biol. 1997, 9: 70-5.
65. Yonemura S, et al. Ezrin/radixin/moesin (ERM) proteins bind to a positively charged amino acid cluster in the juxta-membrane cytoplasmic domain of CD44, CD43, and ICAM-2. J. Cell Biol. 1998, 140: 885-95.
66. Kobayashi H, et al. Clinical significance of cellular distribution of moesin in patients with oral squamous cell carcinoma. Clin. Cancer Res. 2004, 10: 572-80.
67. Moriyama-Kita M, et al. Correlation of S100A4 expression with invasion and metastasis in oral squamous cell carcinoma. Oral Oncol. 2004, 40: 496-500.
68. Imanishi Y, et al. Clinical significance of expression of membrane type 1 matrix metalloproteinase and matrix metalloproteinase-2 in human head and neck squamous cell carcinoma. Hum. Pathol. 2000, 31: 895-904.
69. Magary SP, et al. Expression of matrix metalloproteinases and tissue inhibitor of metalloproteinases in laryngeal and pharyngeal squamous cell carcinoma: A quantitative analysis. Otolaryngol. Head Neck Surg. 2000, 122: 712-6.
70. Tomita T, et al. Granulocyte-macrophage colony-stimulating factor upregulates matrix metalloproteinase-2 (MMP-2) and membrane type-1 MMP (MT1-MMP) in human head and neck cancer cells. Cancer Lett. 2000, 156: 83-91.
71. Kawashiri S, et al. Development of a new invasion and metastasis model of human oral squamous cell carcinomas. Eur. J. Cancer B. Oral Oncol. 1995, 31: 216-21.

72. Zhang X, et al. A lymph node metastatic mouse model reveals alterations of metastasis-related gene expression in metastatic human oral carcinoma sublines selected from a poorly metastatic parental cell line. Cancer 2002, 95: 1663-72.
73. O'Malley BW, Jr., et al. A new immunocompetent murine model for oral cancer. Arch. Otolaryngol. Head Neck Surg. 1997, 123: 20-4.
74. Chung CH, et al. Molecular classification of head and neck squamous cell carcinomas using patterns of gene expression. Cancer Cell 2004, 5: 489-500.
75. Schmalbach CE, et al. Molecular profiling and the identification of genes associated with metastatic oral cavity/pharynx squamous cell carcinoma. Arch. Otolaryngol. Head Neck Surg. 2004, 130: 295-302.
76. Lin Y, et al. Transfusion of ABO-nonidentical platelets is not associated with adverse clinical outcomes in cardiovascular surgery patients. Transfusion 2002, 42: 166-72.
77. Nemunaitis J, et al. Combined analysis of studies of the effects of the matrix metalloproteinase inhibitor marimastat on serum tumor markers in advanced cancer: selection of a biologically active and tolerable dose for longer-term studies. Clin. Cancer Res. 1998, 4: 1101-9.
78. O-charoenrat P, Rhys-Evans P, Eccles S. A synthetic matrix metalloproteinase inhibitor prevents squamous carcinoma cell proliferation by interfering with epidermal growth factor receptor autocrine loops. Int. J. Cancer 2002, 100: 527-33.
79. Maekawa K, et al. Inhibition of cervical lymph node metastasis by marimastat (BB-2516) in an orthotopic oral squamous cell carcinoma implantation model. Clin. Exp. Metastasis 2002, 19: 513-8.
80. Wu Y, et al. Inhibition of head and neck squamous cell carcinoma growth and invasion by the calcium influx inhibitor carboxyamido-triazole. Clin. Cancer Res. 3: 1915-21.
81. Chikamatsu K, et al. Immunotherapy with effector cells and IL-2 of lymph node metastases of human squamous-cell carcinoma of the head and neck established in nude mice. Int. J. Cancer 1999, 82: 532-7.

Chapter 12

MOTILITY IN MELANOMA PROGRESSION

Anja Katrin Bosserhoff
Institute of Pathology, University of Regensburg, Germany

Abstract: This chapter provides an overview on the role of cell motility of melanoma progression. Mechanistically, several molecules are known to be involved in deregulation of adhesive interaction of tumor cells with each other and with extracellular matrices, in synthesis and activation of proteases and other enzymes, and in locomotion of tumor cells and organization of the cytoskeleton. Examples for these gene families and their role in melanoma motility are discussed.

Key words: invasion, extracellular proteases, extracellular matrix, cytoskeleton

1. INTRODUCTION

Conventional therapy of malignant tumors is frequently limited by acquisition of an invasive and metastatic phenotype and progression to a systemic disease. Thus, understanding and manipulating molecular events leading to systemic spread with high cellular motility represent major challenges for current cancer research.

Motility involves a highly regulated and coordinated cascade of complex molecular processes including cell attachment, cell detachment, secretion of proteases, cell migration, and exchanging signals with other cells or matrix in the local milieu (Figure 1). While invasion is regarded as a key signature of malignant tumors, it is also found as part of the normal behavior of inflammatory blood cells, in tissues engaged in morphogenetic movements of normal embryogenesis, and in a number of instances of normal and pathological tissue remodeling in the adult. Numbers of different proteins involved in cancer cell motility have been identified during the past years.

A. Wells (ed.), Cell Motility in Cancer Invasion and Metastasis, 265-300.

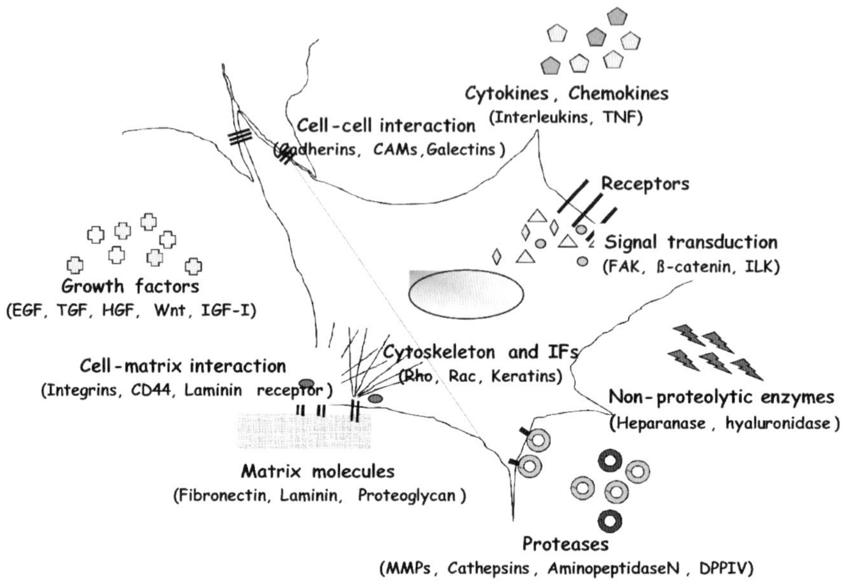

Figure 1. Schematic overview on relations between molecules discussed in the review. Matrix molecules, cytokines / growth factors, receptors and signaling pathways documented in this review are placed in direct connection. Changes observed in malignant tumors can not be understood without analyzing the effects on the whole cellular system.

Although a few, such as the family of matrix metalloproteases (MMPs), have been analyzed in some detail, most of them have been identified only recently and still have to be put in a greater context. Moreover, probably many of them still have to be discovered. In light of this largely preliminary knowledge, this chapter summarizes data on cellular and molecular mechanisms underlying the phenotypic characteristics of melanoma cells that determine their mobile capability.

Some of the genes listed in table 1 are reviewed intensively in other chapters of this book and will therefore not be described here. The broad family of MMPs, their inhibitors and activators are covered in Chapter 7, and the growth factor signaling in Chapter 5. The impact of the cytoskeleton is discussed in Chapter 9. Chapter 8 details various aspects of integrin-mediated cell-matrix interaction. Here, the other genes listed in table 1 and their roles in cell motility of malignant melanoma are discussed in further detail.

Table -1. Overview on gene families involved in melanoma cell motility and examples.

Protein function	Members involved in motility of melanoma
Cell-cell adhesion molecules	Cadherins, ICAM-1, MCAM, Galectin
Cell-matrix adhesion molecules	Integrins, CD44, Laminin receptor
Matrix molecules	Fibronectin, Laminin, Collagens, Thrombospondin
Cytoskeleton and intermediary filaments	Rho, Rac, Cdc 42, keratins, actins
Growth factors	EGF, HGF (scatter factor), TGF-beta, wnt
Cytokines, chemokines	IL-1, TNF-alpha, IL-6
Proteases	MMPs, Cathepsins, tPA, uPA, Transmembranous proteases
Non-proteolytic enzymes	Heparanase, Hyaluronidase
Signal transduction molecules	PI3-K, PLC-gamma, ras, FAK, ß-catenin
Recently identified molecules with complex functions	MIA, TIAM, Amphoterin, SPARC

2. CELL-CELL ADHESION MOLECULES

2.1 Cadherins

E-Cadherin and its associated catenin complexes have been identified as key molecules in cell adhesion. The transmembrane E-Cadherin proteins form calcium-dependent homodimers via extracellular binding and concomitantly transduce signals from the micro-environment to other molecular complexes implicated in invasion. Changes in functional E-Cadherin expression have been described to play important roles in human cancer (Figure 2). E-Cadherin inactivating mutants were found in breast, colon, and gastric carcinomas, whereas downregulation of E-Cadherin expression was reported to occur in malignant melanomas (1, 2). This downregulation was shown to be due to the transcriptional repressor snail (3). Overexpression of E-Cadherin significantly decreased the metastatic capacity of tumor cells as measured in *in vitro* as well as *in vivo* experiments (4) (5), indicating that levels of E-Cadherin expression may represent critical parameters for the invasive and metastatic tumor phenotype. Recent studies clearly showed that also important downstream signaling pathways are regulated by E-cadherin. Qian et al. revealed that re-transfection of E-cadherin into melanoma cell lines conferred negative receptor tyrosine kinase regulation to human melanoma (6). Further, a study of Kuphal et al. presented a negative regulation of NFkappaB by E-cadherin (Oncogene in press). To illustrate the complexity of regulation, a study described that the transfection of GnT-lll (beta 1,4-N-acetylglucosaminyltransferase III) catalyzes the addition of an N-acetylglucosamine (GlcNAc) residue on glycoproteins into mouse melanoma cells and results in the enhanced expression of E-cadherin In turn, this leads to the suppression of lung metastasis (7).

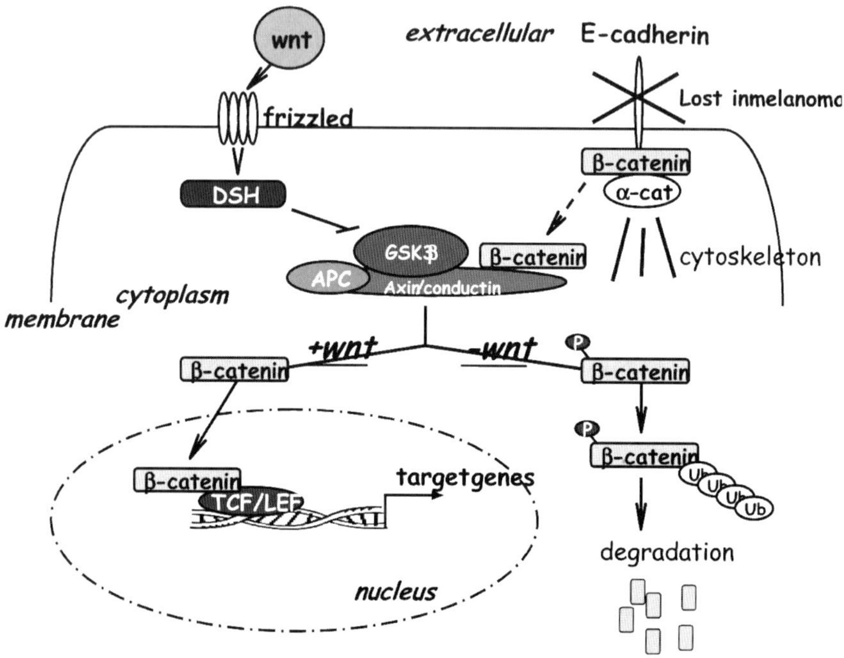

Figure 2. Schematic overview of interactions between molecules involved in E-cadherin/ wnt/ ß-catenin signaling pathways. All molecules discussed in this review have to be seen in the context of interactions. See 2.1, 6.4 and 10.2 for details. (Adapted from Poser and Bosserhoff, Histol. Histopath. 2004).

Controversially, N-cadherin expression is upregulated when E-cadherin expression is lost. Until today details of this regulation are not understood, but the functional importance of this process was determined in a study of Li et al. (8). N-Cadherin is known to mediate homotypic aggregation among melanoma cells as well as heterotypic adhesion of melanoma cells to dermal fibroblasts and vascular endothelial cells. Blocking of N-cadherin-mediated intercellular interaction by N-cadherin-specific antibodies increased the number of cells undergoing apoptosis. Furthermore, N-cadherin promoted migration of melanocytic cells over dermal fibroblasts, suggesting that N-cadherin may also play a role in metastasis.

2.2 Cellular Adhesion Molecules (CAMs)

Members of the immunoglobulin superfamily of adhesion molecules, including ICAM-1, ICAM-2, ICAM-3, VCAM-1, L1 (CD171), and MelCAM (MCAM, Muc-18) bind to integrins on leukocytes and mediate their flattening onto the blood vessel wall. This interaction is critical for subsequent extravasation and the further course of inflammatory reactions in the surrounding tissue. Interestingly, tumor cells have been shown to use similar mechanisms for extravasation during metastasis (9, 10, 11, 12). Alteration in the expression of invasion/metastasis-related melanoma cell adhesion molecule (MelCAM) is strongly associated with the acquisition of malignancy by human melanomas. MelCAM is expressed only sporadically in benign melanocytic nevi and thin primary melanomas with a low probability to metastasize. However, with increasing tumor thickness, MelCAM expression becomes more frequent and it is found in 80% of advanced primary tumors and metastases (13). ICAM is known to be strongly expressed by malignant melanoma cells and detection of soluble ICAM in the serum of melanoma patients can serve as a diagnostic marker. The results of Slattery and Dong suggest that transmigration of melanoma cells through endothelial barriers is mediated by ICAM-1 adhesive interactions (14).

Fogel et al. found an increase in L1 protein expression in malignant melanomas and metastases of malignant melanomas as compared to acquired melanocytic nevi that was statistically significant (15). They assumed that L1 plays a role in metastasis of malignant melanoma.

2.3 Galectins

Galectins represent a family of mammalian lectins with specificity to ß-galactosides, e.g. the poly-N-acetyllactosamine residue of laminin. Galectins are involved in growth-regulatory mechanisms but also in cell-cell and cell-matrix adhesion. It has been suggested that the level of galectin expression, mainly galectin-3, parallels malignancy in different cancer types (16, 17, 18). Galectin-3 was shown to bind specifically the poly-N-acetyllactosamine residues of laminin and has been implicated in tumor invasion and metastasis. Furthermore, galectin-1 might participate in melanoma cell adhesion to laminin and therefore could be a modulator of invasion and metastasis (19). *In vitro* studies suggested that galectin-3 protects cancer cells from apoptosis by promoting cell adhesion properties (20, 21) and also modulates their invasive capacity. Iurisci et al. assessed the circulating levels of galectin-3 in sera from cancer patients as well as from healthy controls. Low serum levels of galectin-3 were detected in healthy individuals whereas galectin-3 serum levels in patients with melanoma were significantly elevated (22). Galectin-9 protein is strongly and homogeneously expressed in melanocytic nevi, but is down-regulated in melanoma cells especially in metastatic lesions. In accordance, high galectin-9 expression was inversely correlated with the progression of this disease (23).

3. CELL-MATRIX ADHESION MOLECULES

3.1 CD44

CD44 is a receptor for hyaluronan and mediates signaling that regulates complex cell behavior including cancer cell migration and invasion. The main form of CD44 (CD44H), a 80 kD transmembrane glycoprotein, is widely expressed in a variety of lymphoid and epithelial cells and also in malignant tumors. CD44 has many variant forms (at least 45 are currently known), which are generated by alternative splicing. During the past years, expression of certain variant CD44 isoforms has been correlated with the degree of tumor differentiation, tumor cell invasion, and metastatic potential (24, 25). In model systems, interaction of CD44 and hyaluronan was a critical determinant for cell adhesion and transendothelial invasion (9). Other studies provided evidence for proteolytic cleavage of CD44 on the surface of cancer cells at the extracellular domain through membrane-associated metalloproteases. It has been suggested that CD44 cleavage plays an important role in CD44-mediated tumor cell migration (26) and in efficient cell-detachment from the hyaluronate substrate within extracellular matrix. In melanoma , overexpression of various CD44 splice variants was found to be correlated to metastasis (27, 28, 29). Yoshinari et al. assumed that CD44v10 expression is functionally related to melanoma cell migration and suggested that interaction between CD44v10 and HA plays a role in the variable tissue invasion and aggressiveness of different melanoma clones (30).

3.2 Laminin receptor

The 67kD laminin receptor belongs to the group of non-integrin adhesion molecules and interacts with the YIGSR sequence in the ß1 chain of laminin. Expression of the receptor correlates with invasive potential of malignant melanoma cells, which can be inhibited by the YIGSR amino acid peptide *in vitro* (31, 32). Givant-Horwitz et al. revealed the involvement of the laminin receptor in different mechanisms related to tumor dissemination in malignant melanoma. Furthermore, they provided first evidence of the involvement of MAPK and dual-specificity phosphatases in its signal transduction pathway (33).

4. MATRIX MOLECULES

4.1 Fibronectin

The 400-500kD homodimeric fibronectin protein represents an important component of extracellular matrix and basement membranes. *In vitro*, fibronectin is chemoattractive for many different types of cancer cells (34), and overexpression of fibronectin is frequently detected in invading areas of malignant tumors *in vivo* (35). Further, an effect of fibronectin expression on enhanced secretion of proteases, including uPA (urokinase-type plasminogen activator), was observed *in vitro* (36). As this effect was depended on RGD sequences and ß1-integrin expression, cell-matrix interactions seem to trigger enhanced protease secretion. Fibronectin was also suggested to be involved in cohort migration, a phenomenon relating to the observation that carcinoma cells frequently invade the surrounding tissue as coherent clusters or nests of cells (37).

4.2 Laminin

The basement membrane molecule laminin forms a family of proteins. Laminin is a cross shaped multifunctional glycoprotein formed by the multimeric assembly of subunits which result from activation of several genes. *In vivo*, depending on its location, and due to its adhesive properties and multivalent affinities, laminin is associated with compounds of the plasma, the basement membrane, and the cell coat as a part of supramolecular complexes. In the basement membrane laminin has structural and functional roles. It may also be adsorbed on the cell coat or secreted. Laminin-5 was identified as a key protein in the anchoring filaments of the basement membrane. The anchoring filaments connect the basement membrane to the epithelial cells and together form the epithelial adhesion complex. Laminin has been implicated in a number of stages in tumor invasion and metastasis. In addition to its roles in cell adhesion and migration, laminin may be important in mediating interactions of tumor cells with the immune system and may have further, more subtle roles in controlling metastatic behavior. Tumor invasion through the basement membrane has been correlated with a decreased expression of laminin. Laminin stimulates migration of several tumor cells tested (38) and laminin-5 is specifically degraded by MMP-2 to produce a soluble chemotactic fragment (39). Purified laminin-5 was capable of potentiating melanoma cell migration as measured in a chemotaxis assay with a soluble form of laminin-5 or a haptotaxis assay with membranes coated with a mixture of laminin-5 and Matrigel, respectively. Furthermore, immobilized laminin-5 induced A375 melanoma cells to secrete matrix metalloproteinase-9 (type IV collagenase) into the culture medium (40).

4.3 Proteoglycans

Abnormal expression of proteoglycans, highly glycosylated matrix molecules, has been implicated in cancer and metastasis primarily because these macromolecules are involved in control of cell growth and matrix assembly. Significant upregulation of perlecan expression, a major heparan sulfate proteoglycan of basement membranes and pericellular matrices, was detected in melanomas and other malignant tumors (41). Perlecan has also ben shown to be upregulated by the growth factor neutrophin (see VIII), suggesting a further role in early steps of invasion. Lumican is a member of the small leucine-rich proteoglycan (SLRP) family. It contributes to the organization of the collagen network and plays an important role in cell migration and tissue repair. The results of Vuillermoz B et al. (42) suggest that lumican is involved in the control of melanoma growth and invasion and may be considered, like decorin, as an anti-tumor factor from the extracellular matrix.

5. CYTOSKELETON

Over the past years, an increasing number of genes have been identified that modify the cytoskeleton and play an active role in cell motility and invasion. Oncogenic signaling can also be significantly influenced by cytoskeletal regulation. Thus, the cytoskeleton has evolved from a static structure to a dynamic modulator of signal transduction pathways. Intercellular communication and the active movement of malignant cells into and through host tissue barriers play a critical role during the complex process of tumor invasion. Motile activity, cytoskeletal actin and vinculin organization as well as gap junctional communication of *in vivo* benign and malignant melanocytes are related to *in vitro* invasiveness (43).

5.1 Keratins

The cytokeratins are intermediate filament proteins characteristic for epithelial cells. In human cells, some 20 different cytokeratin isotypes have been identified. Epithelial cells express between two and ten cytokeratin isotypes and the consequent profile which reflects both epithelial type and differentiation status may be useful in tumor diagnosis. In some cancers, particularly malignant melanoma and breast carcinoma, there is a strong indication that vimentin and keratin intermediate filaments are coexpressed, thus presenting as a dedifferentiated or interconverted (between epithelial and mesenchymal) cellular phenotype (44, 45, 46, 47). These studies have provided direct evidence linking overexpression of keratin IFs in human melanoma and breast carcinoma with increased migratory and invasive activity *in vitro* (48), which can be down-regulated by substituting dominant-negative keratin mutants (49, 50).

5.2 Rho, Rac

The Ras superfamily of GTPases comprises several subfamilies of small GTP-binding proteins whose functions include control of proliferation, differentiation, apoptosis, but also cytoskeletal organization. As the multistep process of invasion requires coordinate activation of migration, motility and cytoskeletal reorganization, the small GTPases Rho, Cdc42 and Rac (Rho-like GTPases) play a central role in this ability. Rho-like GTPases have been implicated in the orchestration of changes in the actin cytoskeleton in response to receptor stimulation, but have also been shown to be involved in transcriptional activation and cell cycle regulation (51, 52, 53). Moreover, they can induce oncogenic transformation in fibroblasts. Recently, RhoC has been shown to enhance metastasis when overexpressed in cells whereas dominant-negative Rho inhibits metastasis (54). Consistently, members of the Rho family are frequently overexpressed in breast cancer and malignant melanoma. Recent studies using the RhoC inhibitor Atorvastatin resulted in reversion of the metastatic phenotype of melanoma cells *in vitro* and consequently inhibited metastasis *in vivo* (55).

Among several proteins isolated as putative target molecules of Rho, the ROCK (or ROK) family of Rho-associated serine-threonine protein kinases is thought to participate in the induction of focal adhesions and stress fibers in cultured cells (56). As a specific ROCK inhibitor (Y-27632) blocked both Rho-mediated activation of actomyosin and invasive activity of cells ROCK also seems to play a role in tumor cell invasion (57). In addition, also other inhibitors like Wf-536 have been developed and shown to inhibit metastasis in *in vivo* melanoma models (58).

6. GROWTH FACTORS

Extracellular signaling via growth factors has been demonstrated to elicit deregulated cell motility in invasive tumor cells. Recent findings suggest that growth factor receptor-mediated motility is one of the most common alterations in malignant tumors, which is causally involved in acquisition of an invasive cell phenotype and represents a biological phenomenon distinct from adhesion-related haptokinetic and haptotactic migration (59). In addition, extracellular growth factor signaling was also shown to elicit upregulation of MMP expression in malignant tumors.

6.1 Epidermal Growth Factor (EGF) and Transforming Growth Factor alpha (TGF-alpha)

Transforming growth factor alpha (TGF-alpha) is a polypeptide growth factor sharing significant structural and functional homology with epidermal growth factor (EGF). TGF-alpha is synthesized as a 160 amino acid transmembrane precursor molecule from which the mature 50 amino acid TGF-alpha peptide is released after proteolytic cleavage by specific elastase-like enzymes. All biological properties of TGF-alpha and EGF are thought to be elicited by binding of the proteolytically processed ligands to transmembrane tyrosine kinase-coupled receptors of the EGF receptor family. Thus, the biochemical cascade following activation of EGF receptor tyrosine kinase is likely to be identical in response to either ligand. However, heparin-binding EGF-like growth factor (HB-EGF), an additional EGF family member, represents a much more potent mitogen for smooth muscle cells than either EGF or TGF, despite the fact that all three ligands bind to the same receptor. EGF was shown to stimulate tumor cell migration in nanomolare concentrations. This effect involved upregulation of MMP-2 and uPA expression (60, 61), which seems to be mediated by a phosphatidylinositol 3'-kinase and phospholipase C-dependent mechanism (62). Signaling of EGF to migratory pathways appears to be dissociated, at least in part, from the proliferative pathway. Therefore, EGF can elicit different biological responses depending on the cellular context and further downstream signal transduction cascades.

6.2 Hepatocyte Growth Factor (HGF)

Hepatocyte growth factor (HGF), also referred to as scatter factor (SC) or hepatopoietin, was originally identified as a pleiotropic growth factor from rat platelets. Mature HGF is an 82 kD heterodimeric glycoprotein of 674 amino acids that lacks significant homology with other known growth factors. The transmembrane tyrosine kinase-coupled receptor c-Met functions as the HGF-receptor. HGF is synthesized as a 728 amino acid precursor protein and is proteolytically processed into the mature growth factor consisting of a disulfide-linked 69 kD chain and a 34 kD chain. Signaling through the HGF/c-Met system elicits a cellular response linked to invasive growth in several ways. HGF was shown to promote cell adhesion to laminin, fibronectin and vitronectin through a PI3-K-dependent mechanism (63). Subsequently, increased adhesion induced by HGF is followed by increased invasiveness through these matrix proteins. This phenomenon seems to be regulated by triggering integrin-clustering (64) as well as by upregulation of MMP and uPA expression (65). Furthermore, elevation of tyrosine-phosphorylated ß-catenin after HGF treatment has been observed (66).

6.3 Transforming Growth Factor beta (TGF-ß)

The TGF superfamily comprehends a large number of polypeptide growth and differentiation factors, including transforming growth factors beta (TGF-ß) -1, -2, and -3 and further growth/differentiation factors (GDFs), Mullerian inhibiting substance (MIS), bone morphogenic proteins (BMPs), glial cell line-derived neurotrophic factor (GDNF), inhibins or activins, Lefty and Nodal. Members of the TGF superfamily are involved in embryonic development and adult tissue homeostasis. Growth regulatory factors of the TGF-ß family inhibit proliferation of epithelial, endothelial and hematopoietic cells and stimulate synthesis of extracellular matrix components. Recent evidence suggests that acquisition of resistance to TGF-ß-conferred growth inhibition plays a significant role in progression of malignant epithelial and hematopoietic tumors. Under these circumstances secretion of TGF-ß by tumor cells may promote invasion rather than inhibit growth (67). Consistently, upregulation of MMP-2 and MMP-9 by TGF-ß1 has been observed in different kinds of cancers (61).

6.4 Wnt

Products of the highly conserved Wnt gene family, including Wnt-1 through Wnt-10, play key roles in regulating cellular growth and differentiation (Figure 2). Wnt-1 is a cysteine-rich, secreted glycoprotein that associates with cell membranes and likely functions as a key regulator of cellular adhesion. Wnt-1, which is essential for normal development of the embryonic nervous system, contributes to hyperplasia and tumorigenic progression when improperly expressed in mammary tissue (68, 69).

Furthermore, Wnt-3 is involved in tumorigenesis and Wnt-2, Wnt-4 and Wnt-5a may be associated with abnormal proliferation in human malignancies (70, 71).

Based on microarray studies, Wnt-5a expression best determined the *in vitro* invasive behavior of malignant melanoma cells (72). Melanoma cells overexpressing Wnt-5a after transfection showed changes in actin reorganization and increased cell adhesion. In direct correlation with Wnt-5a expression and PKC activation, there was an increase in melanoma cell invasion. Furthermore, Wnt-5a expression in human melanoma biopsies directly correlated with increasing tumor grade. These observations indicate a critical role of Wnt-5a in human melanoma progression.

6.5 Insulin-like Growth Factor-1 (IGF-1)

Deregulated signaling contributes to altered cellular growth, motility, and survival during cancer progression. Neudauer and McCarthy (73) have evaluated the ability of several factors to stimulate migration in WM1341D, a cell line derived from an invasive human vertical growth phase melanoma. Basic fibroblast growth factor (bFGF), hepatocyte growth factor (HGF), interleukin-8, and CCL27 each slightly increased migration. Insulin-like growth factor I (IGF-I), however, elevated cell migration by 15-fold. This response required the IGF-I receptor, which activates phosphoinositide 3-kinase (PI3K) and mitogen-activated protein kinase/extracellular signal-regulated kinase (MAPK/ERK) pathways. Both pathways have been implicated in migration in a variety of cell types, but the signaling required for IGF-I-induced melanoma cell migration is not well defined. IGF-I-stimulated activation of MAPK/ERK signaling in WM1341D cells was inhibited by U0126 (a specific ERK inhibitor), but a 33-fold higher dose of U0126 was needed to inhibit IGF-I-stimulated cellular migration. In contrast, similar concentrations of either wortmannin or LY294002 (specific PI-3-kinase (PI3K) inhibitors) were required to inhibit both IGF-I-induced PI3K activation and migration. These results indicate that IGF-I-stimulated migration of WM1341D cells requires PI3K activation but is independent of MAPK/ERK signaling. Further inverstiogations analyzing the contributions of IGF-I signaling pathways to migration will help to understand melanoma progression and may identify new therapeutic targets of this highly metastatic cancer (73).

7. CYTOKINES / CHEMOKINES

Similar to their response to growth factors, invasive tumor cells have been demonstrated to reveal deregulated cell motility in response to extracellular signals from cytokines and chemokines. Since they represent polypeptide, frequently soluble molecules that interact with specific receptors on the cell surface the discrimination between growth factors and cytokines or chemokines is largely arbitrary and reflects the discovery of cytokines as modulators of immune responses.

7.1 Interleukins (IL)

Interleukins comprise a broad family of well characterized cytokines, primarily of hematopoietic cell origin. For example, IL-2 is secreted primarily by mitogen-activated T helper lymphocytes and functions as an autocrine growth factor, driving the clonal expansion of antigen specific cells. In contrast, IL-7 primarily stimulates proliferation of pre-B cells while IL-13 is a potent regulator of inflammatory and immune responses and acts to ensure the rapid onset of a Th2-like response. IL-15 is produced by a broad range of tissues and cell types, and shares many of its biological properties with IL-2. Four distinct interleukins, IL-2, IL-4, IL-7 and IL-15, functionally interact with the common IL-2R receptor subunit. Additional members of the interleukin family include IL-16, IL-17 and IL-18 (also designated IGIF). IL-1 receptor antagonist (IL-1ra) is a cytokine that inhibits IL-1 binding to interleukin receptors.

Cytokines such as IL-1 induce the synthesis and surface expression of adhesion molecules like ICAM-1 and ELAM-1 and contribute to autocrine and paracrine induction of pro-metastatic genes in cancer (74). Furthermore, *in vitro* the invasiveness of tumor cells can be enhanced by addition of recombinant IL-1 and reduced by adding inhibitory anti IL-1 antibodies (75). Moreover, IL-1 alpha induces enhanced migration of melanoma cells on fibronectin and has been related to an upward shift in the alpha 4 and alpha 5 integrin subunit expression (76). Interleukin 8, a chemokine, expression was shown to be enhanced in many different solid tumors including malignant melanomas (77) and to promote tumorigenicity and metastasis in prostate (78) and breast cancer (79). Specific upregulation of MMP-9 and MMP-2 expression and activation was described as a consequence of IL-8 overexpression resulting in increased invasion through Matrigel (78, 80). Recent studies using a fully human anti-IL-8 antibody in a mouse *in vivo* attempt showed a potential therapeutic strategy to control angiogenesis, growth, and metastasis of melanoma (81). Also enhanced levels of IL-6 expression correlate with tumor progression (82, 83) and upregulation of

MMP-2, MMP-9 and TIMP-1 (84). In the case of IL-10 a dose-dependent stimulating effect on glioma invasiveness in Boyden chamber assays was measured. However, this effect is mechanistically poorly understood since MMP expression levels remained unchanged after IL-10 treatment, but Marimastat, a synthetic MMP-inhibitor, markedly reduced IL-10 stimulated invasiveness (85).

7.2 Tumor Necrosis Factors (TNFs)

The proinflammatory polypeptides tumor necrosis factor alpha (TNF-alpha) (also designated cachectin) and TNF beta (also designated lymphotoxin) exhibit approximately 30% sequence homology and bind to a common receptor complex. TNF is produced as a precursor membrane-bound form, from which the soluble mature factor is derived by cleavage of the extracellular domain. The biologically active forms of both TNF-alpha and TNF-beta are non-covalently linked trimers. TNF-alpha may be involved in the enhancement of tumor invasion and metastasis in part by upregulating proteolytic enzymes such as MMPs and u-PA in tumor tissues (86). TNF-alpha tightly regulates Gelatinase B secretion in glioma cells, an enzyme which is believed to play an important role in the local invasion of brain tissue by tumor cells (87). These effects on tumor cell invasion have been shown for several types of cancer. TNF-alpha, similar as shown for IL-1, induces migration on fibronectin of melanoma cells and has been correlated to an induction of alpha4 and alpha5 integrin subunit expression (76). Zhu et al. (88) recently reported that TNF-alpha can upregulate integrin expression, cell attachment and invasion of cells through fibronectin in a human melanoma cell line (HBL). These data support the hypothesis that an inflammatory environment promotes melanoma cell migration (88). In line with this finding TNF autovaccination is a cheap and highly efficient alternative to block TNF and reduce melanoma metastasis in a murine *in vivo* model. Due to these findings, trials with TNF autovaccination are already underway in patients with metastatic cancer (89).

7.3 Chemokines

Chemokines are secreted into the tumor microenvironment by tumor-infiltrating inflammatory cells as well as by tumor cells themselves. Chemokine receptors mediate agonist-dependent cell responses, including migration and activation of several signaling pathways. Chemokines binding to their receptors provide chemotactic cues guiding cells to specific tissues and organs; they therefore could potentially participate in tumor cell dissemination. Melanoma cells express CXCR3 and CXCR4, the receptors for the chemokine stromal cell-derived factor-1alpha (SDF-1alpha) (90). The identification of SDF-1alpha as a potential stimulatory molecule for MT1-MMP as well as for RhoA and Rac1 activities during melanoma cell invasion, associated with an up-regulation in CXCR4 expression by interaction with basement membrane factors, could contribute to better knowledge of mechanisms stimulating melanoma cell dissemination (91). Furthermore, the chemokine SDF-1alpha, the ligand of CXCR4, triggeres modulation of beta1 integrin-dependent melanoma cell adhesion to fibronectin. The chemokine Mig, a ligand for CXCR3, activates the small GTPases RhoA and Rac1, induces a reorganization of the actin cytoskeleton, and triggers cell chemotaxis and modulation of integrin VLA-5- and VLA-4-dependent cell adhesion to fibronectin. Additionally, Mig and SDF-1alpha activated MAPKs p44/42 and p38 in melanoma cells. Expression of functional CXCR3 and CXCR4 receptors on melanoma cells indicate that both chemokines contribute to cell motility during invasion as well as to regulation of cell proliferation and survival (90).

8. PROTEASES

8.1 Cathepsins

Cathepsin A, B, H , D and L, all belonging to the family of cysteine endopeptidases, have been shown to participate in processes of tumor growth, vascularization, invasion and metastasis. In several different types of cancer, including malignant melanomas, cathepsins were found to be upregulated and useful as prognostic markers (92, 93, 94). Cathepsin B was shown to degrade extracellular matrix proteins like collagen IV and laminin, and to activate the precursor form of urokinase plasminogen activator (uPA), eventually thereby initiating an extracellular proteolytic cascade (95). It has been shown that the switch from a nonmetastatic to a highly metastatic phenotype of human melanoma cells is directly related to secretion of procathepsin L form (96). Overexpression of procathepsin-L in melanoma cells increases their tumorigenicity and switches their phenotype from nonmetastatic to highly metastatic cells. Additionally, the use of an anti-cathepsin L antibody inhibits the tumorigenic and metastatic phenotype of human melanoma and may therefore be used as a molecular tool in a therapeutic cellular approach (97). Further, Dennhofer et al. have shown that cysteine protease inhibitors (E-64 and leupeptin) significantly reduce invasion of melanoma cells into basement membranes and dermal composits. These results indicate an important role of cysteine proteases for tumor invasion (98)

8.2 Transmembranous proteases

Aminopeptidase N/CD13 (143 kD) and dipeptidyl peptidase IV/CD26 are Zn^{2+}-dependent ectopeptidases localized on the cell surface of a wide variety of cells. They are involved in tumor cell invasion and the formation of metastases. Both aminopeptidase N and dipeptidyl peptidase play an active role in degradation and invasion of ECM and may be involved in the molecular mechanisms of blood-borne metastasis (99, 100). Cellular migration correlates highly with aminopeptidase N activity. Further, the up-regulated invasion of cancer cells is inhibited by bestatin, a specific inhibitor of aminopeptidase N. These findings suggest that aminopeptidase N expression contributes to the invasive potential of human cancer cells (101).

Dipeptidyl peptidase IV (DPPIV) is a 110-kD, trans-membrane ectoenzyme, with ubiquitous expression. DPPIV has numerous functions including involvement in T-cell activation, cell adhesion, digestion of proline-containing peptides in the kidney and intestines, HIV infection and apoptosis, and regulation of tumorigenicity in certain melanoma cells. Dipeptidyl peptidase IV is a cell surface peptidase expressed by several normal cells including melanocytes. In contrast, malignant cells, including melanomas, frequently loose or alter DPPIV cell surface expression. Loss of DPPIV expression occurs during melanoma progression at a stage where transformed melanocytes become independent of exogenous growth factors for survival. DPPIV expression leads to a loss of tumorigenicity, anchorage-independent growth, a reversal in a block in differentiation, and an acquired dependence on exogenous growth factors for cell survival (102). Pethiyagoda et al. have demonstrated that expression of a proteolytically active form of the DPPIV protein inhibits the invasiveness of malignant melanoma cell lines lacking endogenous DPPIV expression. Furthermore, they have shown that neither the protease activity nor the cytoplasmic domain of DPPIV is required for its anti-invasive activity (103).

9. NON-PROTEOLYTIC ENZYMES

9.1 Heparanase

Endo-ß-D-glucuronidase is commonly referred to as heparanase. Enzymatic targets of the 50kD heparanases (543 aa) are heparan sulfate proteoglycans (HSPGs) in the extracellular matrix. Expression of heparanase correlates with the invasive potential of tumor cell lines (104, 105). Further, treatment with heparanase inhibitors markedly reduces the frequency of metastases in experimental animal models of tumor metastasis (106, 107). It was demonstrated that cell-surface syndecan-1 and ECM perlecan are degradative targets of heparanase, and syndecan-1 regulates heparanase biological activity. This suggests that expression of syndecan-1 on the melanoma cell surface and its degradation by heparanase are important determinants in the control of tumor cell invasion and metastasis (108). Elevated levels of heparanase are associated with the metastatic potential of melanoma cells. Treatment of murine and human melanoma cells with the prototypic neurotrophin nerve growth factor (NGF) increases the production of heparanase by melanoma cells. Some groups suggest that development of brain metastasis in malignant melanoma is due to this relation (109, 110, 111, 112).

9.2 Hyaluronidase

Hyaluronidases are broadly distributed enzymes with varying substrate specificities, a wide range of pH optima and different catalytic mechanisms (113). One of their substrates, hyaluronic acid, represents a major component of the extracellular matrix in brain (114) and soft tissues and has been identified as the ligand of CD44. Disruption of basement membrane integrity by hyaluronidase during cell invasion has been implicated in the development of metastatic carcinoma. The amount of hyaluronidase expression correlates with prognosis in a variety of different cancer types (115, 116, 117). Furthermore, experimental overexpression of hyaluronidase in tumor cells leads to accelerated tumor growth and formation of highly vascularized and more invasive tumors also shown for malignant melanomas (118, 119, 120).

10. INTRACELLULAR MOLECULES OF SIGNAL TRANSDUCTION PATHWAYS

10.1 Focal Adhesion Kinase (FAK)

Integrins are cell surface molecules which promote adhesion of cells to the extracellular matrix. In addition to providing a molecular "glue" essential for tissue organization and survival, integrins are dynamic signaling molecules. They allow cells to sense adhesion to the extracellular matrix and the nature of the specific matrix composition (for details see chapters 7 and 8). Thereby, a cell survival signal is provided preventing a particular form of cell death in detached cells which has been referred to as anoikis (121). An important mediator of this signal is the focal adhesion kinase (FAK), which becomes phosphorylated and activated during integrin-mediated cell adhesion. Initially, FAK was identified as a major 125 kD substrate for the intrinsic protein tyrosine kinase activity of Src encoded pp60. Visualization of p125 by immunofluorescence revealed its preferred subcellular localization in focal adhesions, leading to its designation as focal adhesion kinase (FAK). FAK is concentrated at the basal edge of focal contacts of cells that are actively migrating and spreading.

FAK has been shown to be frequently overexpressed in many different kinds of malignant tumors as colon and breast cancer as well as in malignant melanoma (122, 123, 124). FAK overexpression causes increased cell migration whereas cells derived from FAK -/- mice exhibit reduced migratory capabilities as compared to wild type cells. FAK is thought to exert a dual function with respect to invasion: Overexpression leads to increased cell migration and increased cell survival under conditions of anchorage-independent growth (125). FAK also seems to be important for enhanced MMP-9 expression after exposure of ovarian cancer cells to fibronectin (126).

10.2 ß-catenin

The catenins (alpha, beta and gamma) are ubiquitously expressed, cytoplasmic proteins that associate with E-cadherin at cellular junctions (Figure 2). The p85 kD protein ß-catenin was initially described as an E-cadherin associated protein. However, subsequent studies have shown that ß-catenin can also bind P-cadherin and N-cadherin. In addition to its ability to bind to cadherins, ß-catenin has been shown to co-immunoprecipitate with APC. ß-catenin is a major component of desmosomes, where it is complexed with desmoglein. Increased tyrosine phosphorylation can disrupt catenin-cadherin complexes and thereby attenuate cellular adhesion. ß-catenin also belongs to the important downstream effectors of the wnt signaling pathway (127) and has been implicated in two major biological processes of vertebrates: early embryonic development and tumorigenesis (128, 129). GSK-3ß and other kinases destabilize ß-catenin by phosphorylation at serine 33, 37, 45 and threonine 41 (130). Stabilizing mutations of these phosphorylation sites in ß-catenin were detected in many tumor cell lines *in vitro* (131) and subsequently in malignant tumors *in vivo* (132). In addition to binding to cadherins, beta-catenin also interacts with transcription factors of the TCF-subfamily of HMG box proteins and regulates their activity. Furthermore, ß-catenin has been shown to be involved in MMP-7 upregulation in colorectal cancer (133). Elevated levels of ß-catenin, sometimes concentrated in the invasive front, have been described in several kinds of cancer (134). Also in melanoma, the induction of Nr-CAM transcription by beta-catenin or plakoglobin seems to play a role in tumorigenesis, probably by promoting cell growth and motility (135).

10.3 Integrin-linked Kinase (ILK)

Integrin-linked kinase (ILK), a key component of the extracellular matrix adhesion, has been studied extensively in recent years. Overexpression of ILK in epithelial cells results in anchorage-independent cell growth with increased cell cycle progression. Furthermore, increased ILK expression is correlated with progression of several human tumor types, including breast, prostate, and colon carcinomas. Also in melanoma, ILK expression increases dramatically with invasion and progression and has been shown to be inversely correlated with patient survival (136).

11. RECENTLY IDENTIFIED GENES WITH FUNCTION IN MOTILITY

11.1 Melanoma Inhibitory Activity (MIA)

The melanoma inhibitory activity (MIA) protein was identified within growth-inhibiting activities purified from tissue culture supernatant of the human melanoma cell line HTZ-19d (137). It is expressed as a 131 amino acid precursor molecule and processed into a mature 107 amino acid protein after cleavage of a secretion signal. MIA provides a clinically useful parameter in patients with metastatic melanoma stages III and IV (138,139,140,141). Expression of the wild-type MIA protein gene is not detected in normal skin and melanocytes but has been shown to be associated with progression of melanocytic tumors (138,142). More recently, it has been suggested that the MIA protein specifically inhibits attachment of melanoma cells to fibronectin and laminin, thereby masking the binding site of integrins to these extracellular matrix (ECM) components and promoting migration *in vitro*, and invasion and metastasis *in vivo* (Figure 3) (138, 143, 144, 145).

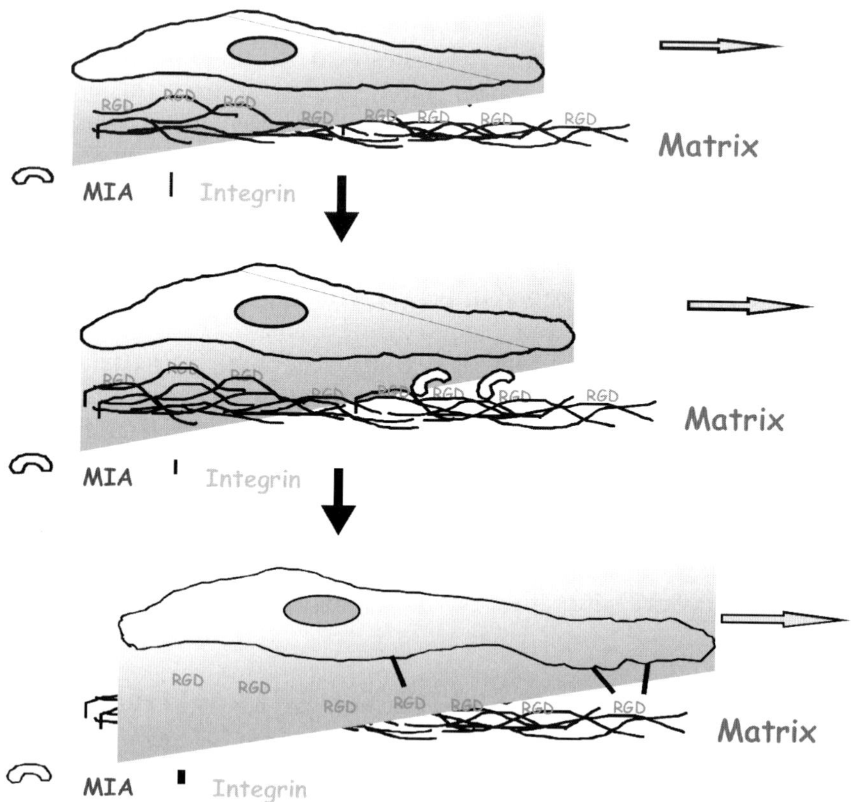

Figure 3. Function of the molecule MIA in migration of melanoma cells. MIA represents the first extracellular protein adopting a SH3 domain-like structure. This structure is stabilized by two disulfide bonds. MIA binds specifically to fibronectin and laminin. By blocking the integrin binding sites on these matrix molecules it leads to cell detachment. In consequence, melanoma cells have a higher ability to migrate (see 11.1 for details).

Interestingly, analysis of the 3D structure of recombinant human MIA by multidimensional NMR spectroscopy revealed that MIA is the first extracellular protein known to adopt an SH3 domain-like fold (146). These studies also provide evidence of specific interaction between a binding fold of MIA and a partial fibronectin peptide that has been implicated in integrin binding. Therefore, it can be concluded that MIA may play a role in tumor progression and spread of malignant melanomas via mediating active detachment of cells from extracellular matrix molecules within their local milieu. Inhibiting MIA functions *in vivo* may provide a novel therapeutic strategy for metastatic melanoma disease (145).

11.2 T-cell lymphoma Invasion and Metastasis (TIAM)

Studying an *in vitro* model system of ***T***-cell-lymphoma ***i***nvasion ***a***nd ***m***etastasis identified a novel protein, Tiam1, which shares a Dbl-homology domain with GDP dissociation stimulator (GDS) proteins. *In vitro* TIAM enhances invasion and metastasis of mouse T-cell-lymphoma cells (51). TIAM functions as a guanine nucleotide exchange factor for the Rho-like GTPase Rac1, a member of the ras superfamily of small GTP-binding proteins (see XI.3), and thereby induces Rac1 signaling (147). Binding to the ankyrin repeat domain of ankyrin activates GDP/GTP exchange on Rho GTPases (147). Via ankyrin binding and Rac1 activation tumor cell invasion and migration are activated *in vitro*. However, so far *in vivo* studies are merely descriptive showing TIAM expression in several tumors including malignant melanomas (148, 149, 150). Further studies have to elucidate the functional role of TIAM in tumorigenesis *in vivo*.

11.3 Secreted Protein, Acidic and Rich in Cysteine (SPARC)

SPARC (secreted protein, acidic and rich in cysteine, also known as osteonectin or BM40) is a matricellular protein that modulates cell adhesion, migration and growth and is thought to play an important role in tissue remodeling and angiogenesis. Alterations of SPARC expression have been observed in a variety of solid tumors including malignant melanomas (151, 152, 153, 154). Overexpression *in vitro* indicated that SPARC contributes to increased motility, invasion and altered adhesion (155). Mechanistically, binding of SPARC to collagen type IV has been proposed (156) but other investigators have postulated activation of MMP-2 (157). A further study reported that the suppression of SPARC expression by human melanoma cells using a SPARC antisense expression vector results in a significant decrease in the *in vitro* adhesive and invasive capacities of tumor cells, completely abolishing their *in vivo* tumorigenicity. This finding is the first evidence that SPARC plays a functional role in the development of a invasive/metastatic phenotype in human melanoma (158, 159). Already before, it has been shown that SPARC expression is a predictor of clinical outcome in thin cutaneous melanomas (158).

11.4 Amphoterin / RAGE

Amphoterin (HMG1) is a 30-kD heparin-binding protein which is functionally associated with the outgrowth of cytoplasmic processes in developing neurons. Amphoterin has been shown to mediate adhesive and proteolytic interactions at the leading edge of motile cells. Recently, it has been shown that inhibition of amphoterin interactions with its cell surface receptor (RAGE) suppresses tumor growth and metastasis. Furthermore, enhanced expression of the high-mobility-group protein amphoterin/HMG1 correlates with increasing invasive potential and progression of tumors (160).

RAGE, the receptor for advanced glycation end products, is a multiligand member of the immunoglobulin superfamily of cell surface molecules. RAGE interacts with distinct molecules implicated in homeostasis, development and inflammation. Engagement of RAGE by a ligand triggers activations of key signaling pathways including ras-signaling to MAP kinases, NF-kappaB and cdc42/rac activation (161). RAGE functions also as a cell surface receptor for amphoterin. Inhibition of amphoterin-RAGE interaction suppresses activation of p44/p42, p38 and SAP/JNK MAP kinases and decreases growth of metastases as assayed in *in vivo* models (161). Moereover, intracellular signaling pathways regulated by RAGE are important molecular effector mechanisms linked to tumor proliferation, invasion and expression of MMPs.

A c-terminal motif in amphoterin (amino acids 150-183) is responsible for RAGE binding (162). This part of amphoterin is sufficient to induce RAGE-dependent process extension, suggesting a role in the regulation of cell motility. When applied in solution, the RAGE-binding COOH-terminal motif of amphoterin efficiently inhibits process extension and transendothelial migration of tumor cells (162). Specific decrease of amphoterin mRNA and protein, using antisense oligonucleotides transfected into cells, inhibits cell migration to laminin in a transfilter assay. Moreover, affinity-purified anti-amphoterin antibodies inhibited cell migration to laminin, supporting an extracellular role for the endogenous amphoterin in cell motility. The finding that amphoterin expression is more pronounced in cells with a motile phenotype as compared to cells of dense cultures, is consistent with the results of the cell migration assays. These results strongly suggested that amphoterin is a key player in the migration of immature and transformed cells (163).

11.5 Eph receptors / Ephrins

Eph receptor tyrosine kinases and ephrins (ligand) regulate morphogenesis in the developing embryo where they effect adhesion and motility of interacting cells (Figure 4). The ligands belonging to the ephrin A family are bound to the cell menbrane via GPI-anchor whereas ephrin Bs are linked via transmembrane domain. The Eph receptors are classified into two subfamilies, EphA and EphB, based on sequence homology and binding affinity for the respective ligands. Receptors and ligands are known to participate in bi-diectional signaling. Although scarcely expressed in adult tissues, Eph receptors and ephrins are overexpressed in a range of tumors. In malignant melanoma, increased Eph and ephrin expression levels correlate with metastatic progression (164, 165). Moreover, strong expression of ephrin-A1 and EphA2 in tumor cells is associated with increased melanoma thickness, and ephrin-A1 staining is related to decreased survival (166). Expression of EphA2 in tumor cells has been shown to be associated with increased tumor cell proliferation (Ki-67 positivity), indicating possible autocrine growth stimulation.

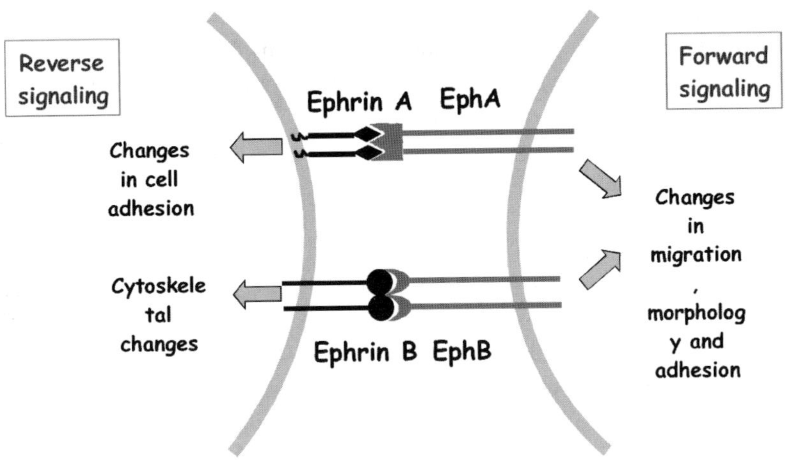

Figure 4. Schematic model of Eph/Ephrins signaling. Model of forward and reverse signaling of Eph receptors and their ligands Ephrins (see 11.5 for details).

11.6 Endothelins / Endothelin receptors

Endothelin receptors are G-protein-coupled receptors with seven membrane-spanning domains and are involved in various physiological processes in adults. The conjunction of recent studies in mammals focused on the action of endothelin 3 in controlling both the emergence and the maintenance of the neural crest-derived melanocyte phenotype (167) and proved the endothelin system to play an important role. Furthermore, ET-1 has been found to be a weak mitogen for melanoma cells, however, melanoma cell chemokinesis was significantly increased by ET-1. These data suggest that ET-1 may be involved in providing a chemokinetic and growth factor environment that enhances perivascular proliferation and invasiveness of melanoma cells (168). Phenotypic and genotypic analyses of cutaneous melanoma have identified the endothelin B receptor (ET(B)R) as tumor progression marker, thus representing a potential therapeutic target (169). Activation of ET(B)R by endothelin-1 (ET-1) and ET-3 leads to loss of expression of the cell adhesion molecule E-cadherin and associated catenin proteins and gain of N-cadherin expression. Exposure of melanoma cells to ET-1 leads to a 60% inhibition in intercellular communication by inducing phosphorylation of gap junctional protein connexin 43. Additionally, activation of the ET(B)R pathway increases alphav beta3 and alpha2 beta1 integrin expression, matrix metalloproteinase (MMP)-2, MMP-9, and membrane type-1-MMP activation, as well as tissue inhibitor of MMP-2 secretion. The ET(B)R pathway results in downstream activation of focal adhesion kinase and extracellular signal-regulated kinase 1/2 signaling pathways, which lead to enhanced cell proliferation, adhesion, migration, and MMP-dependent invasion (170, 171) (see also 10.1).

12. DISCUSSION

Although many pieces of evidence linking molecular pathways to melanoma cell motility have been gathered, still much of the information remains isolated. Nonetheless, a network of interactions emerges linking extracellular and cell surface molecules to cytoplasmic signal transduction pathways and nuclear regulation of gene expression patterns (Figure 1).

This network is constantly subject of modification and further detailing by current application of microarrays and other novel techniques which lead to detection of a wide variety of new molecules relevant for invasion. After completion of the human genome project and with the new dimension of proteomics, even more new genes and the respective proteins will be isolated by methods like 2-dimensional electrophoresis, peptide sequencing and mass-spectrometry. However, much effort will be needed to understand the biological significance of this novel information and to unravel the precise interactions and molecular pathways. As Jules Henri Poincare said more than a century ago:

> "Science is built up with facts, as a house is with stones. But a collection of facts is no more science than a heap of stones is a house".

Cell motility is not a separated tumor cell function but closely integrated in mechanisms like proliferation, immune response, apoptosis, cell-cell-contact und others. Therefore, a precise understanding of the metastatic process will lead to fundamental basal knowledge in tumor cell biology. This will offer new possibilities for therapeutic option in tumor treatment.

Some of the proteins discussed in this review are already used as marker molecules in tumor pathology. In future they could also serve as prognostic markers to distinguish tumors which are highly likely to metastasize from those that are not. Furthermore, these proteins provide potential targets for novel anti-tumor therapies.

ACKNOWLEDGEMENTS

I am indebted to Claus Hellerbrand and Marion Schuierer for critical discussions and proofreading of the manuscript.

REFERENCES

1. Saeki Y, Hazeki K, Matsumoto M, Toyoshima K, Akedo H, Seya T. Correlation between metastatic potency and the down-regulation of E-cadherin in the mouse hepatoma cell lines G-1 and G-5. Oncol. Rep. 2000, 7: 731-735.
2. Mialhe A, Levacher G, Champelovier P, Martel V, Serres M, Knudsen K, Seigneurin D. Expression of E-, P-, n-cadherins and catenins in human bladder carcinoma cell lines. J. Urol. 2000, 164: 826-835.
3. Poser I, Dominguez D, de Herreros AG, Varnai A, Buettner R, Bosserhoff AK. Loss of E-cadherin expression in melanoma cells involves up-regulation of the transcriptional repressor Snail. J. Biol. Chem. 2001, 276: 24661-24666.
4. Luo J, Lubaroff DM, Hendrix MJ. Suppression of prostate cancer invasive potential and matrix metalloproteinase activity by E-cadherin transfection. Cancer Res. 1999, 59: 3552-3556.
5. Hsu MY, Meier FE, Nesbit M, Hsu JY, Van Belle P, Elder DE, Herlyn M. E-cadherin expression in melanoma cells restores keratinocyte-mediated growth control and down-regulates expression of invasion-related adhesion receptors. Am. J. Pathol. 2000, 156: 1515-1525.

6. Qian X, Karpova T, Sheppard AM, McNally J, Lowy DR. E-cadherin-mediated adhesion inhibits ligand-dependent activation of diverse receptor tyrosine kinases. EMBO J. 2004, 23: 1739-1784.
7. Yoshimura M, Ihara Y, Matsuzawa Y, Taniguchi N. Aberrant glycosylation of E-cadherin enhances cell-cell binding to suppress metastasis. J. Biol. Chem. 1996, 271: 13811-13815.
8. Li G, Satyamoorthy K, Herlyn M. N-cadherin-mediated intercellular interactions promote survival and migration of melanoma cells. Cancer Res. 2001, 61: 3819-3825.
9. Okada T, Hawley RG, Kodaka M, Okuno H. Significance of VLA-4-VCAM-1 interaction and CD44 for transendothelial invasion in a bone marrow metastatic myeloma model. Clin. Exp. Metastasis 1999, 17: 623-629.
10. Sun JJ, Zhou XD, Liu YK, Tang ZY, Feng JX, Zhou G, Xue Q, Chen J. Invasion and metastasis of liver cancer: expression of intercellular adhesion molecule 1. J. Cancer Res. Clin. Oncol. 1999, 125: 28-34.
11. Umansky V, Schirrmacher V, Rocha M. New insights into tumor-host interactions in lymphoma metastasis. J. Mol. Med. 1996, 74: 353-363.
12. Koyama S, Ebihara T, Fukao K. Expression of intercellular adhesion molecule 1 (ICAM-1) during the development of invasion and/or metastasis of gastric carcinoma. J. Cancer Res. Clin. Oncol. 1992, 118: 609-614.
13. Lehmann JM, Riethmuller G, Johnson JP. MUC18, a marker of tumor progression in human melanoma, shows sequence similarity to the neural cell adhesion molecules of the immunoglobulin superfamily. Proc. Natl. Acad. Sci. USA 1989, 86: 9891-9895.
14. Slattery MJ, Dong C. Neutrophils influence melanoma adhesion and migration under flow conditions. Int. J. Cancer 2003, 106: 713-722.
15. Fogel M, Mechtersheimer S, Huszar M, Smirnov A, Abu-Dahi A, Tilgen W, Reichrath J, Georg T, Altevogt P, Gutwein P. L1 adhesion molecule (CD 171) in development and progression of human malignant melanoma. Cancer Lett. 2003, 189: 237-247.
16. Camby I, Belot N, Rorive S, Lefranc F, Maurage CA, Lahm H, Kaltner H, Hadari Y, Ruchoux MM, Brotchi J, Zick Y, Salmon I, Gabius HJ, Kiss R. Galectins are differentially expressed in supratentorial pilocytic astrocytomas, astrocytomas, anaplastic astrocytomas and glioblastomas, and significantly modulate tumor astrocyte migration. Brain Pathol. 2001, 11: 12-26.
17. Kawachi K, Matsushita Y, Yonezawa S, Nakano S, Shirao K, Natsugoe S, Sueyoshi K, Aikou T, Sato E. Galectin-3 expression in various thyroid neoplasms and its possible role in metastasis formation. Hum. Pathol. 2000, 31: 428-433.
18. Nakamura M, Inufusa H, Adachi T, Aga M, Kurimoto M, Nakatani Y, Wakano T, Nakajima A, Hida JI, Miyake M, Shindo K, Yasutomi M. Involvement of galectin-3 expression in colorectal cancer progression and metastasis. Int. J. Oncol. 1999, 15: 143-148.
19. van den Brule FA, Buicu C, Baldet M, Sobel ME, Cooper DN, Marschal P, Castronovo V. Galectin-1 modulates human melanoma cell adhesion to laminin. Biochem. Biophys. Res. Commun. 1995, 209: 760-767.
20. Matarresea P, Tinari N, Semeraroa ML, Natolib C, Iacobelli S, Malorni W. Galectin-3 overexpression protects from cell damage and death by influencing mitochondrial homeostasis. FEBS Lett. 2000, 473: 311-315.

21. Matarrese P, Fusco O, Tinari N, Natoli C, Liu FT, Semeraro ML, Malorni W, Iacobelli S. Galectin-3 overexpression protects from apoptosis by improving cell adhesion properties. Int. J. Cancer 2000, 85: 545-554.
22. Iurisci I, Tinari N, Natoli C, Angelucci D, Cianchetti E, Iacobelli S. Concentrations of galectin-3 in the sera of normal controls and cancer patients. Clin. Cancer Res. 2000, 6: 1389-1393.
23. Kageshita T, Kashio Y, Yamauchi A, Seki M, Abedin MJ, Nishi N, Shoji H, Nakamura T, Ono T, Hirashima M. Possible role of galectin-9 in cell aggregation and apoptosis of human melanoma cell lines and its clinical significance. Int. J. Cancer 2002, 99: 809-816.
24. Kuryu M, Ozaki T, Nishida K, Shibahara M, Kawai A, Inoue H. Expression of CD44 variants in osteosarcoma. J. Cancer Res. Clin. Oncol. 1999, 125: 646-652.
25. Ladeda V, Aguirre GJ, Bal dKJ. Function and expression of CD44 during spreading, migration, and invasion of murine carcinoma cells. Exp. Cell Res. 1998, 242: 515-527.
26. Okamoto I, Kawano Y, Tsuiki H, Sasaki J, Nakao M, Matsumoto M, Suga M, Ando M, Nakajima M, Saya H. CD44 cleavage induced by a membrane-associated metalloprotease plays a critical role in tumor cell migration. Oncogene 1999, 18: 1435-1446.
27. Dome B, Somlai B, Ladanyi A, Fazekas K, Zoller M, Timar J. Expression of CD44v3 splice variant is associated with the visceral metastatic phenotype of human melanoma. Virchows Arch. 2001, 439: 628-635.
28. Ranuncolo SM, Ladeda V, Gorostidy S, Morandi A, Varela M, Lastiri J, Loria D, Del Aguila R, Joffe EB, Pallotta G, Puricelli L. Expression of CD44s and CD44 splice variants in human melanoma. Oncol. Rep. 2002, 9: 51-56.
29. Seiter S, Schadendorf D, Herrmann K, Schneider M, Rosel M, Arch R, Tilgen W, Zoller M. Expression of CD44 variant isoforms in malignant melanoma. Clin. Cancer Res. 1996, 2: 447-456.
30. Yoshinari C, Mizusawa N, Byers HR, Akasaka T. CD44 variant isoform CD44v10 expression of human melanoma cell lines is upregulated by hyaluronate and correlates with migration. Melanoma Res. 1999, 9: 223-231.
31. Iwamoto Y, Nomizu M, Yamada Y, Ito Y, Tanaka K, Sugioka Y. Inhibition of angiogenesis, tumour growth and experimental metastasis of human fibrosarcoma cells HT1080 by a multimeric form of the laminin sequence Tyr-Ile-Gly-Ser-Arg (YIGSR). Br. J. Cancer 1996, 73: 589-595.
32. Graf J, Iwamoto Y, Sasaki M, Martín GR, Kleinman HK, Robey FA, Yamada Y. Identification of an amino acid sequence in laminin mediating cell attachment, chemotaxis, and receptor binding. Cell 1987, 48: 989-996.
33. Givant-Horwitz V, Davidson B, Reich R. Laminin-induced signaling in tumor cells: the role of the M(r) 67,000 laminin receptor. Cancer Res. 2004, 64: 3572-3579.
34. Ito H, Miyazaki M, Nishimura F, Nakajima N. Secretion of extracellular matrix (fibronectin), growth factor (transforming growth factor beta) and protease (cathepsin D) by hepatoma cells. Oncology 2000, 58: 261-270.
35. Ryu S, Jimi S, Eura Y, Kato T, Takebayashi S. Strong intracellular and negative peripheral expression of fibronectin in tumor cells contribute to invasion and metastasis in papillary thyroid carcinoma. Cancer Lett. 1999, 146: 103-109.
36. Urtreger AJ, Aguirre GJ, Werbajh SE, Puricelli LI, Muro AF, Bal dKJ. Involvement of fibronectin in the regulation of urokinase production and binding in murine mammary tumor cells. Int. J. Cancer 1999, 82: 748-753.

37. Shimao Y, Nabeshima K, Inoue T, Koono M. Role of fibroblasts in HGF/SF-induced cohort migration of human colorectal carcinoma cells: fibroblasts stimulate migration associated with increased fibronectin production via upregulated TGF-beta1. Int. J. Cancer 1999, 82: 449-458.
38. Hunt G, Sherbet GV. Effects of laminin on the attachment of glioma cells to type IV collagen. Clin. Exp. Metastasis 1989, 7: 353-359.
39. Kleiner DE, Stetler-Stevenson WG. Matrix metalloproteinases and metastasis. Cancer Chemother. Pharmacol. 1999, 43 Suppl:S42-51: S42-S51.
40. Tsuji T, Kawada Y, Kai-Murozono M, Komatsu S, Han SA, Takeuchi K, Mizushima H, Miyazaki K, Irimura T. Regulation of melanoma cell migration and invasion by laminin-5 and alpha3beta1 integrin (VLA-3). Clin. Exp. Metastasis 19: 127-134.
41. Cohen IR, Murdoch AD, Naso MF, Marchetti D, Berd D, Iozzo RV. Abnormal expression of perlecan proteoglycan in metastatic melanomas. Cancer Res. 1994, 54: 5771-5774.
42. Vuillermoz B, Khoruzhenko A, D'Onofrio MF, Ramont L, Venteo L, Perreau C, Antonicelli F, Maquart FX, Wegrowski Y. The small leucine-rich proteoglycan lumican inhibits melanoma progression. Exp. Cell Res. 2004, 296: 294-306.
43. Helige C, Zellnig G, Hofmann-Wellenhof R, Fink-Puches R, Smolle J, Tritthart HA. Interrelation of motility, cytoskeletal organization and gap junctional communication with invasiveness of melanocytic cells in vitro. Invasion Metastasis 1997, 17: 26-41.
44. Kirschmann DA, Seftor EA, Nieva DR, Mariano EA, Hendrix MJ. Differentially expressed genes associated with the metastatic phenotype in breast cancer. Breast Cancer Res. Treat. 1999, 55: 127-136.
45. Shinohara M, Hiraki A, Ikebe T, Nakamura S, Kurahara S, Shirasuna K, Garrod DR. Immunohistochemical study of desmosomes in oral squamous cell carcinoma: correlation with cytokeratin and E-cadherin staining, and with tumour behaviour. J. Pathol. 1998, 184: 369-381.
46. Nakayama S, Sasaki A, Mese H, Alcalde RE, Matsumura T. Establishment of high and low metastasis cell lines derived from a human tongue squamous cell carcinoma. Invasion Metastasis 1998, 18: 219-228.
47. Chu YW, Yang PC, Yang SC, Shyu YC, Hendrix MJ, Wu R, Wu CW. Selection of invasive and metastatic subpopulations from a human lung adenocarcinoma cell line. Am. J. Respir. Cell Mol. Biol. 1997, 17: 353-360.
48. Hembrough TA, Kralovich KR, Li L, Gonias SL. Cytokeratin 8 released by breast carcinoma cells in vitro binds plasminogen and tissue-type plasminogen activator and promotes plasminogen activation. Biochem. J. 1996, 317: 763-769.
49. Hendrix MJ, Seftor EA, Seftor RE, Trevor KT. Experimental co-expression of vimentin and keratin intermediate filaments in human breast cancer cells results in phenotypic interconversion and increased invasive behavior. Am. J. Pathol. 1997, 150: 483-495.
50. Hendrix MJ, Seftor EA, Chu YW, Trevor KT, Seftor RE. Role of intermediate filaments in migration, invasion and metastasis. Cancer Metastasis Rev. 1996, 15: 507-525.
51. Michiels F, Collard JG. Rho-like GTPases: their role in cell adhesion and invasion. Biochem. Soc. Symp. 1999, 65:125-46.
52. Evers EE, van der Kammen RA, ten Klooster JP, Collard JG. Rho-like GTPases in tumor cell invasion. Methods Enzymol. 2000, 325: 403-415.
53. Evers EE, Zondag GC, Malliri A, Price LS, ten Klooster JP, van der Kammen RA, Collard JG. Rho family proteins in cell adhesion and cell migration. Eur. J. Cancer 36: 1269-1274.

54. Clark EA, Golub TR, Lander ES, Hynes RO. Genomic analysis of metastasis reveals an essential role for RhoC. Nature 2000, 406: 532-535.
55. Collisson EA, Kleer C, Wu M, De A, Gambhir SS, Merajver SD, Kolodney MS. Atorvastatin prevents RhoC isoprenylation, invasion, and metastasis in human melanoma cells. Mol. Cancer Ther. 2003, 2: 941-948.
56. Suyama E, Kawasaki H, Kasaoka T, Taira K. Identification of genes responsible for cell migration by a library of randomized ribozymes. Cancer Res. 2003, 63: 119-124.
57. Itoh K, Yoshioka K, Akedo H, Uehata M, Ishizaki T, Narumiya S. An essential part for Rho-associated kinase in the transcellular invasion of tumor cells. Nat. Med. 1999, 5: 221-225.
58. Nakajima M, Hayashi K, Egi Y, Katayama K, Amano Y, Uehata M, Ohtsuki M, Fujii A, Oshita K, Kataoka H, Chiba K, Goto N, Kondo T. Effect of Wf-536, a novel ROCK inhibitor, against metastasis of B16 melanoma. Cancer Chemother. Pharmacol. 2003, 52: 319-324.
59. Ueda M, Ueki M, Terai Y, Ueki K, Kumagai K, Fujii H, Yoshizawa K, Nakajima M. Biological implications of growth factors on the mechanism of invasion in gynecological tumor cells. Gynecol. Obstet. Invest. 1999, 48: 221-228.
60. Ueda M, Fujii H, Yoshizawa K, Terai Y, Kumagai K, Ueki K, Ueki M. Effects of EGF and TGF-alpha on invasion and proteinase expression of uterine cervical adenocarcinoma OMC-4 cells. Invasion Metastasis 1998, 18: 176-183.
61. Rooprai HK, Rucklidge GJ, Panou C, Pilkington GJ. The effects of exogenous growth factors on matrix metalloproteinase secretion by human brain tumour cells. Br. J. Cancer 2000, 82: 52-55.
62. Price JT, Tiganis T, Agarwal A, Djakiew D, Thompson EW. Epidermal growth factor promotes MDA-MB-231 breast cancer cell migration through a phosphatidylinositol 3′-kinase and phospholipase C-dependent mechanism. Cancer Res. 1999, 59: 5475-5478.
63. Jiang W, Hiscox S, Matsumoto K, Nakamura T. Hepatocyte growth factor/scatter factor, its molecular, cellular and clinical implications in cancer. Crit. Rev. Oncol. Hematol. 1999, 29: 209-248.
64. Trusolino L, Cavassa S, Angelini P, Ando M, Bertotti A, Comoglio PM, Boccaccio C. HGF/scatter factor selectively promotes cell invasion by increasing integrin avidity. FASEB J. 2000, 14: 1629-1640.
65. Rosenthal EL, Johnson TM, Allen ED, Apel IJ, Punturieri A, Weiss SJ. Role of the plasminogen activator and matrix metalloproteinase systems in epidermal growth factor- and scatter factor-stimulated invasion of carcinoma cells. Cancer Res. 1998, 58: 5221-5230.
66. Hiscox S, Jiang WG. Association of the HGF/SF receptor, c-met, with the cell-surface adhesion molecule, E-cadherin, and catenins in human tumor cells. Biochem. Biophys. Res. Commun. 1999, 261: 406-411.
67. Taipale J, Saharinen J, Keski-Oja J. Extracellular matrix-associated transforming growth factor-beta: role in cancer cell growth and invasion. Adv. Cancer Res. 1998, 75:87-134: 87-134.
68. Naylor S, Smalley MJ, Robertson D, Gusterson BA, Edwards PA, Dale TC. Retroviral expression of Wnt-1 and Wnt-7b produces different effects in mouse mammary epithelium. J. Cell Sci. 2000, 113: 2129-2138.

69. Vider BZ, Zimber A, Chastre E, Prevot S, Gespach C, Estlein D, Wolloch Y, Tronick SR, Gazit A, Yaniv A. Evidence for the involvement of the Wnt 2 gene in human colorectal cancer. Oncogene 1996, 12: 153-158.
70. Iozzo RV, Eichstetter I, Danielson KG. Aberrant expression of the growth factor Wnt-5A in human malignancy. Cancer Res. 1995, 55: 3495-3499.
71. Lejeune S, Huguet EL, Hamby A, Poulsom R, Harris AL. Wnt5a cloning, expression, and up-regulation in human primary breast cancers. Clin. Cancer Res. 1995, 1: 215-222.
72. Weeraratna AT, Jiang Y, Hostetter G, Rosenblatt K, Duray P, Bittner M, Trent JM. Wnt5a signaling directly affects cell motility and invasion of metastatic melanoma. Cancer Cell 2002, 1: 279-288.
73. Neudauer CL, McCarthy JB. Insulin-like growth factor I-stimulated melanoma cell migration requires phosphoinositide 3-kinase but not extracellular-regulated kinase activation. Exp. Cell Res. 2003, 286: 128-137.
74. Nozaki S, Sledge GWJ, Nakshatri H. Cancer cell-derived interleukin 1alpha contributes to autocrine and paracrine induction of pro-metastatic genes in breast cancer. Biochem. Biophys. Res. Commun. 2000, 275: 60-62.
75. Yasuda T, Matsui H, Kanamori M, Yudoh K, Ohmori K, Aoki M, Tsuji H. Effects of tumor cell-derived interleukin 1 alpha on invasiveness of metastatic clones of murine RCT sarcoma through endothelial cells. Tumour Biol. 1999, 20: 105-116.
76. Dekker SK, Vink J, Vermeer BJ, Bruijn JA, Mihm MC, Jr., Byers HR. Differential effects of interleukin 1-alpha (IL-1 alpha) or tumor necrosis factor-alpha (TNF-alpha) on motility of human melanoma cell lines on fibronectin. J. Invest. Dermatol. 1994, 102: 898-905.
77. Bar-Eli M. Role of interleukin-8 in tumor growth and metastasis of human melanoma. Pathobiology 1999, 67: 12-18.
78. Inoue K, Slaton JW, Eve BY, Kim SJ, Perrotte P, Balbay MD, Yano S, Bar-Eli M, Radinsky R, Pettaway CA, Dinney CP. Interleukin 8 expression regulates tumorigenicity and metastases in androgen-independent prostate cancer. Clin. Cancer Res. 2000, 6: 2104-2119.
79. Youngs SJ, Ali SA, Taub DD, Rees RC. Chemokines induce migrational responses in human breast carcinoma cell lines. Int. J. Cancer 1997, 71: 257-266.
80. Luca M, Huang S, Gershenwald JE, Singh RK, Reich R, Bar-Eli M. Expression of interleukin-8 by human melanoma cells up-regulates MMP-2 activity and increases tumor growth and metastasis. Am. J. Pathol. 1997, 151: 1105-1113.
81. Huang S, Mills L, Mian B, Tellez C, McCarty M, Yang XD, Gudas JM, Bar-Eli M. Fully humanized neutralizing antibodies to interleukin-8 (ABX-IL8) inhibit angiogenesis, tumor growth, and metastasis of human melanoma. Am. J. Pathol. 2002, 161: 125-134.
82. Nishino H, Miyata M, Kitamura K. The effect of interleukin-6 on enhancing the invasiveness of head and neck cancer cells in vitro. Eur. Arch. Otorhinolaryngol. 1998, 255: 468-472.
83. Obata NH, Tamakoshi K, Shibata K, Kikkawa F, Tomoda Y. Effects of interleukin-6 on in vitro cell attachment, migration and invasion of human ovarian carcinoma. Anticancer. Res. 1997, 17: 337-342.
84. Kossakowska AE, Edwards DR, Prusinkiewicz C, Zhang MC, Guo D, Urbanski SJ, Grogan T, Marquez LA, Janowska-Wieczorek A. Interleukin-6 regulation of matrix metalloproteinase (MMP-2 and MMP-9) and tissue inhibitor of metalloproteinase (TIMP-1) expression in malignant non-Hodgkin's lymphomas. Blood 1999, 94: 2080-2089.

85. Wagner S, Stegen C, Bouterfa H, Huettner C, Kerkau S, Roggendorf W, Roosen K, Tonn JC. Expression of matrix metalloproteinases in human glioma cell lines in the presence of IL-10. J. Neurooncol. 1998, 40: 113-122.
86. Wu W, Yamaura T, Murakami K, Ogasawara M, Hayashi K, Murata J, Saiki I. Involvement of TNF-alpha in enhancement of invasion and metastasis of colon 26-L5 carcinoma cells in mice by social isolation stress. Oncol. Res. 1999, 11: 461-469.
87. Esteve PO, Tremblay P, Houde M, St-Pierre Y, Mandeville R. In vitro expression of MMP-2 and MMP-9 in glioma cells following exposure to inflammatory mediators. Biochim. Biophys. Acta. 1998, 1403: 85-96.
88. Zhu N, Lalla R, Eves P, Brown TL, King A, Kemp EH, Haycock JW, MacNeil S. Melanoma cell migration is upregulated by tumour necrosis factor-alpha and suppressed by alpha-melanocyte-stimulating hormone. Br. J. Cancer 2004, 90: 1457-1463.
89. Waterston AM, Salway F, Andreakos E, Butler DM, Feldmann M, Coombes RC. TNF autovaccination induces self anti-TNF antibodies and inhibits metastasis in a murine melanoma model. Br. J. Cancer 2004, 90: 1279-1284.
90. Robledo MM, Bartolome RA, Longo N, Rodriguez-Frade JM, Mellado M, Longo I, Van Muijen GN, Sanchez-Mateos P, Teixido J. Expression of functional chemokine receptors CXCR3 and CXCR4 on human melanoma cells. J. Biol. Chem. 2001, 276: 45098-45105.
91. Bartolome RA, Galvez BG, Longo N, Baleux F, Van Muijen GN, Sanchez-Mateos P, Arroyo AG, Teixido J. Stromal cell-derived factor-1alpha promotes melanoma cell invasion across basement membranes involving stimulation of membrane-type 1 matrix metalloproteinase and Rho GTPase activities. Cancer Res. 2004, 64: 2534-2543.
92. Bartenjev I, Rudolf Z, Stabuc B, Vrhovec I, Perkovic T, Kansky A. Cathepsin D expression in early cutaneous malignant melanoma. Int. J. Dermatol. 2000, 39: 599-602.
93. Frohlich E, Schlagenhauff B, Mohrle M, Weber E, Klessen C, Rassner G. Activity, expression, and transcription rate of the cathepsins B, D, H, and L in cutaneous malignant melanoma. Cancer 2001, 91: 972-982.
94. Kozlowski L, Wojtukiewicz MZ, Ostrowska H. Cathepsin A activity in primary and metastatic human melanocytic tumors. Arch. Dermatol. Res. 2000, 292: 68-71.
95. Mai J, Waisman DM, Sloane BF. Cell surface complex of cathepsin B/annexin II tetramer in malignant progression. Biochim. Biophys. Acta. 2000, 1477: 215-230.
96. Frade R, Rodrigues-Lima F, Huang S, Xie K, Guillaume N, Bar-Eli M. Procathepsin-L, a proteinase that cleaves human C3 (the third component of complement), confers high tumorigenic and metastatic properties to human melanoma cells. Cancer Res. 1998, 58: 2733-2736.
97. Rousselet N, Mills L, Jean D, Tellez C, Bar-Eli M, Frade R. Inhibition of tumorigenicity and metastasis of human melanoma cells by anti-cathepsin L single chain variable fragment. Cancer Res. 2004, 64: 146-151.
98. Dennhofer R, Kurschat P, Zigrino P, Klose A, Bosserhoff A, van Muijen G, Krieg T, Mauch C, Hunzelmann N. Invasion of melanoma cells into dermal connective tissue in vitro: evidence for an important role of cysteine proteases. Int. J. Cancer 2003, 106: 316-323.
99. Fujii H, Nakajima M, Saiki I, Yoneda J, Azuma I, Tsuruo T. Human melanoma invasion and metastasis enhancement by high expression of aminopeptidase N/CD13. Clin. Exp. Metastasis 1995, 13: 337-344.

100. Saiki I, Fujii H, Yoneda J, Abe F, Nakajima M, Tsuruo T, Azuma I. Role of aminopeptidase N (CD13) in tumor-cell invasion and extracellular matrix degradation. Int. J. Cancer 1993, 54: 137-143.
101. Kido A, Krueger S, Haeckel C, Roessner A. Possible contribution of aminopeptidase N (APN/CD13) to invasive potential enhanced by interleukin-6 and soluble interleukin-6 receptor in human osteosarcoma cell lines [In Process Citation]. Clin. Exp. Metastasis 1999, 17: 857-863.
102. Wesley UV, Albino AP, Tiwari S, Houghton AN. A role for dipeptidyl peptidase IV in suppressing the malignant phenotype of melanocytic cells. J. Exp. Med. 190: 311-322.
103. Pethiyagoda CL, Welch DR, Fleming TP. Dipeptidyl peptidase IV (DPPIV) inhibits cellular invasion of melanoma cells. Clin. Exp. Metastasis 2000, 18: 391-400.
104. Kosir MA, Quinn CC. Sorting of heparan sulfate proteoglycan into matrix compartments of prostate adenocarcinoma cells. J. Surg. Res. 1995, 58: 46-52.
105. Friedmann Y, Vlodavsky I, Aingorn H, Aviv A, Peretz T, Pecker I, Pappo O. Expression of heparanase in normal, dysplastic, and neoplastic human colonic mucosa and stroma : evidence for its role in colonic tumorigenesis [In Process Citation]. Am. J. Pathol. 2000, 157: 1167-1175.
106. Vlodavsky I, Elkin M, Pappo O, Aingorn H, Atzmon R, Ishai-Michaeli R, Aviv A, Pecker I, Friedmann Y. Mammalian heparanase as mediator of tumor metastasis and angiogenesis. Isr. Med. Assoc. J. 2000, Suppl: 37-45.
107. Vlodavsky I, Friedmann Y, Elkin M, Aingorn H, Atzmon R, Ishai-Michaeli R, Bitan M, Pappo O, Peretz T, Michal I, Spector L, Pecker I. Mammalian heparanase: gene cloning, expression and function in tumor progression and metastasis. Nat. Med. 1999, 5: 793-802.
108. Reiland J, Sanderson RD, Waguespack M, Barker SA, Long R, Carson DD, Marchetti D. Heparanase degrades syndecan-1 and perlecan heparan sulfate: functional implications for tumor cell invasion. J. Biol. Chem. 2004, 279: 8047-8055.
109. Denkins Y, Reiland J, Roy M, Sinnappah-Kang ND, Galjour J, Murry BP, Blust J, Aucoin R, Marchetti D. Brain metastases in melanoma: roles of neurotrophins. Neuro – oncol. 2004, 6: 154-165.
110. Marchetti D, Aucoin R, Blust J, Murry B, Greiter-Wilke A. p75 neurotrophin receptor functions as a survival receptor in brain-metastatic melanoma cells. J. Cell Biochem. 2004, 91: 206-215.
111. Marchetti D, Nicolson GL. Human heparanase: a molecular determinant of brain metastasis. Adv. Enzyme Regul. 2001, 41:343-59: 343-359.
112. Marchetti D, McQuillan DJ, Spohn WC, Carson DD, Nicolson GL. Neurotrophin stimulation of human melanoma cell invasion: selected enhancement of heparanase activity and heparanase degradation of specific heparan sulfate subpopulations. Cancer Res. 1996, 56: 2856-2863.
113. Csoka TB, Frost GI, Stern R. Hyaluronidases in tissue invasion. Invasion Metastasis 1997, 17: 297-311.
114. Maleski M, Hockfield S. Glial cells assemble hyaluronan-based pericellular matrices in vitro. Glia 1997, 20: 193-202.
115. Madan AK, Pang Y, Wilkiemeyer MB, Yu D, Beech DJ. Increased hyaluronidase expression in more aggressive prostate adenocarcinoma. Oncol. Rep. 1999, 6: 1431-1433.
116. Madan AK, Yu K, Dhurandhar N, Cullinane C, Pang Y, Beech DJ. Association of hyaluronidase and breast adenocarcinoma invasiveness. Oncol. Rep. 1999, 6: 607-609.

117. Pham HT, Block NL, Lokeshwar VB. Tumor-derived hyaluronidase: a diagnostic urine marker for high-grade bladder cancer. Cancer Res. 1997, 57: 778-783.
118. Novak U, Stylli SS, Kaye AH, Lepperdinger G. Hyaluronidase-2 overexpression accelerates intracerebral but not subcutaneous tumor formation of murine astrocytoma cells. Cancer Res. 1999, 59: 6246-6250.
119. Victor R, Chauzy C, Girard N, Gioanni J, d'Anjou J, Stora DN, Delpech B. Human breast-cancer metastasis formation in a nude-mouse model: studies of hyaluronidase, hyaluronan and hyaluronan-binding sites in metastatic cells. Int. J. Cancer 1999, 82: 77-83.
120. Liu D, Pearlman E, Diaconu E, Guo K, Mori H, Haqqi T, Markowitz S, Willson J, Sy MS. Expression of hyaluronidase by tumor cells induces angiogenesis in vivo. Proc. Natl. Acad. Sci. USA 1996, 93: 7832-7837.
121. Frisch SM, Ruoslahti E. Integrins and anoikis. Curr. Opin. Cell Biol. 1997, 9: 701-706.
122. Judson PL, He X, Cance WG, Van Le L. Overexpression of focal adhesion kinase, a protein tyrosine kinase, in ovarian carcinoma. Cancer 1999, 86: 1551-1556.
123. Owens LV, Xu L, Craven RJ, Dent GA, Weiner TM, Kornberg L, Liu ET, Cance WG. Overexpression of the focal adhesion kinase (p125FAK) in invasive human tumors. Cancer Res. 1995, 55: 2752-2755.
124. Weiner TM, Liu ET, Craven RJ, Cance WG. Expression of focal adhesion kinase gene and invasive cancer. Lancet. 1993, 342: 1024-1025.
125. Kornberg LJ. Focal adhesion kinase and its potential involvement in tumor invasion and metastasis. Head Neck 1998, 20: 745-752.
126. Shibata K, Kikkawa F, Nawa A, Thant AA, Naruse K, Mizutani S, Hamaguchi M. Both focal adhesion kinase and c-Ras are required for the enhanced matrix metalloproteinase 9 secretion by fibronectin in ovarian cancer cells. Cancer Res. 1998, 58: 900-903.
127. Cadigan KM, Nusse R. Wnt signaling: a common theme in animal development. Genes Dev. 1997, 11: 3286-3305.
128. Polakis P. The oncogenic activation of beta-catenin. Curr. Opin. Genet. Dev. 1999, 9: 15-21.
129. Wodarz A, Nusse R. Mechanisms of Wnt signaling in development. Annu. Rev. Cell Dev. Biol. 1998, 14:59-88.: 59-88.
130. Yost C, Torres M, Miller JR, Huang E, Kimelman D, Moon RT. The axis-inducing activity, stability, and subcellular distribution of beta-catenin is regulated in Xenopus embryos by glycogen synthase kinase 3. Genes Dev. 10: 1443-1454.
131. Morin PJ, Sparks AB, Korinek V, Barker N, Clevers H, Vogelstein B, Kinzler KW. Activation of beta-catenin-Tcf signaling in colon cancer by mutations in beta-catenin or APC. Science 1997, 275: 1787-1790.
132. Wijnhoven BP, Nollet F, De Both NJ, Tilanus HW, Dinjens WN. Genetic alterations involving exon 3 of the beta-catenin gene do not play a role in adenocarcinomas of the esophagus. Int. J. Cancer 2000, 86: 533-537.
133. Brabletz T, Jung A, Hermann K, Gunther K, Hohenberger W, Kirchner T. Nuclear overexpression of the oncoprotein beta-catenin in colorectal cancer is localized predominantly at the invasion front. Pathol. Res. Pract. 1998, 194: 701-704.
134. Brabletz T, Jung A, Dag S, Hlubek F, Kirchner T. beta-catenin regulates the expression of the matrix metalloproteinase-7 in human colorectal cancer. Am. J. Pathol. 1999, 155: 1033-1038.

135. Conacci-Sorrell ME, Ben Yedidia T, Shtutman M, Feinstein E, Einat P, Ben Ze'ev A. Nr-CAM is a target gene of the beta-catenin/LEF-1 pathway in melanoma and colon cancer and its expression enhances motility and confers tumorigenesis. Genes Dev. 2002, 16: 2058-2072.
136. Dai DL, Makretsov N, Campos EI, Huang C, Zhou Y, Huntsman D, Martinka M, Li G. Increased expression of integrin-linked kinase is correlated with melanoma progression and poor patient survival. Clin. Cancer Res. 2003, 9: 4409-4414.
137. Blesch A, Bosserhoff AK, Apfel R, Behl C, Hessdoerfer B, Schmitt A, Jachimczak P, Lottspeich F, Buettner R, Bogdahn U. Cloning of a novel malignant melanoma-derived growth-regulatory protein, MIA. Cancer Res. 1994, 54: 5695-5701.
138. Bosserhoff AK, Kaufmann M, Kaluza B, Bartke I, Zirngibl H, Hein R, Stolz W, Buettner R. Melanoma-inhibiting activity, a novel serum marker for progression of malignant melanoma. Cancer Res. 1997, 57: 3149-3153.
139. Bosserhoff AK, Buettner R. Expression, function and clinical relevance of MIA (melanoma inhibitory activity). Histol. Histopath. 2002, 17: 289-300.
140. Dreau D, Bosserhoff AK, White RL, Buettner R, Holder WD. Melanoma-inhibitory activity protein concentrations in blood of melanoma patients treated with immunotherapy. Oncol. Res. 1999, 11: 55-61.
141. Deichmann M, Benner A, Bock M, Jackel A, Uhl K, Waldmann V, Naher H. S100-Beta, melanoma-inhibiting activity, and lactate dehydrogenase discriminate progressive from nonprogressive American Joint Committee on Cancer stage IV melanoma. J. Clin. Oncol. 1999, 17: 1891-1896.
142. van Groningen JJ, Bloemers HP, Swart GW. Identification of melanoma inhibitory activity and other differentially expressed messenger RNAs in human melanoma cell lines with different metastatic capacity by messenger RNA differential display. Cancer Res. 1995, 55: 6237-6243.
143. Bosserhoff AK, Moser M, Hein R, Landthaler M, Buettner R. In situ expression patterns of melanoma-inhibiting activity (MIA) in melanomas and breast cancers. J. Pathol. 1999, 187: 446-454.
144. Guba M, Bosserhoff AK, Steinbauer M, Abels C, Anthuber M, Buettner R, Jauch KW. Overexpression of melanoma inhibitory activity (MIA) enhances extravasation and metastasis of A-mel 3 melanoma cells in vivo. Br. J. Cancer 2000, 83: 1216-1222.
145. Bosserhoff AK, Stoll R, Sleeman JP, Bataille F, Buettner R, Holak TA. Active detachment involves inhibition of cell-matrix contacts of malignant melanoma cells by secretion of melanoma inhibitory activity. Lab. Invest. 2003, 83: 1583-1594.
146. Stoll R, Renner C, Ambrosius D, Golob M, Voelter W, Buettner R, Bosserhoff AK, Holak TA. Letter to the editor: Sequence-specific 1H, 13C, and 15N assignment of the human melanoma inhibitory activity (MIA) protein. J. Biomol. NMR 2000, 17: 87-88.
147. Bourguignon LY, Zhu H, Shao L, Chen YW. Ankyrin-Tiam1 interaction promotes Rac1 signaling and metastatic breast tumor cell invasion and migration. J. Cell Biol. 2000, 150: 177-191.
148. Habets GG, Scholtes EH, Zuydgeest D, van der Kammen RA, Stam JC, Berns A, Collard JG. Identification of an invasion-inducing gene, Tiam-1, that encodes a protein with homology to GDP-GTP exchangers for Rho-like proteins. Cell 1994, 77: 537-549.
149. Michiels F, Habets GG, Stam JC, van der Kammen RA, Collard JG. A role for Rac in Tiam1-induced membrane ruffling and invasion. Nature 1995, 375: 338-340.

150. Habets GG, van der Kammen RA, Stam JC, Michiels F, Collard JG. Sequence of the human invasion-inducing TIAM1 gene, its conservation in evolution and its expression in tumor cell lines of different tissue origin. Oncogene 1995, 10: 1371-1376.
151. Paley PJ, Goff BA, Gown AM, Greer BE, Sage EH. Alterations in SPARC and VEGF immunoreactivity in epithelial ovarian cancer. Gynecol. Oncol. 2000, 78: 336-341.
152. Thomas R, True LD, Bassuk JA, Lange PH, Vessella RL. Differential expression of osteonectin/SPARC during human prostate cancer progression. Clin. Cancer Res. 2000, 6: 1140-1149.
153. Massi D, Franchi A, Borgognoni L, Reali UM, Santucci M. Osteonectin expression correlates with clinical outcome in thin cutaneous malignant melanomas. Hum. Pathol. 1999, 30: 339-344.
154. Porte H, Triboulet JP, Kotelevets L, Carrat F, Prevot S, Nordlinger B, DiGioia Y, Wurtz A, Comoglio P, Gespach C, Chastre E. Overexpression of stromelysin-3, BM-40/SPARC, and MET genes in human esophageal carcinoma: implications for prognosis. Clin. Cancer Res. 1998, 4: 1375-1382.
155. Golembieski WA, Ge S, Nelson K, Mikkelsen T, Rempel SA. Increased SPARC expression promotes U87 glioblastoma invasion in vitro. Int. J. Dev. Neurosci. 1999, 17: 463-472.
156. Kato Y, Frankenne F, Noel A, Sakai N, Nagashima Y, Koshika S, Miyazaki K, Foidart JM. High production of SPARC/osteonectin/BM-40 in mouse metastatic B16 melanoma cell lines. Pathol. Oncol. Res. 2000, 6: 24-26.
157. Gilles C, Bassuk JA, Pulyaeva H, Sage EH, Foidart JM, Thompson EW. SPARC/osteonectin induces matrix metalloproteinase 2 activation in human breast cancer cell lines. Cancer Res. 1998, 58: 5529-5536.
158. Ledda MF, Adris S, Bravo AI, Kairiyama C, Bover L, Chernajovsky Y, Mordoh J, Podhajcer OL. Suppression of SPARC expression by antisense RNA abrogates the tumorigenicity of human melanoma cells. Nat. Med. 1997, 3: 171-176.
159. Kato Y, Frankenne F, Noel A, Sakai N, Nagashima Y, Koshika S, Miyazaki K, Foidart JM. High production of SPARC/osteonectin/BM-40 in mouse metastatic B16 melanoma cell lines. Pathol. Oncol. Res. 2000, 6: 24-26.
160. Xiang YY, Wang DY, Tanaka M, Suzuki M, Kiyokawa E, Igarashi H, Naito Y, Shen Q, Sugimura H. Expression of high-mobility group-1 mRNA in human gastrointestinal adenocarcinoma and corresponding non-cancerous mucosa. Int. J. Cancer 1997, 74: 1-6.
161. Taguchi A, Blood DC, del Toro G, Canet A, Lee DC, Qu W, Tanji N, Lu Y, Lalla E, Fu C, Hofmann MA, Kislinger T, Ingram M, Lu A, Tanaka H, Hori O, Ogawa S, Stern DM, Schmidt AM. Blockade of RAGE-amphoterin signalling suppresses tumour growth and metastases. Nature 2000, 405: 354-360.
162. Huttunen HJ, Fages C, Kuja-Panula J, Ridley AJ, Rauvala H. Receptor for advanced glycation end products-binding COOH-terminal motif of amphoterin inhibits invasive migration and metastasis. Cancer Res. 2002, 62: 4805-4811.
163. Fages C, Nolo R, Huttunen HJ, Eskelinen E, Rauvala H. Regulation of cell migration by amphoterin. J. Cell Sci. 2000, 113: 611-620.
164. Hendrix MJ, Seftor EA, Hess AR, Seftor RE. Molecular plasticity of human melanoma cells. Oncogene 2003, %19; 22: 3070-3075.
165. Easty DJ, Bennett DC. Protein tyrosine kinases in malignant melanoma. Melanoma. Res. 2000, 10: 401-411.
166. Straume O, Akslen LA. Importance of vascular phenotype by basic fibroblast growth factor, and influence of the angiogenic factors basic fibroblast growth factor/fibroblast

growth factor receptor-1 and ephrin-A1/EphA2 on melanoma progression. Am. J. Pathol. 2002, 160: 1009-1019.
167. Dupin E, Le Douarin NM. Development of melanocyte precursors from the vertebrate neural crest. Oncogene 2003, 22: 3016-3023.
168. Yohn JJ, Smith C, Stevens T, Hoffman TA, Morelli JG, Hurt DL, Yanagisawa M, Kane MA, Zamora MR. Human melanoma cells express functional endothelin-1 receptors. Biochem. Biophys. Res. Commun. 1994, 201: 449-457.
169. Demunter A, Wolf-Peeters C, Degreef H, Stas M, van den Oord JJ. Expression of the endothelin-B receptor in pigment cell lesions of the skin. Evidence for its role as tumor progression marker in malignant melanoma. Virchows Arch. 2001, 438: 485-491.
170. Bagnato A, Rosano L, Spinella F, Di C, V, Tecce R, Natali PG. Endothelin B receptor blockade inhibits dynamics of cell interactions and communications in melanoma cell progression. Cancer Res. 2004, 64: 1436-1443.
171. Jamal S, Schneider RJ. UV-induction of keratinocyte endothelin-1 downregulates E-cadherin in melanocytes and melanoma cells. J. Clin. Invest. 2002, 110: 443-452.

Chapter 13

CELL MOTILITY IN PROSTATE TUMOR INVASION AND METASTASIS

Sourabh Kharait, Kien Tran, Clayton Yates and Alan Wells
Department of Pathology, University of Pittsburgh School of Medicine, Pittsburgh, PA, USA

Abstract: The mortality and morbidity from prostate carcinomas result primarily from their progression to invasive and metastatic state. Recent studies have uncovered the crucial role of cell motility in dictating these highly orchestrated biological events. This review will focus on the intracellular and extracellular factors that govern cell motility during prostate tumor cell invasion. The role of receptor signaling systems, including those for epidermal growth factor and hepatocyte growth factor, chemokine receptors, and sex steroid hormones (estrogen and testosterone) will be discussed with reference to their modulation of cellular invasive potential by regulating motility. Lastly, detailed cellular biophysical events and the underlying biochemical molecules that govern these distinct events will be discussed with the goal that targeting these individual events using the emerging genetic and pharmacological interventions can lead to more specific and successful therapies against prostate cancer.

Key words: EGF receptor, HGF, met, chemokine receptors, EMT

1. INTRODUCTION

Prostate cancer mortality results mainly from progression to the invasion and metastatic state. Significant morbidity ensues from surgical or radiological ablation of non-invasive, organ-confined carcinomas, and these adverse effects are greatly amplified once the tumor breaches the prostate capsule or migrates out along neurovascular conduits (1). The approach to organ-confined carcinomas is still uncertain as development of clinically insignificant prostate adenocarcinoma appears to be a consequence of aging

A. Wells (ed.), Cell Motility in Cancer Invasion and Metastasis, 301-338.

with small carcinomas found upon autopsy examination of most men dying in their 90s. Still, the current practice is to remove aggressive carcinomas as prostate cancer responds poorly to the current generation of therapies (2). This is due in large part to these agents being optimized to kill growing cells while prostate cancer present mitogenic indices far lower than many normal tissues. As such, future interventions need to focus on the cell properties that enable prostate carcinomas to migrate from an organ-confined situation to invade adnexia and metastasize to distant organs. Recent studies have highlighted a central role for cell motility in acquiring the ability of prostate tumor cells to progress to a metastatic focus.

1.1 Prostate Cancer Epidemiology

Cancer is the second leading cause of death in the United States, second only to heart disease. Prostate cancer is the most commonly diagnosed cancer in men, with more than 70 % of prostate cancers being diagnosed in men over 65. In the year 2003, approximately 220,990 new prostate cancer cases were diagnosed (ACS: Cancer facts and figures, 2003). It is the second leading cause of cancer deaths in men after lung cancer. Family history, age and ethnicity are the only well-established risk factors in prostate cancer. Prostate cancer is rare in young men and the probability rises steeply with increasing age with the risk being 1 in 7 after the age of 60 as compared to a risk of 1 in 44 between the age 40 and 59 years. Organ-confined prostate cancer advances slowly with survival lasting over a decade even in untreated cases. This creates a therapeutic dilemma in older patients, as surgical and radiological ablation of the tumor (and the prostate) carry significant morbidity and even subsequent mortality (1).

Recent studies on prostate cancer genetics suggest that only about 5-10 % of prostate cancers have a familial predisposition (3, 4). Also, prostate cancer is 2.4 times more common in African-American men as compared to caucasians (Surveillance, Epidemiology and End-Results program, 1975-2000, Division of Cancer control and population sciences, NCI, 2003). In the last few years, research has focused on studying the role of various other risk factors such as diet, sexually transmitted diseases, obesity and smoking as being a risk factor for prostate cancer. At present, conclusive evidence for any of the above being contributory in many patients is lacking.

The current therapeutic management of prostate cancer involves chemical or physical castration to induce an androgen-withdrawal apoptosis of the tumor cells. However, this prolongs survival by only about 10 % on average with the relapse being androgen-independent. The progression to invasion and metastasis is thus only slowed and not blocked.

Local extracapsular invasion of prostate cancer continues to be a significant problem especially in elderly patients, with limited options of radiotherapy and combination chemotherapy (5, 6). Invasion of the adnexia results in compromised function of the renal and genital systems with significant physiological and psychological sequelae. Additionally, metastatic spread carries a high mortality burden. Bone is the most common site of metastasis in prostate cancers (approximately 90%) with 85% of the lesions affecting the vertebral column. Once in bone, the prostate carcinoma cells induce an osteoblastic response that is responsible for debilitating bone pain. Other frequent sites of metastasis include the lung (50%) and liver (25%). Brain metastasis is rare (1-2%) but carries the worst prognosis with a maximum life expectancy of 6 – 7 months.

1.2 Tumor Invasion and Metastasis

Aggressiveness of tumors is directly defined by their ability to invade surrounding tissues and their ability to grow in ectopic environments. The process of invasion is a highly coordinated one (7). As compared to normal cells, tumor cells have an increased capability to loosen their connections to the substratum and break cell-cell adhesions. Integrin receptors that mediate the cell-substratum adhesion and cadherins that mediate cell-cell cohesion, are of pivotal in this behavior. Integrins serve a double role, as they interact as adhesion receptors with the substratum during tumor cell migration. Migration through the extracellular matrix (ECM) barriers, while mainly occuring via natural cleavage planes, also requires matrix remodeling accomplished by various proteases including matrix metalloproteinases. This motile strategy is used both to invade local adnexia and gain access to conduits for distant dissemination.

Prostate tumor cells disseminate mainly via bloodstream or lymphatics. During this process, the tumor cells must survive the de-adhesion induced anoikes either by cell intrinsic changes rendering them resistant or by forming clumps to recreate the 'attached' cell signals. These individual cells or clusters of cells reach target organs, extravasate, and migrate to appropriate sites within the tissues. If the target tissue provides the missing extrinsic signals, the cells will proliferate to form a metastasis; the exact combination of signals required depends on the tumor cell intrinsic changes that provide for site autonomous survival and growth (8). It is this final step of cell proliferation that is the least efficient (9) and relates to the low rate of metastatic foci despite significant extravasation of many tumors (9). Still, any of these individual steps may be rate-limiting and represent a potential therapeutic target. Both local invasion and metastasis are significant issues in the early dissemination of prostate carcinoma, as opposed to breast cancer in which invasiveness is a later event or ovarian carcinoma in which peritoneal spread predates demonstrable metastases. Thus, maximum benefit may accrue from deciphering and targeting a tumor cell acquired property that is critical for both, such as motility.

Recent research in tumor biology has bolstered the concept that tumor invasion is the net result of dysregulated cell motility (10). Invasiveness of tumors can directly be attributed to their migratory potential along with their ability to penetrate and remodel the stroma. Inhibiting the motility of tumor cells decreases their invasiveness (11-13).

2. FOUNDATIONAL MODEL FOR CELL MOTILITY

Cell migration is highly orchestrated and comprises many "sub-processes" that are carefully organized temporally to achieve the net response of cell locomotion (14) (Figure 1).

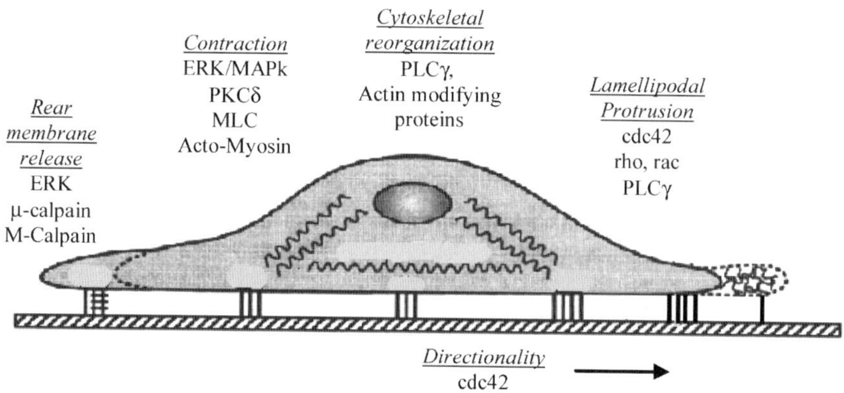

Figure 1. Various biophysical processes (underlined) during cell migration are elucidated. These individual biophysical events during cell motility are controlled by underlying biochemical signals. Adapted from (14).

In order for a cell to move, it first has to be polarized or attain a sense of directionality. Various simplistic models for cell migration have now been elucidated (15, 16). One of these proposes the polarized cell to have a tail or a rear end and a front end upon stimulation with a motogenic agent like epidermal growth factor (EGF). Attainment of polarity is accompanied by extension of dominant lamellipodia at the 'front end'. This is followed by detachment of the cell's rear end and subsequent transcellular contractility that provides the necessary force for locomotion. Each of these processes are controlled by numerous intracellular signaling molecules, some of them being involved in other cellular responses as well. Substantial evidence suggests that disruption of these individual biophysical events like lamellipodal protrusion or rear cell-membrane detachment by targeting the underlying signaling switches offers a very promising approach to novel anti-cancer drug discovery.

Upon stimulation with a motogenic agent like EGF, the first step towards cell locomotion involves extensions of the cell membrane, as lamellipodia, in the direction of movement. Lamellipodia are principally composed of cytoskeletal elements like actin and myosin along with integrin receptors and signaling molecules (17). Phospholipase C-γ (PLCγ) signaling pathway is shown to be crucial in lamellipodial extension in fibroblasts and cancer cells upon activation by growth factors, though it likely plays little if any role in integrin-mediated extension (16). Phosphoinositide 3-OH kinase (PI3K) is also active at the front of the cell during adhesion receptor-induced motility and growth factor-induced motility but in epithelial and not fibroblastoid cells (16, 18). Both of these enzymes alter the phosphoinositide face of the inner membrane, with PI3K creating phosphoinositol 3,4,5-trisphosphate (PIP3) and PLCγ hydrolysing and removing the PIP2. This alters the docking sites for a variety of molecules that impinge on the actin cytoskeleton. New sites are created by PIP3 and the loss of PIP2, particularly by PLCγ hydrolysis which releases prebound molecules including vinculin, gelsolin, cofilin and profilin. All of these molecules then act to disassemble focal adhesions and accelerate actin polymerization, the details of which have been ably reviewed elsewhere (19, 20), enabling protrusion of lamellipodia.

Emergence of a dominant lamellipod marks the state of polarity of the cell (21). The small GTPase cdc42 is required to either establish or maintain a persistent cell polarity and directionality that leads to productive locomotion (22-24). This is tightly balanced in that either under- or over-activation of cdc42 tips the balance and either no lamellipodia are formed or none is established as dominant and the cell 'dances' in place (25). These membrane extensions or lamellipodia are then stabilized by attachment of the protruding lamellipodia to the substratum, thereby enabling new focal contacts between cell and extracellular matrix. Thus, cell-substratum adhesion is a crucial element for migration of tumor cells and is achieved principally by the action of integrins and other adhesion receptors (14, 21).

Sites of cell-substratum adhesion are not merely sites of passive contact between the cell and the extracellular matrix but are active in cell signaling. While this might seem obvious from adhesion receptor-mediated motility, all migratory prostate carcinoma cells present autocrine growth factor signaling systems, and thus one must examine adhesion signaling locally and not globally. Focal adhesions are rich not only in cytoskeletal proteins, integrins, other adhesion receptors and linker proteins, but also various signaling kinases. The non-receptor tyrosine kinases focal adhesion kinase (FAK), src, and integrin-linked kinase (ILK), all modulate cell-adhesion dependent growth, survival and motility by activating numerous signaling pathways (26). These have been shown to interact directly with the adhesion complex and to be activated therein. Interestingly, level of FAK expression and signaling has been correlated with higher invasive and migratory capacity of prostate cancer (27, 28). These kinases are activated by integrin ligation to substratum or even solubilized fragments (see Chapter 8). For instance, within prostate cancer cells and various other cell lines, expression of $\alpha v\beta 3$ has been correlated with increased invasion and tumorigenicity by modulating cell-adhesion and motility associated signaling pathways (29, 30) (and Chapter 8). Expression of $\alpha v\beta 3$ within prostate cancer cells results in upregulation of cdc2 (cdk1), with colocalization of cdc2 and caldesmon within the membrane ruffles of motile cells (29). Cell migration was enhanced in cells expressing this integrin and limited upon inhibition of Cdc2. Thus, the signals elicited within the contact complexes not only alter the adhesivity of the site but also are transmitted to distal parts of the cells.

The cell then contracts and generates sufficient force needed for translocation. Adhesion sites are vital as a fulcrum for mechanical forces. Transcellular contractility is achieved via a consorted action of actin-myosin cytoskeletal machinery. In fibroblasts and prostate cancer cells at least, growth factors can activate the regulatory element myosin light chain (MLC) proteins of myosin II via Protein kinase Cδ (PKCδ) (31). Such contractility is needed by tumor cells for motility and by fibroblasts during contraction of the wound edges during later stages of wound healing. Unpublished studies from our laboratory have shown that invasion of DU145 and PC3 prostate cancer cells through a thin layer of Matrigel is substantially reduced by pharmacological and molecular inhibition of the PKCδ > MLC signaling pathway. Overall, the phosphorylation/activation state of MLC is controlled by a balance between the activating MLC Kinase (MLCK) and deactivating MLC Phosphatase (MLCP) enzymes. A second pathway regulating this occurs via Rho Kinase directly activating MLCK and inhibiting MLCP, thereby stimulating cell contractility and motility; this pathway is likely used by adhesion receptor signaling (32, 33). Indeed, abrogation of Rho Kinase in

PC3 prostate carcinoma cells substantially inhibited their migration and invasion *in vitro* and *in vivo* (13).

Adhesion of the lamellipodia to the substratum at the front end is accompanied by the detachment of the cell membrane at the rear that enables the cell to move forwards. Rear detachment is both passive in response to transcellular tension and active in that adhesion sites in the rear are weakened through active signaling. Transcellular tension is sufficient only under regimes of low adhesivity (34), under which overall locomotion of fibroblasts is actually decreased (35). Active deadhesion is achieved largely by the intracellular protease calpain, two isoforms of which are found within motile fibroblasts and prostate cancer cells (12, 16, 36); Calpain-I (μ-calpain) and calpain-II (M-calpain) activated *in vitro* by micromolar and millimolar levels calcium respectively. Calpain II has been shown to be activated by direct phosphorylation by ERK/MAP Kinase downstream of signaling from EGFR and other receptor tyrosine kinases (37). In the absence of receptor tyrosine kinase activity, it is possible that calpain I subsumes this role being activated by calcium influx upon triggering stretch-activated calcium channels (38). Calpain cleaves the cytoplasmic tails of integrins and/or linker proteins such as talin, loosening the attachment to substratum. Coupled with transcellular contractility this leads to membrane de-adhesion at the rear end of the cell. The importance of this biochemical event in prostate cancer progression was shown by abrogation of calpain using molecular and pharmacological agents substantially inhibiting migration and invasion of DU145 prostate cancer cells (12). In addition, calpains may be involved in lessening the cell-cell adhesions required for tumor cell dissemination from the primary localized mass by targeting E-cadherins (39).

Productive locomotion of the cell involves repetitive cycling of these key biophysical processes. However, although these biophysical processes are described individually, they are tightly coordinated temporally, since each of these is necessary for the others to be executed in an orderly manner. By conceptually segregating these individual events, we can begin to understand the key elements of cell motility that together dictate the final response. Each of these individual events is controlled by numerous intracellular biochemical signaling molecules, some of which are being explored as operative targets in limiting cell motility in tumor invasion and metastasis.

3. PROSTATE TUMOR CELL MOTILITY

Most solid tumors possess multiple genetic perturbations that lead to inactivation or downregulation of tumor suppressor genes and/or activation and amplification of proto-oncogenes. These are critical for initial oncogenesis and may contribute to further progression. However, at least a major part of the transition to progression appears to involve epigenetic signaling. Interestingly, some of the epigenetic signaling pathways may be linked to the initial genetic alterations but later induce additional cell behaviors. Carcinomas of breast, lung, prostate and urinary bladder are associated with upregulated activation of various growth factor receptors including the Epidermal Growth Factor receptor (EGFR), Hepatocyte growth factor receptor c-Met and many more. This ensues not only from increased receptor expression but also ectopic autocrine signaling. Increased expression of these receptors increases the efficiency of ligand capture at the cell surface stimulating migratory and proliferation associated intracellular signaling pathways. Reports suggest that invasive tumors secrete numerous growth factors and chemokines; commonly implicated are the epidermal growth factor (EGF), hepatocyte growth factor (HGF), fibroblast growth factor-1 (FGF-1, aFGF), keratinocyte growth factor (KGF, FGF-7), insulin like growth factor 1 (IGF1), interleukins (IL)-6, IL-8, and vascular endothelial growth factor (VEGF). Such upregulation of receptors and autonomous secretion of ligands create self-stimulatory autocrine signaling loops. The increased intracellular signaling stimulates tumor cell migration resulting in a more aggressive phenotype. Thus, growth factor receptors that "drive" motility of tumor cells offer a unique therapeutic target in addition to conventional surgical and endocrinological modalities.

There is an extensive list of growth factors receptors that have been reported as altered in their expression levels and signaling characteristics in at least some prostate cancers. Here, we will highlight signaling and describe the signaling characteristics of the two most widely implicated in tumor cell motility and invasion, the EGF receptor and HGF receptor (c-Met).

3.1 The Epidermal Growth Factor Receptor (EGFR) Family

The type 1 growth factor receptor tyrosine kinase (RTK) family, also known as erbB or HER (human epidermal growth factor receptor) family, consists of four well characterised members (40). These include c-erbB1 (or HER-1) or EGF receptor (EGFR), c-erbB2 (or HER-2/neu), c-erbB3 (HER-3) and c-erbB4 (or HER-4). Several ligands have been identified for these receptors except for HER-2, for which a ligand is yet to be found. However, much of the signaling from this family of receptors involves cross activation and hetero-aggregation of members. The amino acid homology between these receptor subtypes is about 40-60 % in the extracellular domain and about 60-80 % in the intracellular domain (most of the divergence is in the long, protein interactive tail). Specific ligands to EGFR (HER-1) include epidermal growth factor, transforming growth factor-α, heparin-binding EGF and amphiregulin; the high affnity ligands for HER-3 and HER-4 are heregulins, with the first member also known as neu differentiation factor (41). Upon binding of the ligand to the extracellular domain, these RTKs are activated by homo- or hetero-dimerization thereby leading to tyrosine phosphorylation at multiple residues within the long cytoplasmic tails. These phosphotyrosine residues then serve as multiple docking sites for various SH2- or PTB-containing adaptor proteins. Direct activation of these docking proteins often by phosphorylation, or indirect activation of their downstream effectors elicits the multiple cell responses.

EGFR was the first receptor identified as a proto-oncogene and, thus, the first to be implicated by upregulation in numerous human cancers. Interestingly, the level of EGFR activation correlated with tumor progression and not carcinogenesis (7). Various genetic alterations such as gene amplification or alternatively spliced variants have been found in glioblastomas and other carcinomas. However, in prostate cancer the relationship appears primarily epigenetic with a vigorous upregulation of ligand production and autocrine stimulation (42).

ErbB2 was initially invoked as a proto-oncogene when a chemically-induced tumor model system was found due to an activating mutation in the transmembrane domain of this receptor (43). While this mutation has yet to be defined in human carcinomas, increased levels of erbB2 have been noted in many human carcinomas, breast cancer in particular. The upregulation of erbB2 accentuates signaling from EGFR ligands as the heterodimer is less efficiently attenuated by receptor internalization and degradation (44). However, ErbB2 has been reported as upregulated only in a fraction of prostate cancers and this correlates with aggressive and invasive nature of the tumors (45, 46). Reportedly, erbB2 gene amplification seems to be very uncommon in androgen independent prostate cancers (47).

The precise role of erbB3 and erbB4 in prostate cancer is still very unclear. ErbB3 at least must function via transactivation and aggregation with other members of the family, as it lacks intrinsic kinase activity. Interestingly, it is the only member of the erbB family to possess canonical PI3K phosphotyrosine docking sites, and thus involvement of this receptor would be expected to enhance this particular signaling pathway (48).

The primary role of these receptors in cancer progression in general is highlighted by the fact that two new therapies specifically target EGFR (Iressa by AstraZeneca and Tarceva by OSI Pharmaceuticals) and erbB2 (Herceptin by Genentech). These are the first biologicals to target signaling receptors and the first, along with Gleevec (anti-abl), to inhibit tyrosine kinase function. While these therapies have been approved for other carcinomas, they are being explored in prostate carcinomas due to the upregulation of the receptors in this cancer.

3.1.1 EGFR signaling in prostate cancer

EGFR signaling is upregulated in prostate cancers due to autocrine production of activating ligands (42). In normal prostate functioning, EGFR is presented on the basolateral surfaces of the prostate acinar cells to respond to low levels of TGF-α produced by stromal cells. At the same time, these acinar cells produce copious EGF secreted into the lumen, resulting in the highest concentration of EGF in any body fluid. The function of this EGF remains undetermined at present. During the epithelial dedifferentiation that accompanies carcinogenic transformation, the physiologically tight cellular asymmetry is lost secondary to cadherin downregulation and a potential autocrine loop becomes an operative one. In many of these prostate carcinoma cells, the ligand production switches from EGF to TGF-α. While the mechanism behind this is yet undefined, the outcome is stimulation by a ligand whose receptor-binding characteristics spares receptor downregulation and maintains a higher level of signaling (49) (see Chapter 5).

That EGFR is involved in prostate tumor invasiveness has been shown in cell systems. EGFR overexpression in DU145 prostate tumor cells substantially increases their invasive potential *in vitro* and *in vivo* (50, 51). This increase in invasion was substantially ameliorated using a blocking antibody against EGFR or inhibitor of EGFR kinase activity.

EGFR activation triggers a myriad of downstream signaling pathways (Figure 2) that overlap in initiating diverse cell responses including cell proliferation, survival, and migration. It appears that the cell chooses among these sometimes mutually exclusive responses depending on the current cell proteome, other signaling pathway extant, and the temporospatial aspects of EGFR and secondary effector signaling. This systems biological issue is poorly understood at present, but a flurry of investigations is likely to provide insights in the near future.

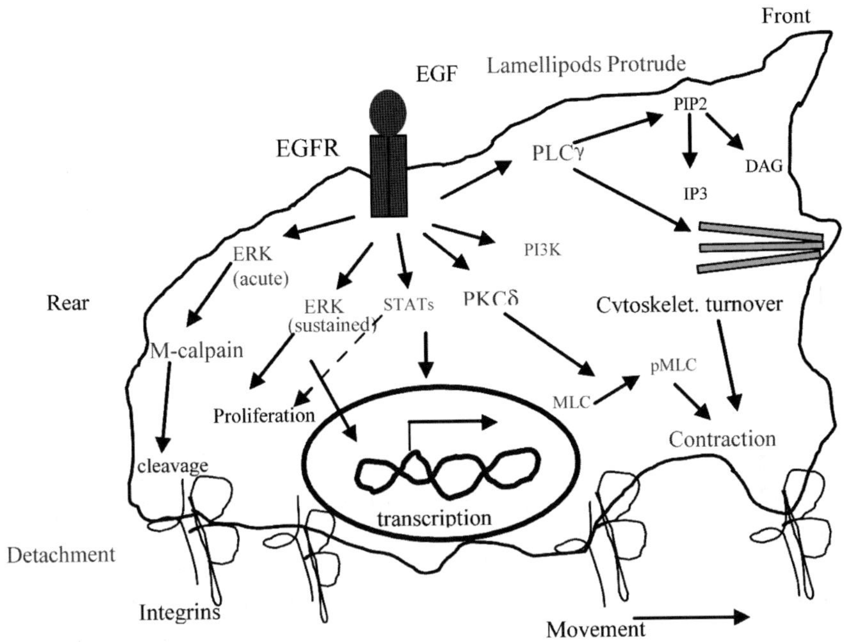

Figure 2. EGF receptor activation by its ligand results in stimulation of a plethora of intracellular signaling cascades that participate in diverse cell responses like cell proliferation and migration. EGFR induced motility associated signaling pathways are highlighted in red. Pathways involved in cell responses other than migration are highlighted in blue. Of these, PI3Kinase is involved in cell survival as well as motility. Similarly, sustained ERK activation is involved in cell proliferation as well as migration.

Upon ligand occupancy, activated EGFR is proposed to undergo a conformational change which leads to receptor dimerization and auto (cross) phosphorylation of multiple tyrosine residues within the cytoplasmic domains. Binding of multiple adapting molecules, including Grb2, Shc and Sos, to these phosphotyrosine residues activates the well characterized Ras > Raf > Mek signaling pathway that finally activates ERK/MAP Kinase pathway. Sustained and tonic activation of ERK1 and 2 (p44/42) is required for cell proliferation and gene expression, while transient and acute activation leads to its membrane translocation that is required for cell motility by causing membrane detachment at the rear end of the cell.

Another signaling pathway activated by EGFR that is by and large specific for growth factor induced motility (and not for proliferation) is the phospholipase C-γ (PLCγ) signaling pathway that ultimately leads to reorganization of cellular cytoskeleton. EGFR also activates signal transducer and activator of transcription (STAT) group of transcription factors, small GTPases including rac, rho, and cdc42, Phosphatidylinositol 3-kinase (PI3K), certain Protein kinase C isoforms (PKCδ) (31), and to a lesser extent phospholipase D and tyrosine kinase Src (52). Thus it is evident, that EGFR participates in diverse cell responses by activation of different downstream signaling pathways in different cell types.

Specifically the PLCγ, ERK/MAP Kinase, and PI3K pathways have been well characterized to govern cell motility downstream of EGFR signaling and these will be discussed in detail. In addition, the functional relevance and role of these pathways will be discussed in the context of prostate tumor cell migration and invasion.

3.1.1.1 Phospholipase C-γ signaling pathway

Signaling through phospholipase C-γ may be a rate-limiting step in EGF-induced cell motility (16). Since EGF receptor activation stimulates proliferation as well as migration, divergent signaling pathways are identified immediately downstream of EGFR. While ERK1/2 activation is required for EGF induced proliferation as well as motility, PLCγ is motility specific and is not required for proliferation (53). Such selective parsing of signals offers novel signaling nodes for therapeutic purposes. Indeed, it has been shown that pharmacological and genetic abrogation of PLCγ signaling pathway significantly decreased migration and invasion of EGFR expressing DU145 and other prostate tumor cells both *in vitro* and *in vivo* (11). Similarly, continual activation of PLCγ signaling pathway via overexpression of EGFR in prostate tumor cells was associated with their significantly increased invasive capability as compared to nontransfected cells (51). This would partly explain the increased motility of certain other tumor cells that express high levels of EGFR on the cell surface.

The net effect of PLCγ signaling is to reorganize the cellular cytoskeleton and "prepare" the cells for movement. Activation of PLCγ leads to hydrolysis of membrane associated phospholipids, predominantly, phosphoinositide 4,5-bisphosphate (PIP2) thereby yielding inositol 1,4,5-triphosphate (IP3) and diacylglycerol (DAG). IP3 stimulates calcium release from endoplasmic reticulum potentially activating classical protein kinase C isoforms. However, the consequence of such PKC signaling has not been demonstrated in terms of tumor progression or carcinogenesis. Interestingly, more impact likely results from the hydrolysis of PIP2 substrate. PLCγ removal of this membrane docking site leads to mobilization of actin binding proteins (ABP) like gelsolin, profilin and cofilin that help in actin polymerization and cytoskeletal reorganization (54). These ABPs are bound to membrane PIP2 in an 'inactive' status as they are removed from the actin target. Upon hydrolysis by PLCγ these ABPs shuttle from the plasma membrane to severe, cap, and polymerize actin molecules enabling that submembrane cytoskeleton to participate in a new burst of polymerization that characterizes the leading lamellipod.

A number of other PIP2 binding proteins have been implicated in cell attachment to the substratum that needs to be loosened for forward protrusion. Vinculin, talin and α-actinin can bind to membrane PIP2 and serve as an anchoring bridge between focal adhesions and actin cytoskeleton. These proteins can localize and sequester actin domains to sites of contact upon PIP2 formation and release them upon PIP2 hydrolysis, thereby spatio-temporally controlling the location of anchor points. Thus, activation of PLCγ pathway can not only disrupt preformed actin filaments to enable new extensions but also loosen the underlying points of attachment for ruffling to occur.

Cytoskeletal reorganization is central to cell motility. Along these lines, some of ABP have been implicated in tumor cell invasion. Recent evidence shows that profilin 1 (PFN1), plays a tumor suppressive role in breast cancer cells (55). Immunohistochemical analysis has revealed intermediate and low levels of PFN1 in different human breast cancers (55). Indeed overexpression of PFN1 significantly reduced *in vitro* invasiveness of breast cancer cells and increased cell-substratum adhesion via an increased tyrosine phosphorylation of focal adhesion kinase (56). Mobilization of cofilin is noted at the leading edge of extravasating breast carcinoma cells (18). This promotes activation of the Arp2/3 complex to nucleate new actin branches and generate the protrusive force in the lamellipod (19).

3.1.1.2 ERK/MAP Kinase signaling pathway

EGF is a robust activator of ERK/MAP Kinase (ERK) via the well-described Ras > Raf > Mek signaling pathway. Growth factor receptors exhibit their proliferative effects on tumor cells, at least in part, by activating the ERK/MAP kinase signaling pathway (57, 58). In agreement with these findings, inhibition of ERK signaling in prostate tumor cells leads to apoptosis (59, 60) and growth cessation. Advanced stage prostate cancers possess significantly higher levels of activated ERK 1/2 levels compared with low grade prostate cancer (61). Other investigations have also focused on the role of ERK 1/2 in tumor cell motility and invasion (62-64). One study found that FGF-1 stimulation induced the expression of promatrilysin via ERK/STAT 3 dependent manner thereby contributing to invasion of prostate carcinoma cells (63). In another study, it was found that hypoxia of prostate cancer cells, induced the expression of urokinase type plasminogen activator (uPA) receptor (uPAR) via ERK and p38 MAP Kinase pathways

(62). These findings suggest downstream changes in the cell proteome underlie the mode by which ERK increases tumor progression.

Epigenetic events might contribute to tumor progression also. ERK/MAP Kinase is required for haptokinetic motility of fibroblasts as well as EGF-induced motility. During haptokinesis, ERK1/2 is proposed to stimulate MLC kinase to phosphorylate the regulatory myosin light chain (MLC) protein (65). However, upon EGF stimulation, ERK1/2 signaling activates the intracellular protease m-calpain by direct phosphorylation (37). Calpain is required for tail de-adhesion during migration over moderately to highly adhesive surfaces. To further extend these studies in cancer cells, pharmacological or molecular inhibition of m-calpain significantly decreased prostate tumor cell invasiveness (12). Thus, ERK signaling might play multiple roles in immediate motility via m-calpain, supportive haptokinesis signaled by adhesion receptors, and longer term upregulation of matrix modifying enzymes for penetration through the ECM barriers.

3.1.1.3 Phosphatidylinositol 3-Kinase signaling pathway

The phosphatidylinositol 3-Kinase (PI3K) signaling pathway is likely as contributory as Ras > ERK pathway to cell proliferation and survival. PI3 kinases comprise a family of three subclasses, of which the originally defined class I has been implicated in oncogenesis and tumor progression (66). Each enzyme molecule is a heterodimer consisting of a regulatory p85 subunit and a catalytic p110 subunit. The regulatory subunit is a direct substrate of many receptor tyrosine kinases. Also, p85 can directly associate with phosphotyrosine residues of activated growth factor receptors via its paired SH2 domains, thereby leading to activation secondary to steric changes. In addition, p110 subunit associates with Ras GTPase upon RTK activation, directly activating the catalytic domain of the enzyme. These many nodes of regulation highlight the central role of PI3K.

PI3K's most evident action is to phosphorylate the 3' position of the inositol ring, with preference for the membrane phospholipid PI(4,5)P2 to produce phosphatidylinositol 3,4,5- triphosphate (PIP3). The counterregulatory tumor suppressor PTEN antagonizes PI3K actions by removing this modification. PTEN is downregulated in many prostate tumors, presumably leading to unopposed PI3K actions and high levels of PIP3. The production of PIP3 leads to activation of Akt/Protein Kinase B via intermediary kinase(s) PDK1 (and possibly PDK2). Akt signaling strongly promotes cell survival and proliferation.

PI3K activity has also been linked to increased migration, matrix metalloproteinase production and stromal invasion of certain tumors (67, 68), with some of this occuring through activation of Akt1 (69, 70).

Similarly, overexpression of Akt2/ Protein kinase Bβ, increases invasion of breast and ovarian cancer cells via upregulation of β1 integrins (71). EGFR activation has been shown to increase migration and invasion of breast cancer cells via activation of PI3K and PLC signaling pathways (72). Integrins also activate PI3K, with the PI3Kinase pathway being required for αvβ3 mediated migration of highly invasive PC3 cells (30, 73).

How PI3K (or reduced PTEN) may promote cell motility, independent of gene transcription, relates the creation of unique PH (plekstrin homology) domain docking sites on the membrane. This has been best described in *Dictyostelium* in which chemotactic G protein coupled receptors recruit PI3K to the front lamellipod (74, 75). Breast carcinoma cells chemotaxing to EGF display similar frontal recruitment of PI3K promoting actin nucleation of this site (76). Thus, while PLCγ removes PIP2 docking sites from the front of the cell, PI3K both removes these sites and creates new ones in PIP3. As such, these two molecules may reinforce the actions of the others, at least in this biochemical event.

3.1.1.4 Signal Transducer and activator of transcription (STAT) proteins

EGFR dependent activation of STATs, especially 1, 3, and 5, is an independent signaling mechanism downstream of EGFR activation, and does not require activation of intermediate tyrosine kinases (77). STATs are constitutively activated in squamous carcinoma cells which express high levels of EGFR on the surface and have increased autocrine signaling (78-80). This signaling nexus is required for EGFR dependent proliferation of prostate carcinoma cells (81-84). Unpublished reports from our laboratory have shown that EGFR mediated cell motility can be abrogated using genetic interventions against STATs in both fibroblasts and prostate cancer cells.

3.1.2 Ligand-receptor interactions in modulating EGFR signaling

Upon binding the known, high affinity ligands, EGF receptors undergo internalization and subsequent degradation within the endosomes (52). This serves primarily as an attenuation mechanism (85). However, internalized EGF-EGFR complexes still activate certain signaling pathways that are crucial in cell proliferation and gene expression (86, 87). Once internalized, EGFR occupancy determines receptor and ligand fate. EGF remains tightly bound to the receptor driving both ligand and receptor to degradation. TGF-α, which does not signal from the endosomes since the endosomal acidic pH leads to dissociation of TGF-EGFR complex, results in EGFR recycling, while ligand sorts with the fluid phase in that 2/3 is shunted to the lysosome.

Interestingly, when ligand is replenished in an autocrine fashion, TGF-α produces more prolonged EGFR signaling as the receptor is 'spared' compared to EGF autocrine production (49). Thus, the specific ligand present during autocrine stimulation of prostate carcinomas, predominantly being TGF-α, dictates EGFR signaling in a temporal and spatial manner.

Co-expression of high levels of erbB2 (HER2, c-neu) limits EGFR degradation by retarding the internalization step (88). This results in a perceived more robust EGFR signaling and likely underlies the mechanism by which erbB2 promotes tumor progression.

A second mode of EGFR activation that also limits EGFR degradative attenuation involves transient binding by ultra-low affinity ligands (89, 90). Some of the EGF-like repeats in the ECM components tenascin-C, laminin, and decorin bind to EGFR but in a manner that is sufficiently transient so as not to drive internalization (89-91). This restricts signaling from EGFR to the plasma membrane, and under this restriction motility appears preferential to proliferation at limiting levels of ligand (unpublished observations). These low affinity EGFR ligands can be liberated or unmasked from the extracellular matrix via the action of matrix metalloproteinases during conditions of organogenesis, matrix remodeling or tumor invasion (89). The fact that many of the ECM proteins, tenascin-C in particular, are upregulated (or dysregulated) in solid tumors strengthens the observation that cell-ECM interactions or "sensing of the stroma" by the cell is a vital element during tumor invasion. Also, since these ECM-embedded ligands cannot be internalized, they can stimulate EGFR dependent motility thereby enhancing invasiveness of tumor cells. Elucidation of the role of such "cryptic" ligands in the ECM offers a new avenue for further intensive research in extracellular matrix biology.

3.2 Hepatocyte Growth Factor and c-Met Signaling System

Hepatocyte growth factor (HGF) was initially shown to cause "scattering" of epithelial cells in culture, hence alternative name scatter factor (SF) (92). HGF binds to its receptor, c-Met, a proto-oncogene and receptor tyrosine kinase, with resultant tyrosine autophosphorylation (93). HGF and c-Met activities have been linked to the genesis of various physiological responses like morphogenesis and mitogenesis as well as pathological ones like tumor adhesion and growth, angiogenesis, tumor invasion and metastasis (94-96).

HGF and c-Met play a crucial role in prostate cancer progression and metastasis especially to the bone (97-99). HGF/SF is overexpressed in human prostate adenocarcinomas as compared to normal prostatic tissue

(98). In established prostate cancer cell lines, highly invasive and androgen-independent DU145 and PC3 cells showed significantly higher expression of c-Met as compared to minimal levels seen in non-invasive androgen-dependent LNCaP cells (100). Increasing HGF levels increases the invasive potential of prostate cancer cells (100). HGF production is likely controlled by other RTK systems as EGFR signaling induces this other ligand (101). That the signaling system is critical is shown by similar findings concerning the overexpression of the receptor, c-Met (98). c-Met overexpression is highly associated with bone and lymph node metastasis, as well as increased proliferative potential of prostate tumors (97). Conversely, ribozyme-mediated reduction of c-Met expression inhibited tumorigenic growth and lymph node metastases of PC3-LN4 prostate tumor cells in an orthotopic nude mouse model (102).

The underlying cell behavior promoted by increased c-Met signaling is likely greater migration, as HGF increases haptokinetic migration and invasion of DU145 and PC3 prostate cancer cells through Matrigel. This occurs in conjunction with upregulation of urokinase plasminogen activator (uPA) and its receptor (uPAR); in addition, matrix metalloproteinases (MMP-1 and MMP-9) expression is increased (100, 103, 104).

3.2.1 HGF/SF and cell-cell adhesion

Stable adhesions between normal epithelial cells are maintained by cadherin molecules at adherens junctions. Cadherins comprise a family of transmembrane cell surface glycoproteins that mediate calcium-dependent, homotypic cell–cell interactions through their extracellular domains, and regulate a variety of biological processes during development, morphogenesis, and tumor metastasis (105, 106). E-cadherin's function in cell-cell adhesion critically depends on its ability to interact through its cytoplasmic domain with catenin proteins. A diverse collection of defects alter cadherin-catenin function in cancer cells, including loss-of-function mutations and defects in the expression of E-cadherin and certain catenins, such as α-catenin (107). During prostate cancer progression, epithelial cells transition into a mesenchymal-like phenotype (epithelial-mesenchymal transition or EMT) (108) that facilitates their dissemination and during this process, E-cadherin expression is frequently downregulated or even undetectable (109-111). Abberantly low levels of E-cadherin and α-catenin are associated with poor survival of patients with prostate cancer (112).

Decreased cell-cell adhesion may not be cell autonomous but actually a response to RTK signaling. Both autocrine EGFR signaling and activation of c-Met receptor by HGF downregulate E-cadherin expression with subsequent increased cell proliferation, dedifferentiation, and induction of cell motility (113). EGFR mediated downregulation of E-cadherin has been suggested in the progression of breast carcinoma cells to a more invasive phenotype (113). In fact, blocking autocrine EGFR activation upregulates E-cadherin and catenin levels in prostate cancer cells limiting their migration and invasion [Yates,C; Wells, A; Turner, T; Luteinizing hormone releasing hormone (LHRH) analog reverses the cell adhesion profile of DU145 human prostate carcinoma (submitted)].

Concomitant with the loss of E-cadherin, N-cadherin levels increase during the EMT noted in carcinomas. In line with this, the expression of N-cadherin is increased in invasive prostate cancer cell lines (186). Although not fully characterized in prostate cancer cells, the switch from E-cadherin expression to N-cadherin expression would support an increased metastatic ability of disseminated tumor cells by enabling circulating tumor cells to associate with the stroma and the endothelium at distant sites (108, 114-116).

The cellular basis of the tumor progression is now being deciphered. First, the decreased E-cadherin levels compromise the tight junctions and enable apically-secreted EGF to establish an autocrine loop with the basolaterally sequestered EGFR (42). This further depresses E-cadherin levels and promotes HGF production, in a reinforcing cycle. In addition to disrupting the cell-cell junctions, and enabling a more migratory phenotype (117), HGF/SF upregulates secretion of matrix metalloproteinases that degrade the extracellular matrix aiding in tumor dissemination. HGF upregulates matrilysin (MMP-7) that mediates extracellular cleavage of E-cadherin, thereby further disrupting cell-cell cohesion and switching of prostate cells from a lesser to a highly invasive phenotype (93). Other studies have also confirmed the role of matrilysin in potentiating invasive property of DU145 prostate cancer cells (118). The increase in N-cadherin promotes interactions with endothelial cells needed for extravasation and intravasation during metastases (116). Thus, the EMT conspires to enhance tumor dissemination.

3.3 Urokinase Plasminogen Activator Receptor (uPAR) System

Autocrine growth factor signaling in prostate carcinoma cells is reflected by an altered proteome. Two proteins upregulated to a great degree are the urokinase type plasminogen activator (uPA) and its receptor (uPAR) (119, 120). This is intriguing for a number of reasons. First, the tumor cell-associated urokinase-type plasminogen activator system, consisting of the serine protease uPA, its substrate plasminogen, its membrane-bound receptor uPAR, as well as its inhibitors PAI-1 and PAI-2, plays an important role in the pericellular processes relating to matrix remodeling (121). Plasmin is a serine proteinase generated *in vivo* from the zymogen plasminogen by the action of uPA and tissue plasminogen activator (tPA). Plasmin facilitates activation of metalloproteinases, like MMP-7 and MMP-2 (122, 123). Recent studies demonstrate that cell-associated uPA plays a critical role in regulating the amount of plasmin present at the surface of prostatic carcinoma cells and that differential production of uPA directly correlates with the capacity to bind and activate plasminogen (123). In other words, presence of uPA-uPAR complex on the cell surface characterizes the capacity of prostate cells to activate plasmin and remodel matrix. The uPA-uPAR system has also been shown to be active in other solid tumors like those of breast, brain and urinary bladder (124, 125). Second, uPAR acts as a motogenic receptor though whether the actual cell signaling events occur directly from uPAR or through the EGFR or integrins is still unsettled (126, 127). Third, uPA, bound to uPAR, cleaves pro-HGF generating active ligand that would then trigger HGF receptors on the cell (128, 129). Thus, uPA/uPAR would promote carcinoma progression by inducing multiple cell properties.

This potential role has been confirmed at least in principle. Pharmacological inhibitors, blocking antibodies, and molecular decoys to uPA decrease *in vitro* invasion of DU145 and PC3 prostate tumor cells along with a subsequent inhibition of cell proliferation (123, 126). Molecular downregulation of uPAR using specific antisense oligonucleotides also decreases prostate tumor invasion, thereby fortifying one of the growing concepts in tumor biology that upregulation of growth factors in tumor cells facilitates tumor invasion by upregulating genes that encode for proteins required for increased cell motility as well as those involved in ECM remodeling. Concurrently, other serine proteases have also been implicated in prostate tumor invasion by mediating ECM degradation. Specifically, Hepsin, has been shown to be upregulated in almost 90% of prostate cancers, and is expressed almost exclusively by tumor cells. Overexpression of Hepsin was shown to modulate basement membrane disorganization of prostate epithelial cells and promote prostate tumor invasion and metastasis to lung, liver and bone in a non-metastasizing mouse model of prostate cancer (187).

3.4 Adhesion Receptors

Cell locomotion requires adhesion receptors to provide the interface with the surfaces. These adhesion receptors actively signal during this process, enabling or preventing the RTK-mediated responses, or even, themselves, driving the motility. Extensive evidence suggests that unregulated growth and migration of tumor cells is due in part to alterations of integrin expression accompanied with a loss of cell-cell adhesion molecules. In order for the tumor cells to invade through the stroma, the cells need to "understand" the cues from the extracellular matrix. Integrins, a family of transmembrane receptors composed of heterodimers of one alpha (α) and one beta (β) subunit, are of primal importance in their ability to transmit signals from the remodeled ECM and mediate cell-cytoskeletal-ECM attachments needed for cell migration. In addition, changes in integrin receptor expression alter the intracellular signaling through various signal transduction pathways including the above-described PI3K and ERK pathways. It must be noted that while the intracellular signaling cascades activated might share the same molecular members as those derived from RTK, the temporospatial aspects of signaling are quantitatively and likely qualitatively different, and thus drive distinct cell behaviors.

As prostate cancer progresses, the repertoire of integrin subunit expression is altered, along with the surrounding ECM. The most notable switch of integrins in normal prostate tissue to prostate cancer tissue as determined by immunohistochemistry is the switch from α6β4 to α3β1 and/or α6β1 (130, 131). Changes like these have been correlated with progression from carcinoma *in situ* to invasive prostate carcinoma (132, 133). In some other studies, however, the α6β4 has been shown to be responsible for the highly invasive phenotypic behavior of PC3 prostate carcinoma cells (134). Thus, the localization and function of the α6β4 integrin changes with the carcinoma progression (see Chapter 8).

It is reasonable to contend that tumor cells utilize different integrin receptor complexes depending on the cell type and the site of metastasis. It has been reported that well established prostate cancer cells LNCaP (lymph node metastasis) and PC3 prostate cells (bone marrow metastasis) express different integrins at the metastastic organ, respectively. LNCaP cells express α6β4 and do not migrate on vitronectin (VN) while highly invasive PC3 cells primarily present αvβ3 and migrate on VN which is expressed in the mature bone marrow matrix (135, 136). Integrin αvβ3 was present only on primary prostatic adenocarcinoma cells, but not on cells obtained from normal prostate tissue. Forced expression of αvβ3 in LNCaP cells increased their migratory property and mediated attachment to VN (135). In prostate cancer a novel form of the α6 integrin, which lacks a large portion of extracellular domain (α6p), is found paired with β1 (137). This altered integrin heterodimer (α6pβ1) maintains cell-cytoskeletal linkages whereas altering cell-matrix adhesion, giving tumor cells selective advantages for metastases. Other studies have shown that αII(b)β3 plays a crucial role in invasiveness of DU145 prostate tumor cells in SCID mice; a monoclonal antibody against αII(b)β3 inhibited the metastasis of these cells to the lungs (138). Multiple integrins are likely involved in that DU145 also express

$\alpha 6\beta 1$ and $\alpha v\beta 3$ (139).The selection against the $\alpha 4$ subunit and the persistence of the $\beta 1$ subunit are a result of the EMT evident in prostate cancer cells, thus providing increased potential for tumor cells to invade and migrate.

The altered expression of integrins mediates migratory properties of tumor cells at least in part by altering intracellular signaling from the ECM in addition to modulating cell–matrix adhesive properties. In this regard, FAK has been shown to mediate $\alpha v\beta 3$ triggered migration of prostate cancer cells (135). Signaling from ERK and PI3K pathways downstream of integrins also affects tumor cell motility, cell adhesion, proliferation, and survival (140, 141). It is sufficient to say that altered expression of integrin receptors on the cell surface is a crucial element for the tumor cells to "understand" the cues from the remodeled extracellular matrix through which the tumor cells invade. For a detailed description of the role of adhesion receptors in prostate tumor metastases, the reader is referred to other excellent reviews (141) (see Chapter 10).

3.5 Rho GTPases and cytoskeletal reorganization

Rho GTPases are members of the Ras superfamily of monomeric 20-30 kDa GTP-binding proteins that act as molecular switches by alternating between active GTP-bound and inactive GDP-bound forms (142) (see Chapter 9). Dysregulated Rho GTPases have been discovered in many human tumors, including colon, breast, lung, prostate and head and neck squamous-cell carcinoma (143). The members of this large family of proteins most often associated with carcinogenesis are RhoA, Cdc42 and Rac1. These GTPases are characterized for their role in cytokinesis via cellular cytoskeletal modulation (144), cell contraction and spreading (145), cell-substratum adhesion, etc. That Rho GTPases have also been implicated in modulating prostate cancer invasiveness by driving cell motility is not unexpected (13, 143).

Integrin ligation and RTK signaling both lead to activation of these GTPases. RhoA is involved in maintaining actin stress fibers, cell adhesion and cell spreading on ECM. Abrogation of RhoA using pharmacological inhibitors have shown to substantially decrease invasion of PC3 prostate cancer cells *in vivo* and *in vitro* (13). In addition, Rho GTPases have shown to be involved in upregulating the transcription of certain genes including uPAR (146) and NFκB (147), both of which have been implicated in prostate cancer cell metastasis. Activity of RhoA was shown to be necessary and sufficient for NFκB mediated invasion and migration of PC3 cells (148).

Cdc42 and Rac1 are crucial in maintaining cell polarity, asymmetry, cytoskeletal remodeling and actin dynamics at the leading edge of the moving cell. Specific pharmacological inhibitor of Rac1 blocked anchorage-independent growth and invasiveness of PC3 prostate cancer cells (149). These proteins act via numerous downstream effectors; e.g. Cdc42 can inhibit MLC phosphorylation by inhibiting MLCK via p21-activated kinase (PAK) (150) thereby affecting transcellular contractility. Also, PAK has shown to activate LIM-kinase which has been shown to be overexpressed in cancers of prostate and breast and is proposed to modulate growth, cell shape and motility of these cancer cells (151, 152). Accordingly, expression of dominant negative LIM-kinase in these cells was shown to decrease their invasion through matrigel (152). Thus, substantial and inundating crosstalk is evident between multiple signaling molecules that modulate various cellular biophysical events during cell motility, thereby highlighting different nodes that control and manipulate this highly orchestrated process.

3.6 CXC Chemokines Receptor Systems

Any discussion of prostate cancer migration would not be complete without a mention of the role of chemokines in invasive cancers. One class of chemokines that has been implicated is the CXC chemokines that are classified into two groups based on the presence or absence of the 'ELR' motif (Glu-Leu-Arg) preceding the first conserved cysteine amino acid in the NH2 terminus. The first group of CXC chemokines contains the ELR motif and is angiogenic and the most important member being studied in context of prostate cancer is Interleukin-8 (IL-8). The second group of CXC chemokines, which lack the ELR motif, include interferon- γ -inducible protein (IP-10), monokine induced by γ-interferon (MIG), platelet factor 4 (PF4), and stromal derived factor (SDF-1). Chemokines activate receptors that are members of the large family of seven-transmembrane domain spanning G protein-coupled receptors and activate adenylate cyclase and/or phospholipase C-β to accomplish the bulk of their signaling.

Clear involvement of (SDF-1 or CXCL12) and it's receptor CXCR4 in invasive cancers have been recently demonstrated (153-155). Using high density tissue microarrays it was shown that the chemokine SDF-1 (or CXCL12) and its receptor (CXCR4) are expressed in highly metastatic prostate cancer (155). The importance of CXCR4 is highlighted in its putative role in "homing" of the prostate cancer cells in bone during metastasis (155). Antibody against CXCL12 inhibited growth and proliferation of bone homing LNCaP, PC3 and C4-2B metastatic prostate cancer cells. Upon SDF-1 stimulation alterations in adherence, migration, and invasion of human prostate cancer cell lines have been observed (156). Further emphazing the importance of chemokines in migration there have been preliminary reports that CX3CR1 and CXCR2 receptors and their ligands, fractalkine and IL-8 promote prostate carcinoma cell migration (157, 158). Adhesion of prostate cancer cells to human bone marrow endothelial cells in flow conditions was significantly reduced by a neutralizing antibody against fractalkine, and they migrate toward a medium conditioned by osteoblast, which secrete the soluble form of the chemokine (158). Also an antibody to IL-8, the secretion of which has been shown to be upregulated during prostate cancer progression (159), limited migration of prostate carcinoma cells (157). It is likely that the CXCR3 ligands will also modulate invasiveness as they regulate cell migration in endothelial cells, fibroblasts, and keratinocytes. Chemokine receptor intervention is an emerging target for therapeutics in the progression of cancer, however more studies are needed to fully characterize the pathways these receptors elicit during their role in overt metastasis.

3.7 Sex Hormone Endocrine Factors Governing Prostate Cancer Progression

3.7.1 Androgens and androgen receptor

Prostate epithelial cells, both normal and cancerous, depend on androgens for survival and normal differentiation (160). This is the basis of significant advances and alternatives in managing advanced cases of prostate cancer using chemical (via Leutinizing hormone releasing hormone or LHRH agonists) or surgical castration often combined with peripheral blockade of androgen action. However, in most cases, androgen deprivation therapy is associated with only a transient regression of prostate cancer growth extending survival only by about 10 %, as it is followed by a relapse and progression to an androgen-independent state. There is little current therapy for these advanced, aggressive carcinomas.

The molecular bases of progression from an androgen-sensitive to an androgen-independent phenotype are largely unclear. Some studies suggest that the androgen-independent state is actually just a very low dependency. Recurrent or secondary tumors frequently exhibit overamplified androgen receptor (AR) that can lead to metastatic tumor proliferation despite very low levels of circulating androgens in plasma (161-163). In corroboration with this observation, overexpression of androgen receptor is associated with aggressive clinicopathologic features and decreased biochemical recurrence-free survival in prostate cancer patients treated with radical prostatectomy (164). Few secondary tumors, but not primary prostate cancers, also express androgen receptor gene mutations that can affect receptor-ligand binding affinity, androgen insensitivity and altered expression levels (165, 166).

A second mechanism for escaping androgen-dependency likely utilizes autocrine RTK stimulation. Actually, the autonomous and uncontrolled proliferation of androgen-independent prostate carcinoma cells occurs via secretion of growth factors like EGF and TGF-α in the medium and is not dependent on androgen (167). Androgen independent cell lines like PC3 and DU145, express higher levels of EGF receptor (autocrine) phosphorylation as compared to the androgen responsive LNCaP cells (168). Some of these signaling might feed back into androgen functioning as ligand independent activation of androgen receptor has also been shown to be mediated via growth factor receptors like EGFR, IGFR-1, etc. in prostate tumor cells (169, 170). Treatment of androgen receptor expressing PC3 cells with dihydrotestosterone has shown to upregulate EGF receptor expression and its binding affinity for EGF, thereby enhancing the mitogenic properties of EGF (171). A small subset of clinical prostate cancer specimens has also demonstrated Her2/Neu overexpression but the clinical activity of an antibody treatment directed against Her2/Neu receptor as a single agent has been disappointing (172). These studies point towards substantial crosstalk between androgen receptor and growth factor receptors in prostate tumor progression. An emerging concept is that after therapeutic androgen ablation, a relapse of prostate cancer is characterized by the ability of tumor cells to overcome their androgen dependence and exhibit autonomous growth via secretion of exogenous growth factors. This addresses an interesting possibility of a new therapeutic approach of utilizing growth factor receptor blockade in addition to conventional therapies in treating prostate cancer.

While the above studies point to a detrimental effect of AR expression on prostate tumor cell behavior, some studies have shown that androgen responsiveness is associated with lower recurrence after therapy and lower metastatic potential. Introduction of human AR into the invasive human prostate cancer ARCaP cell line restored its androgen-regulated cell growth, decreased the rate of tumor growth, and selectively activated AR target gene expression (173). Similarly, androgens have also shown to modulate selective integrin expression on cell surface. Increased AR expression in highly invasive PC3 cells resulted in downregulation of α6β4 integrin expression, reduced adhesion on laminin and decreased Matrigel invasion *in vitro* (134).

3.7.2 Estrogens and estrogen receptors

After the discovery of dependence of prostate cancer on androgens but before the advent of LHRH agonists, diethylstilbestrol (DES) emerged as a sound alternative to androgen deprivation in treatment of prostate cancer, although not entirely free of adverse effects. Estrogen acts via two receptors, namely the classical and well characterized α receptor (ER-α) and the novel β receptor (ER-β), both of which are abundantly expressed in normal prostate tissue. ER-β is known to regulate the growth of normal prostate epithelial cells as well as protect these cells from oxidative stress induced neoplastic transformation.

Estrogen receptor β (ER-β) is frequently lost in prostate cancer as the stage advances, but frequently reappears within bone marrow and lymph node metastatic foci (174). Adenovirus mediated restoration of ER-β expression in DU145 prostate cancer cells inhibited their proliferation and *in vitro* invasion through Matrigel, and induced apoptosis (175), thereby suggesting a possible role of ER-β in tumor suppression. However, the reappearance of ER-β in metastatic tumor foci is shown not to impair the prostate cancer growth at such ectopic sites, and possibly even confers a survival advantage to these tumor cells. In contrast to the proposed tumor suppressive role of ER-β, ER-α activation has been proposed to play a role in tumorigenesis (176). It is possible that estrogens alone, or in synergy with an androgen, induce aberrant growth and neoplastic transformation in the prostate, probably via their action on ER-α. In agreement with this contention, the selective estrogen receptor modulator, Trioxifene, was shown to inhibit metastasis and extend survival in the PAIII rat prostatic carcinoma model (177).

In addition to genetic instability, epigenetic stability too is a hallmark of malignant cells. With the help of epigenetic manipulations, the tumor cell is able to control specific gene expression patterns and express proteins that are required at specific stages in tumor progression. Evidence suggests that ER-β gene is subject to DNA methylation and thereby gene silencing as the prostate cancer progresses (178). Interestingly, treatment of both prostate and breast carcinoma cells with the DNA methyltransferase inhibitor 5-aza-2'deoxycytidine (5-aza-dC) led to reactivation of silenced ER-α and in some cases ER-β gene, further providing evidence that the ER-β gene, similar to the ER-α gene, may be subject to regulation by DNA methylation in these solid tumors. In addition to estrogen receptors, the progesterone receptor has also shown to be upregulated in later stages of the androgen independent prostate cancer tissue specimens, along with a corresponding increase in ER-α mRNA levels (176, 179). Although the precise implication of such a finding is open to further investigations, it is possible that at later stages of tumor progression, androgen-independent prostate cancer cells utilize estrogen dependent signaling pathways for growth and/or metastases. As growth factor receptor overexpression also leads to substantial genetic and epigenetic variations in tumor cells, such research questions aimed at elucidation of the precise role of the two estrogen receptors and their crosstalk with other growth factor receptors during prostate cancer progression offers a fertile ground for future research.

4. CELL MOTILITY AS A NOVEL THERAPEUTIC TARGET IN LIMITING PROSTATE TUMOR PROGRESSION

Aims to curb and cure prostate cancer have been only partially successful. Mortality due to prostate cancer, once it has escaped from the confines of the prostate capsule, remains high. Thus, limiting tumor invasion and metastasis provides a strong rationale in novel cancer therapeutics in that tumors can be rendered less invasive and more amenable for surgical resection with a lower probability of recurrence at best, or slowed in their further dissemination. Targeting tumor invasion needs a precise and detailed knowledge of different underlying phenotypic characteristics that a tumor cell acquires during the process of metastasis. The current knowledge of this comprises this chapter.

Cell motility, specifically, as a critical requirement for tumor invasion, affords an attractive target that can be explored in therapy of advanced and aggressive prostate cancers. Segregation of individual biophysical events of cell motility have elucidated the detailed kinetics of this complicated and well orchestrated cellular process. Laboratory based experiments in tumor biology aimed at disrupting these biophysical events have shown promising results, e.g. inhibition of rear cell membrane detachment by targeting calpain using a specific calpain inhibitor ALLN, decreased invasion of DU145 prostate tumor cells across Matrigel and limited dissemination in a murine xenograft model of prostate cancer progression (12). Similarly, abrogation of PLCγ signal transduction pathway that is crucial in lamellipodal protrusion of migrating tumor cell using a specific pharmacological or molecular agent inhibited migration and invasion of prostate carcinoma cells *in vitro* and *in vivo* (11). Preliminary data from our laboratory suggest that PLCγ targeting also reduced the metastases and invasiveness of *in situ* occuring prostate and breast cancers in transgenic mice. Interestingly, none of the agents used in these studies affected the proliferation of these tumor cells. Myosin II also has been a target to limit prostate tumor invasion (13). In addition, investigators have focused on targeting cell adhesion molecules (CAMs) such as integrin receptors in limiting tumor cell motility. Current anti-adhesion molecule cancer therapy is dominated by the use of neutralizing antibodies, immunotoxins, and soluble proteins (180, 181); e.g.Vitaxin, a humanized antibody, and cilengitide, a cyclic peptide mimicking the RGD

(Arg-Gly-Asp) ligand recognition peptidic domain common to alpha v integrin ligands, are in phase II clinical trials (181). Thus targeting key underlying biochemical molecules that control individual as well as integrated biophysical processes of cell migration offer a very promising avenue in cancer therapeutics.

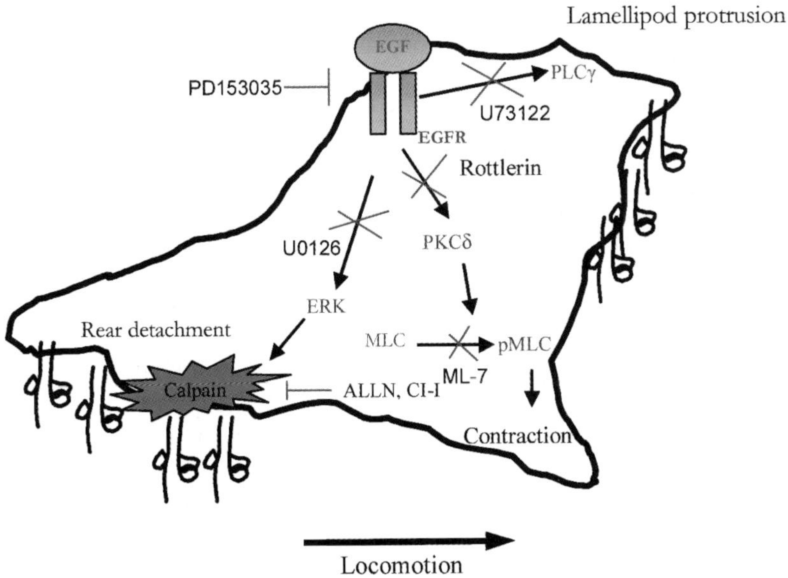

Figure 3. Epidermal growth factor induced motility of tumor cells can be abrogated at multiple signaling nodes that control different biophysical processes. Different commercially available chemical compounds that abrogate specific intracellular signaling cascades in vitro are shown; e.g. inhibition of PLCγ signaling pathway via U73122 can disrupt lamellipodal protrusion as well as cytoskeletal reorganization. Similarly cell contractility can be limited by inhibition of PKCδ by Rottlerin as well as limiting myosin light chain phosphorylation by ML-7. Finally, rear membrane detachment can be inhibited by targeting ERK (by U0126) and limiting calpain activity using specific inhibitors like Calpain inhibitor – I (or CI-I) and ALLN. These compounds can be further explored as potential drugs in the treatment of invasive and metastatic prostate cancers.

This is indeed an era of "biological therapy". Relentless and inundating research investigations in the field of tumor biology have unraveled the roles of countless biologically active molecules and signal transduction pathways in dictating key cellular responses like proliferation, evasion of apoptosis and migration. Indeed, with availability of highly specific and effective tools for genetic interventions like short interfering RNA (siRNA), the prospects of “downregulating” oncogenic proteins / receptors posit a challenging and exciting avenue for cancer researchers. The past few years have seen the clinical use of monoclonal antibodies and specific drugs against growth factor receptors like EGFR, HER-2 and Platelet derived growth factor receptor (PDGFR)/c-kit that are aimed at abrogating tumor proliferation and motility especially in breast, ovarian and lung cancers (182, 183). Pharmacological agents against matrix metalloproteinases (like Batimastat and Marimastat) and the uPA-uPAR system are being exploited for their effectiveness in treating invasive tumors (184, 185). In addition, signal transduction pathways like those of ERK/MAP Kinase and PI3Kinase /Akt pathway that regulate tumor cell proliferation and survival offer new potential targets in anti-cancer drug discovery. Thus anti-cancer therapy aimed at inhibition of these vital signaling molecules and the key biophysical events they control, holds a promise in the management of this deadly disease.

ACKNOWLEDGEMENTS

We thank the members of the Wells's laboratory for their insightful discussions and help in preparing this manuscript. This work was supported by grants from the Veterans Administration Merit Award program and the National Institutes of Health.

REFERENCES

1. Morton, GC. Early prostate cancer. Curr. Probl. Cancer 2000, 24: 5-51.
2. Goodin S, Rao KV, DiPaola RS. State-of-the-art treatment of metastatic hormone-refractory prostate cancer. Oncologist 2002, 7: 360-70.
3. Anwar K, et al. Presence of ras oncogene mutations and human papillomavirus DNA in human prostate carcinomas. Cancer Res. 1992, 52: 5991-6.
4. Carter BS, Epstein JI, Isaacs WB. ras gene mutations in human prostate cancer. Cancer Res. 1990, 50: 6830-2.
5. Porter AT, Williams A, Forman JD. Adjuvant radiation therapy for pathologic T3 prostate cancer. Can. J. Urol. 1997, 4: 57-60.
6. Schild SE. Radiation therapy (RT) after prostatectomy: The case for salvage therapy as opposed to adjuvant therapy. Int. J. Cancer 2001, 96: 94-8.
7. Wells A. Tumor invasion: role of growth factor-induced cell motility. Adv. Cancer Res. 2000, 78: 31-101.

8. Gleave ME, et al. Prostate and bone fibroblasts induce human prostate cancer growth in vivo: implications for bidirectional tumor-stromal cell interaction in prostate carcinoma growth and metastasis. J. Urol. 1992, 147: 1151-9.
9. Luzzi KJ, et al. Multistep nature of metastatic inefficiency: dormancy of solitary cells after successful extravasation and limited survival of early micrometastases. Am. J. Pathol. 1998, 153: 865-73.
10. Kassis J, et al. Tumor invasion as dysregulated cell motility. Semin. Cancer Biol. 2001, 11: 105-17.
11. Kassis J, et al. A role for phospholipase C-gamma-mediated signaling in tumor cell invasion. Clin. Cancer Res. 1999, 5: 2251-60.
12. Mamoune A, et al. Calpain-2 as a target for limiting prostate cancer invasion. Cancer Res. 2003, 63: 4632-40.
13. Somlyo AV, et al. Rho-kinase inhibitor retards migration and in vivo dissemination of human prostate cancer cells. Biochem. Biophys. Res. Commun. 2000, 269: 652-9.
14. Lauffenburger DA, Horwitz AF. Cell migration: a physically integrated molecular process. Cell 1996, 84: 359-69.
15. Wells A, et al. Growth factor-induced cell motility in tumor invasion. Acta. Oncol. 2002, 41: 124-30.
16. Wells A, et al. Epidermal growth factor receptor-mediated motility in fibroblasts. Microsc. Res. Tech. 1998, 43: 395-411.
17. Pollard TD. The cytoskeleton, cellular motility and the reductionist agenda. Nature 2003, 422: 741-5.
18. Condeelis JS, et al. Lamellipodia in invasion. Semin. Cancer Biol. 2001, 11: 119-28.
19. Pollard TD, Borisy GG. Cellular motility driven by assembly and disassembly of actin filaments. Cell 2003, 112: 453-65.
20. Condeelis J. How is actin polymerization nucleated in vivo? Trends Cell Biol. 2001, 11: 288-93.
21. Horwitz R, Webb D. Cell migration. Curr. Biol. 2003, 13: R756-9.
22. Nobes CD, Hall A. Rho, rac, and cdc42 GTPases regulate the assembly of multimolecular focal complexes associated with actin stress fibers, lamellipodia, and filopodia. Cell, 1995, 81: 53-62.
23. Srinivasan S, et al. Rac and Cdc42 play distinct roles in regulating PI(3,4,5)P3 and polarity during neutrophil chemotaxis. J. Cell Biol. 2003, 160: 375-85.
24. Etienne-Manneville S, Hall A. Rho GTPases in cell biology. Nature 2002, 420: 629-35.
25. Chou J, et al. Directional motility induced by epidermal growth factor requires Cdc42. Exp. Cell Res. 2003, 287: 47-56.
26. Shen TL, Guan JL. Differential regulation of cell migration and cell cycle progression by FAK complexes with Src, PI3K, Grb7 and Grb2 in focal contacts. FEBS Lett. 2001, 499: 176-81.
27. Kornberg LJ. Focal adhesion kinase and its potential involvement in tumor invasion and metastasis. Head Neck 1998, 20: 745-52.
28. Slack JK, et al. Alterations in the focal adhesion kinase/Src signal transduction pathway correlate with increased migratory capacity of prostate carcinoma cells. Oncogene 2001, 20: 1152-63.
29. Manes T, et al. Alpha(v)beta3 integrin expression up-regulates cdc2, which modulates cell migration. J. Cell Biol. 2003, 161: 817-26.

30. Zheng DQ, et al. Substrate specificity of alpha(v)beta(3) integrin-mediated cell migration and phosphatidylinositol 3-kinase/AKT pathway activation. J. Biol. Chem. 2000, 275: 24565-74.
31. Iwabu A, et al. Epidermal growth factor induces fibroblast contractility and motility via a protein kinase C delta-dependent pathway. J. Biol. Chem. 2004, 279: 14551-60.
32. Totsukawa G, et al. Distinct roles of ROCK (Rho-kinase) and MLCK in spatial regulation of MLC phosphorylation for assembly of stress fibers and focal adhesions in 3T3 fibroblasts. J. Cell Biol. 2000, 150: 797-806.
33. Riento K, Ridley AJ. Rocks: multifunctional kinases in cell behaviour. Nat. Rev. Mol. Cell Biol. 2003, 4: 446-56.
34. Palecek SP, et al. Physical and biochemical regulation of integrin release during rear detachment of migrating cells. J Cell Sci. 1998, 111: 929-40.
35. DiMilla PA, et al. Maximal migration of human smooth muscle cells on fibronectin and type IV collagen occurs at an intermediate attachment strength. J. Cell Biol. 1993, 122: 729-37.
36. Glading A, et al. Epidermal growth factor receptor activation of calpain is required for fibroblast motility and occurs via an ERK/MAP kinase signaling pathway. J. Biol. Chem. 2000, 275: 2390-8.
37. Glading A, et al. Epidermal growth factor activates m-calpain (calpain II), at least in part, by extracellular signal-regulated kinase-mediated phosphorylation. Mol. Cell Biol, 2004, 24: 2499-512.
38. Lee J, et al. Regulation of cell movement is mediated by stretch-activated calcium channels. Nature 1999, 400: 382-6.
39. Rios-Doria J, et al. The role of calpain in the proteolytic cleavage of E-cadherin in prostate and mammary epithelial cells. J. Biol. Chem. 2003, 278: 1372-9.
40. Aaronson SA, Growth factors and cancer. Science 1991, 254: 1146-53.
41. Tzahar E, et al. Bivalence of EGF-like ligands drives the ErbB signaling network. Embo. J. 1997, 16: 4938-50.
42. Kim HG, et al. EGF receptor signaling in prostate morphogenesis and tumorigenesis. Histol. Histopathol. 1999, 14: 1175-82.
43. Bargmann CI, Weinberg RA. Increased tyrosine kinase activity associated with the protein encoded by the activated neu oncogene. Proc. Natl. Acad. Sci. USA 1988, 85: 5394-8.
44. Worthylake R, Opresko LK, Wiley HS. ErbB-2 amplification inhibits down-regulation and induces constitutive activation of both ErbB-2 and epidermal growth factor receptors. J. Biol. Chem. 1999, 274: 8865-74.
45. Liu HL, et al. Detection of low level HER-2/neu gene amplification in prostate cancer by fluorescence in situ hybridization. Cancer J. 2001, 7: 395-403.
46. Zhau HY, et al. Transfected neu oncogene induces human prostate cancer metastasis. Prostate 1996, 28: 73-83.
47. Reese DM, et al. HER2 protein expression and gene amplification in androgen-independent prostate cancer. Am. J. Clin. Pathol. 2001, 116: 234-9.
48. Grasso AW, et al. ErbB kinases and NDF signaling in human prostate cancer cells. Oncogene 1997, 15: 2705-16.
49. Reddy CC, et al. Engineering epidermal growth factor for enhanced mitogenic potency. Nat. Biotechnol. 1996, 14: 1696-9.

50. Xie H, et al. In vitro invasiveness of DU-145 human prostate carcinoma cells is modulated by EGF receptor-mediated signals. Clin. Exp. Metastasis, 1995, 13: 407-19.
51. Turner T, et al. EGF receptor signaling enhances in vivo invasiveness of DU-145 human prostate carcinoma cells. Clin. Exp. Metastasis 1996, 14: 409-18.
52. Wells A. EGF receptor. Int. J. Biochem. Cell Biol. 1999, 31: 637-43.
53. Chen P, et al. Epidermal growth factor receptor-mediated cell motility: phospholipase C activity is required, but mitogen-activated protein kinase activity is not sufficient for induced cell movement. J. Cell Biol. 1994, 127: 847-57.
54. Chen, P, Murphy-Ullrich JE, Wells A. A role for gelsolin in actuating epidermal growth factor receptor-mediated cell motility. J. Cell Biol. 1996, 134: 689-98.
55. Janke J, et al. Suppression of tumorigenicity in breast cancer cells by the microfilament protein profilin 1. J. Exp. Med. 2000, 191: 1675-86.
56. Roy P, Jacobson K. Overexpression of profilin reduces the migration of invasive breast cancer cells. Cell Motil. Cytoskeleton 2004, 57: 84-95.
57. Hoshino R, et al. Constitutive activation of the 41-/43-kDa mitogen-activated protein kinase signaling pathway in human tumors. Oncogene 1999, 18: 813-22.
58. Wang J, Eltoum IE, Lamartiniere CA. Genistein alters growth factor signaling in transgenic prostate model (TRAMP). Mol. Cell Endocrinol. 2004, 219: 171-80.
59. Hagan M, et al. Ionizing radiation-induced mitogen-activated protein (MAP) kinase activation in DU145 prostate carcinoma cells: MAP kinase inhibition enhances radiation-induced cell killing and G2/M-phase arrest. Radiat. Res. 2000, 153: 371-83.
60. McKinstry R, et al. Inhibitors of MEK1/2 interact with UCN-01 to induce apoptosis and reduce colony formation in mammary and prostate carcinoma cells. Cancer Biol. Ther. 2002, 1: 243-53.
61. Gioeli D, et al. Activation of mitogen-activated protein kinase associated with prostate cancer progression. Cancer Res. 1999, 59: 279-84.
62. Lee KH, et al. Involvement of MAPK pathway in hypoxia-induced up-regulation of urokinase plasminogen activator receptor in a human prostatic cancer cell line, PC3MLN4. Exp. Mol. Med. 2004, 36: 57-64.
63. Udayakumar TS, et al. Fibroblast growth factor-1 induced promatrilysin expression through the activation of extracellular-regulated kinases and STAT3. Neoplasia 2002, 4: 60-7.
64. Suthiphongchai T, et al. Involvement of ERK1/2 in invasiveness and metastatic development of rat prostatic adenocarcinoma. Oncol. Res. 2003, 13: 253-9.
65. Klemke RL, et al. Regulation of cell motility by mitogen-activated protein kinase. J. Cell Biol. 1997, 137: 481-92.
66. Vivanco I, Sawyers CL. The phosphatidylinositol 3-Kinase AKT pathway in human cancer. Nat. Rev. Cancer 2002, 2: 489-501.
67. Kubiatowski T, et al. Association of increased phosphatidylinositol 3-kinase signaling with increased invasiveness and gelatinase activity in malignant gliomas. J. Neurosurg. 2001, 95: 480-8.
68. Ellerbroek SM, et al. Phosphatidylinositol 3-kinase activity in epidermal growth factor-stimulated matrix metalloproteinase-9 production and cell surface association. Cancer Res. 2001, 61: 1855-61.

69. Park BK, Zeng X, Glazer RI. Akt1 induces extracellular matrix invasion and matrix metalloproteinase-2 activity in mouse mammary epithelial cells. Cancer Res. 2001, 61: 7647-53.
70. Kim D, et al. Akt/PKB promotes cancer cell invasion via increased motility and metalloproteinase production. Faseb. J. 2001, 15: 1953-62.
71. Arboleda MJ, et al. Overexpression of AKT2/protein kinase Bbeta leads to up-regulation of beta1 integrins, increased invasion, and metastasis of human breast and ovarian cancer cells. Cancer Res. 2003, 63: 196-206.
72. Price JT, et al. Epidermal growth factor promotes MDA-MB-231 breast cancer cell migration through a phosphatidylinositol 3'-kinase and phospholipase C-dependent mechanism. Cancer Res. 1999, 59: 5475-8.
73. Cooper CR, Chay CH, Pienta KJ. The role of alpha(v)beta(3) in prostate cancer progression. Neoplasia 2002, 4: 191-4.
74. Comer FI, Parent CA. PI 3-kinases and PTEN: how opposites chemoattract. Cell 2002, 109: 541-4.
75. Iijima M, Devreotes P. Tumor suppressor PTEN mediates sensing of chemoattractant gradients. Cell 2002, 109: 599-610.
76. Hill K, et al. Specific requirement for the p85-p110alpha phosphatidylinositol 3-kinase during epidermal growth factor-stimulated actin nucleation in breast cancer cells. J. Biol. Chem. 2000, 275: 3741-4.
77. David M, et al. STAT activation by epidermal growth factor (EGF) and amphiregulin. Requirement for the EGF receptor kinase but not for tyrosine phosphorylation sites or JAK1. J. Biol. Chem. 1996, 271: 9185-8.
78. Grandis JR, et al. Requirement of Stat3 but not Stat1 activation for epidermal growth factor receptor- mediated cell growth In vitro. J. Clin. Invest. 1998, 102: 1385-92.
79. Xi S, et al. Constitutive activation of Stat5b contributes to carcinogenesis in vivo. Cancer Res. 2003, 63: 6763-71.
80. Kijima T, et al. STAT3 activation abrogates growth factor dependence and contributes to head and neck squamous cell carcinoma tumor growth in vivo. Cell Growth Differ. 2002, 13: 355-62.
81. Lou W, et al. Interleukin-6 induces prostate cancer cell growth accompanied by activation of stat3 signaling pathway. Prostate 2000, 42: 239-42.
82. Ok Lee S, et al. RNA interference targeting Stat3 inhibits growth and induces apoptosis of human prostate cancer cells. Prostate 2004, 60: 303-9.
83. Ni Z, et al. Selective activation of members of the signal transducers and activators of transcription family in prostate carcinoma. J. Urol. 2002, 167: 1859-62.
84. Gao B, et al. Constitutive activation of JAK-STAT3 signaling by BRCA1 in human prostate cancer cells. FEBS Lett. 2001, 488: 179-84.
85. Wells A, et al. Ligand-induced transformation by a noninternalizing epidermal growth factor receptor. Science 1990, 247: 962-4.
86. Haugh JM, et al. Effect of epidermal growth factor receptor internalization on regulation of the phospholipase C-gamma1 signaling pathway. J. Biol. Chem. 1999, 274: 8958-65.
87. Wang Y, et al. Endosomal signaling of epidermal growth factor receptor stimulates signal transduction pathways leading to cell survival. Mol. Cell Biol. 2002, 22: 7279-90.
88. Hendriks BS, Wiley HS, Lauffenburger D. HER2-mediated effects on EGFR endosomal sorting: analysis of biophysical mechanisms. Biophys. J. 2003, 85: 2732-45.

89. Schenk S, et al. Binding to EGF receptor of a laminin-5 EGF-like fragment liberated during MMP-dependent mammary gland involution. J. Cell Biol. 2003, 161: 197-209.
90. Swindle CS, et al. Epidermal growth factor (EGF)-like repeats of human tenascin-C as ligands for EGF receptor. J. Cell Biol. 2001, 154: 459-68.
91. Iozzo RV, et al. Decorin is a biological ligand for the epidermal growth factor receptor. J. Biol. Chem. 1999, 274: 4489-92.
92. Stella MC, Comoglio PM. HGF: a multifunctional growth factor controlling cell scattering. Int. J. Biochem. Cell Biol. 1999, 31: 1357-62.
93. Davies G, Jiang WG, Mason MD. Matrilysin mediates extracellular cleavage of E-cadherin from prostate cancer cells: a key mechanism in hepatocyte growth factor/scatter factor-induced cell-cell dissociation and in vitro invasion. Clin. Cancer Res. 2001, 7: 3289-97.
94. Comoglio PM. Structure, biosynthesis and biochemical properties of the HGF receptor in normal and malignant cells. Exs. 1993, 65: 131-65.
95. Comoglio PM, Trusolino L. Invasive growth: from development to metastasis. J. Clin. Invest. 2002, 109: 857-62.
96. Birchmeier C, et al. Met, metastasis, motility and more. Nat. Rev. Mol. Cell Biol. 2003, 4: 915-25.
97. Lang SH, et al. Scatter factor influences the formation of prostate epithelial cell colonies on bone marrow stroma in vitro. Clin. Exp. Metastasis 1999, 17: 333-40.
98. Humphrey PA, et al. Hepatocyte growth factor and its receptor (c-MET) in prostatic carcinoma. Am. J. Pathol. 1995, 147: 386-96.
99. Pisters LL, et al. c-met proto-oncogene expression in benign and malignant human prostate tissues. J. Urol. 1995, 154: 293-8.
100. Nishimura K, et al. Regulation of invasive potential of human prostate cancer cell lines by hepatocyte growth factor. Int. J. Urol. 1998, 5: 276-81.
101. Gohda E, et al. Induction of hepatocyte growth factor in human skin fibroblasts by epidermal growth factor, platelet-derived growth factor and fibroblast growth factor. Cytokine 1994, 6: 633-40.
102. Kim SJ, et al. Reduced c-Met expression by an adenovirus expressing a c-Met ribozyme inhibits tumorigenic growth and lymph node metastases of PC3-LN4 prostate tumor cells in an orthotopic nude mouse model. Clin. Cancer Res. 2003, 9: 5161-70.
103. Fujiuchi Y, et al. Effect of hepatocyte growth factor on invasion of prostate cancer cell lines. Oncol. Rep. 2003, 10: 1001-6.
104. Parr C, Jiang WG. Expression of hepatocyte growth factor/scatter factor, its activator, inhibitors and the c-Met receptor in human cancer cells. Int. J. Oncol. 2001, 19: 857-63.
105. Gumbiner BM. Carcinogenesis: a balance between beta-catenin and APC. Curr. Biol. 1997, 7: R443-6.
106. Fagotto F, Gumbiner BM. Cell contact-dependent signaling. Dev. Biol. 1996, 180: 445-54.
107. Hajra KM, Fearon ER. Cadherin and catenin alterations in human cancer. Genes Chromosomes Cancer 2002, 34: 255-68.
108. Tran NL, et al. N-Cadherin expression in human prostate carcinoma cell lines. An epithelial-mesenchymal transformation mediating adhesion withStromal cells. Am. J. Pathol. 1999, 155: 787-98.

109. Otto T, et al. E-cadherin: a marker for differentiation and invasiveness in prostatic carcinoma. Urol. Res. 1993, 21: 359-62.
110. Bussemakers MJ, et al. Decreased expression of E-cadherin in the progression of rat prostatic cancer. Cancer Res. 1992, 52: 2916-22.
111. Umbas R, et al. Expression of the cellular adhesion molecule E-cadherin is reduced or absent in high-grade prostate cancer. Cancer Res. 1992, 52: 5104-9.
112. Richmond PJ, et al. Aberrant E-cadherin and alpha-catenin expression in prostate cancer: correlation with patient survival. Cancer Res. 1997, 57: 3189-93.
113. Hazan RB, Norton L. The epidermal growth factor receptor modulates the interaction of E-cadherin with the actin cytoskeleton. J. Biol. Chem. 1998, 273: 9078-84.
114. Suyama K, et al. A signaling pathway leading to metastasis is controlled by N-cadherin and the FGF receptor. Cancer Cell 2002, 2: 301-14.
115. Hazan RB, et al. Cadherin switch in tumor progression. Ann. N. Y. Acad. Sci. 2004, 1014: 155-63.
116. Kang Y, Massague J. Epithelial-mesenchymal transitions: twist in development and metastasis. Cell 2004, 118: 277-9.
117. Hiscox S, Jiang WG. Hepatocyte growth factor/scatter factor disrupts epithelial tumour cell-cell adhesion: involvement of beta-catenin. Anticancer Res. 1999, 19: 509-17.
118. Powell WC, et al. Expression of the metalloproteinase matrilysin in DU-145 cells increases their invasive potential in severe combined immunodeficient mice. Cancer Res. 1993, 53: 417-22.
119. Nishimura K, et al. Effects of hepatocyte growth factor on urokinase-type plasminogen activator (uPA) and uPA receptor in DU145 prostate cancer cells. Int. J. Androl. 2003, 26: 175-9.
120. Mamoune A, et al. DU145 human prostate carcinoma invasiveness is modulated by urokinase receptor (uPAR) downstream of epidermal growth factor receptor (EGFR) signaling. Exp. Cell Res. 2004, 299: 91-100.
121. Magdolen V, et al. Inhibition of the tumor-associated urokinase-type plasminogen activation system: effects of high-level synthesis of soluble urokinase receptor in ovarian and breast cancer cells in vitro and in vivo. Recent Results Cancer Res. 2003, 162: 43-63.
122. Andreasen PA, Egelund R, Petersen HH. The plasminogen activation system in tumor growth, invasion, and metastasis. Cell Mol. Life Sci. 2000, 57: 25-40.
123. Festuccia C, et al. Plasminogen activator system modulates invasive capacity and proliferation in prostatic tumor cells. Clin. Exp. Metastasis 1998, 16: 513-28.
124. MacDonald TJ, DeClerck YA, Laug WE. Urokinase induces receptor mediated brain tumor cell migration and invasion. J. Neurooncol. 1998, 40: 215-26.
125. Binder BR. Influence of urokinase on cell proliferation and invasion. Blood Coagul. Fibrinolysis 1990, 1: 717-20.
126. Jo M, et al. Epidermal growth factor receptor-dependent and -independent cell-signaling pathways originating from the urokinase receptor. J. Biol. Chem. 2003, 278: 1642-6.
127. Pai R, et al. Prostaglandins promote colon cancer cell invasion; signaling by cross-talk between two distinct growth factor receptors. Faseb. J. 2003, 17: 1640-7.
128. Mars WM, et al. Immediate early detection of urokinase receptor after partial hepatectomy and its implications for initiation of liver regeneration. Hepatology 1995, 21: 1695-701.

129. Okusa Y, et al. Urokinase type plasminogen activator and its receptor regulate the invasive potential of gastric cancer cell lines. Int. J. Oncol. 2000, 17: 1001-5.
130. Nagle RB, et al. Adhesion molecules, extracellular matrix, and proteases in prostate carcinoma. J. Cell Biochem. Suppl. 1994, 19: 232-7.
131. Knox JD, et al. Differential expression of extracellular matrix molecules and the alpha 6-integrins in the normal and neoplastic prostate. Am. J. Pathol. 1994, 145: 167-74.
132. Cress AE, et al. The alpha 6 beta 1 and alpha 6 beta 4 integrins in human prostate cancer progression. Cancer Metastasis Rev. 1995, 14: 219-28.
133. Rabinovitz I, Nagle RB, Cress AE. Integrin alpha 6 expression in human prostate carcinoma cells is associated with a migratory and invasive phenotype in vitro and in vivo. Clin. Exp. Metastasis 1995, 13: 481-91.
134. Bonaccorsi L, et al. Androgen receptor expression in prostate carcinoma cells suppresses alpha6beta4 integrin-mediated invasive phenotype. Endocrinology 2000, 141: 3172-82.
135. Zheng DQ, et al. Prostatic carcinoma cell migration via alpha(v)beta3 integrin is modulated by a focal adhesion kinase pathway. Cancer Res. 1999, 59: 1655-64.
136. Stewart DA, Cooper CR, Sikes RA. Changes in extracellular matrix (ECM) and ECM-associated proteins in the metastatic progression of prostate cancer. Reprod. Biol. Endocrinol. 2004, 2: 2.
137. Demetriou MC, Cress AE. Integrin clipping: a novel adhesion switch? J. Cell Biochem. 2004, 91: 26-35.
138. Trikha M, et al. Role of alphaII(b)beta3 integrin in prostate cancer metastasis. Prostate 1998, 35: 185-92.
139. Moretti RM, et al. Role of the orphan nuclear receptor ROR alpha in the control of the metastatic behavior of androgen-independent prostate cancer cells. Oncol. Rep. 2002, 9: 1139-43.
140. Fornaro M, et al. Fibronectin protects prostate cancer cells from tumor necrosis factor-alpha-induced apoptosis via the AKT/survivin pathway. J. Biol. Chem. 2003, 278: 50402-11.
141. Fornaro M, Manes T, Languino LR. Integrins and prostate cancer metastases. Cancer Metastasis Rev. 2001, 20: 321-31.
142. Bar-Sagi D, Hall A. Ras and Rho GTPases: a family reunion. Cell 2000, 103: 227-38.
143. Sahai E, Marshall CJ. RHO-GTPases and cancer. Nat. Rev. Cancer 2002, 2: 133-42.
144. Kaibuchi K, Kuroda S, Amano M. Regulation of the cytoskeleton and cell adhesion by the Rho family GTPases in mammalian cells. Annu. Rev. Biochem. 1999, 68: 459-86.
145. Fukata Y, Amano M, Kaibuchi K. Rho-Rho-kinase pathway in smooth muscle contraction and cytoskeletal reorganization of non-muscle cells. Trends Pharmacol. Sci. 2001, 22: 32-9.
146. Muller SM, Okan E, Jones P. Regulation of urokinase receptor transcription by Ras- and Rho-family GTPases. Biochem. Biophys. Res. Commun. 2000, 270: 892-8.
147. Montaner S, et al. Activation of serum response factor by RhoA is mediated by the nuclear factor-kappaB and C/EBP transcription factors. J. Biol. Chem. 1999, 274: 8506-15.

148. Hodge JC, et al. Requirement of RhoA activity for increased nuclear factor kappaB activity and PC-3 human prostate cancer cell invasion. Cancer Res. 2003, 63: 1359-64.
149. Gao Y, et al. Rational design and characterization of a Rac GTPase-specific small molecule inhibitor. Proc. Natl. Acad. Sci. USA 2004, 101: 7618-23.
150. Sanders LC, et al. Inhibition of myosin light chain kinase by p21-activated kinase. Science 1999, 283: 2083-5.
151. Davila M, et al. LIM kinase 1 is essential for the invasive growth of prostate epithelial cells: implications in prostate cancer. J. Biol. Chem. 2003, 278: 36868-75.
152. Yoshioka K, et al. A role for LIM kinase in cancer invasion. Proc. Natl. Acad. Sci. USA 2003, 100: 7247-52.
153. Mochizuki H, et al. Interaction of ligand-receptor system between stromal-cell-derived factor-1 and CXC chemokine receptor 4 in human prostate cancer: a possible predictor of metastasis. Biochem. Biophys. Res. Commun. 2004, 320: 656-63.
154. Darash-Yahana M, et al. Role of high expression levels of CXCR4 in tumor growth, vascularization, and metastasis. Faseb. J. 2004, 18: 1240-2.
155. Sun YX, et al. Expression of CXCR4 and CXCL12 (SDF-1) in human prostate cancers (PCa) in vivo. J. Cell Biochem. 2003, 89: 462-73.
156. Taichman, RS, et al. Use of the stromal cell-derived factor-1/CXCR4 pathway in prostate cancer metastasis to bone. Cancer Res. 2002, 62: 1832-7.
157. Reiland J, Furcht LT, McCarthy JB. CXC-chemokines stimulate invasion and chemotaxis in prostate carcinoma cells through the CXCR2 receptor. Prostate 1999, 41: 78-88.
158. Shulby SA, et al. CX3CR1-fractalkine expression regulates cellular mechanisms involved in adhesion, migration, and survival of human prostate cancer cells. Cancer Res. 2004, 64: 4693-8.
159. Veltri, RW, et al. Interleukin-8 serum levels in patients with benign prostatic hyperplasia and prostate cancer. Urology 1999, 53: 139-47.
160. Tanji N, Aoki K, Yokoyama M. Growth factors: roles in andrology. Arch. Androl. 2001, 47: 1-7.
161. Koivisto P, et al. Androgen receptor gene amplification: a possible molecular mechanism for androgen deprivation therapy failure in prostate cancer. Cancer Res. 1997, 57: 314-9.
162. Koivisto P, et al. Analysis of genetic changes underlying local recurrence of prostate carcinoma during androgen deprivation therapy. Am. J. Pathol. 1995, 147: 1608-14.
163. Latil A, et al. Evaluation of androgen, estrogen (ER alpha and ER beta), and progesterone receptor expression in human prostate cancer by real-time quantitative reverse transcription-polymerase chain reaction assays. Cancer Res. 2001, 61: 1919-26.
164. Li R, et al. High level of androgen receptor is associated with aggressive clinicopathologic features and decreased biochemical recurrence-free survival in prostate: cancer patients treated with radical prostatectomy. Am. J. Surg. Pathol. 2004, 28: 928-34.
165. Hirawat S, Budman DR, Kreis W. The androgen receptor: structure, mutations, and antiandrogens. Cancer Invest. 2003, 21: 400-17.
166. Culig Z, et al. Androgen receptor gene mutations in prostate cancer. Implications for disease progression and therapy. Drugs Aging. 1997, 10: 50-8.

167. Hofer DR, et al. Autonomous growth of androgen-independent human prostatic carcinoma cells: role of transforming growth factor alpha. Cancer Res. 1991, 51: 2780-5.
168. Sherwood ER, et al. Epidermal growth factor receptor activation in androgen-independent but not androgen-stimulated growth of human prostatic carcinoma cells. Br. J. Cancer 1998, 77: 855-61.
169. Heinlein CA, Chang C. Androgen receptor in prostate cancer. Endocr. Rev. 2004, 25: 276-308.
170. Grossmann ME, Huang H, Tindall DJ. Androgen receptor signaling in androgen-refractory prostate cancer. J. Natl. Cancer Inst. 2001, 93: 1687-97.
171. Brass AL, et al. Androgen up-regulates epidermal growth factor receptor expression and binding affinity in PC3 cell lines expressing the human androgen receptor. Cancer Res. 1995, 55: 3197-203.
172. Morris MJ, et al. HER-2 profiling and targeting in prostate carcinoma. Cancer 2002, 94: 980-6.
173. Cinar B, et al. Androgen receptor mediates the reduced tumor growth, enhanced androgen responsiveness, and selected target gene transactivation in a human prostate cancer cell line. Cancer Res. 2001, 61: 7310-7.
174. Horvath LG, et al. Frequent loss of estrogen receptor-beta expression in prostate cancer. Cancer Res. 2001, 61: 5331-5.
175. Cheng J, et al. Expression of estrogen receptor beta in prostate carcinoma cells inhibits invasion and proliferation and triggers apoptosis. FEBS Lett. 2004, 566: 169-72.
176. Bonkhoff H, Motherby H, Fixemer T. [New insights into the role of estogens and their receptors in prostate cancer]. Urologe A. 2003, 42: 1594-601.
177. Neubauer, BL, et al. The selective estrogen receptor modulator trioxifene (LY133314) inhibits metastasis and extends survival in the PAIII rat prostatic carcinoma model. Cancer Res. 2003, 63: 6056-62.
178. Zhu X, et al. Dynamic regulation of estrogen receptor-beta expression by DNA methylation during prostate cancer development and metastasis. Am. J. Pathol. 2004, 164: 2003-12.
179. Bonkhoff H, et al. Progesterone receptor expression in human prostate cancer: correlation with tumor progression. Prostate 2001, 48: 285-91.
180. Yoganathan N, et al. Integrin-linked kinase, a promising cancer therapeutic target: biochemical and biological properties. Pharmacol. Ther. 2002, 93: 233-42.
181. Tucker GC. Alpha v integrin inhibitors and cancer therapy. Curr. Opin. Investig. Drugs 2003, 4: 722-31.
182. Levitzki A, Gazit A. Tyrosine kinase inhibition: an approach to drug development. Science 1995, 267: 1782-8.
183. Bennasroune A, et al. Tyrosine kinase receptors as attractive targets of cancer therapy. Crit. Rev. Oncol. Hematol. 2004, 50: 23-38.
184. Rasmussen, HS, McCann PP. Matrix metalloproteinase inhibition as a novel anticancer strategy: a review with special focus on batimastat and marimastat. Pharmacol. Ther. 1997, 75: 69-75.
185. Muehlenweg B, et al. Interference with the urokinase plasminogen activator system: a promising therapy concept for solid tumours. Expert Opin. Biol. Ther. 2001, 1: 683-91.

186. Tran N, et al. N-Cadherin expression in human prostate carcinoma cell lines. An epithelial-mesenchymal transformation mediating adhesion with Stromal cells. Am. Jour. Pathology. 1999, Vol 155: 787-798.
187. Klezovitch O, et al: Hepsin promotes prostate cancer progression and metastasis. Cancer Cell 2004, Aug;6:185-95.

COLOR PLATES

CHAPTER 2

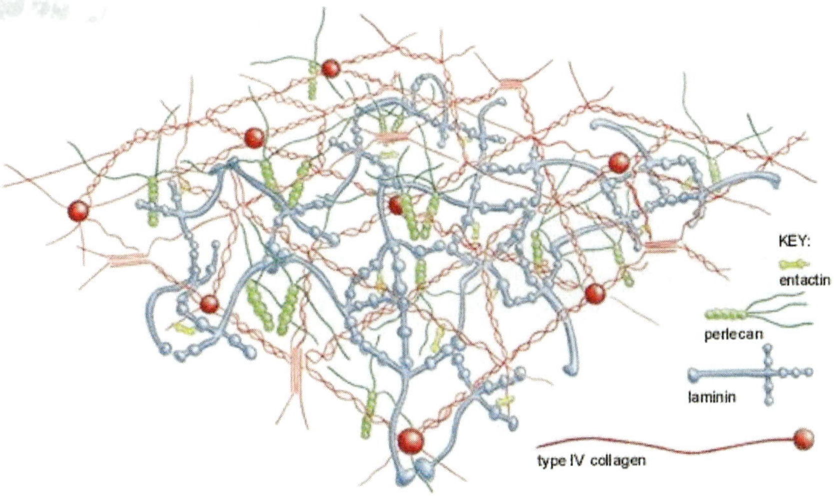

Figure 1. The complicated inter-relationships of the major components of the basement membrane are shown in a diagram to scale based on detailed analyses of molecular interactions by Dr. Peter D. Yurchenco. In the diagram, Collagen Type IV (Col-IV) forms one network. Laminin (Lm) forms another. Entactin (En) and Perlecan (Perl) interact with the two networks. HS is the heparan sulfate glycosaminoglycans linked to the perlecan core protein. (16 adapted with permission).

CHAPTER 3

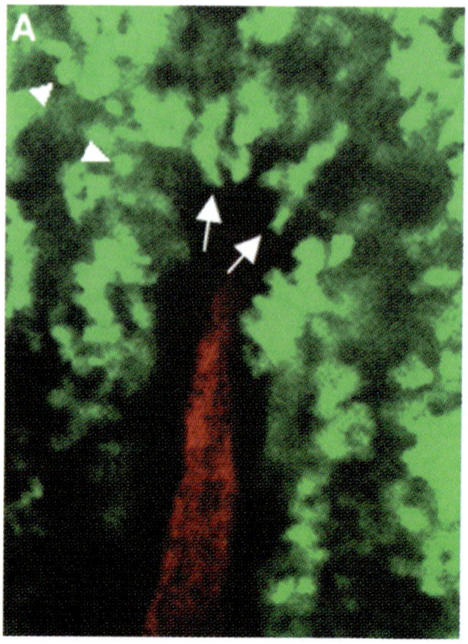

Figure 1. Single photon imaging of a tumor formed by MTLn3 cells expressing GFP and a blood vessel visualized using rhodamine dextran. From (1).

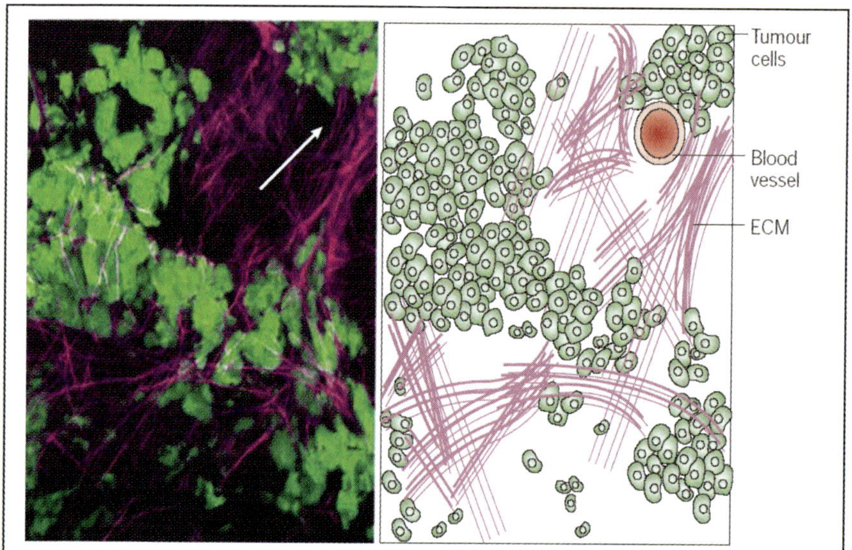

Figure 3. ECM fibres converge on blood vessels in mammary tumours. Multiphoton microscopy shows that carcinoma cells (green) move on extracellular-matrix fibres (purple), some of which converge on blood vessels (arrow). From (18).

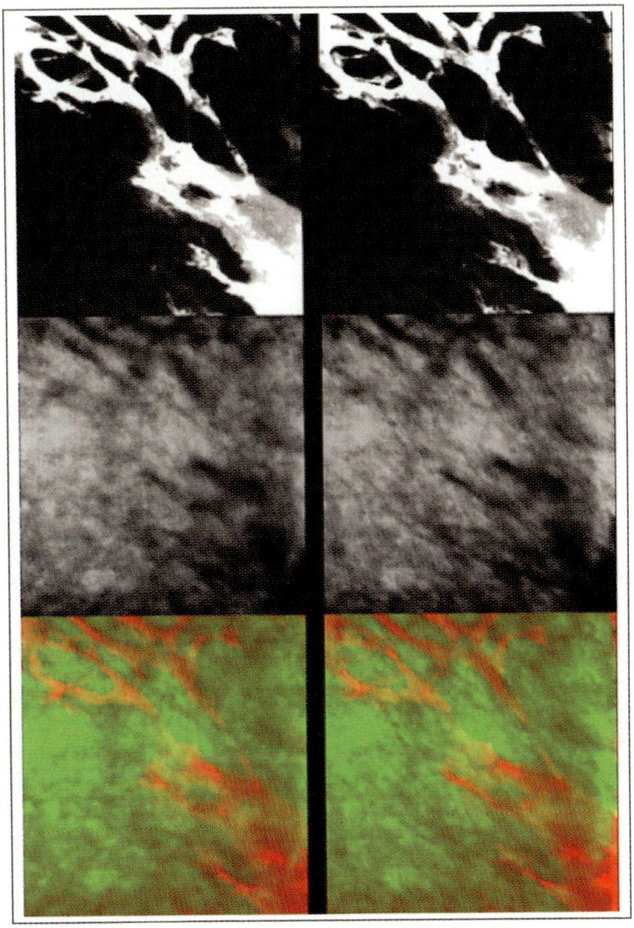

Figure 4. Blood vessel structure determined by multiphoton microscopy. Animals were injected with Texas Red dextran (70,000 MW) and then the tumors were imaged taking a slice every 5 microns. The z series were then projected to generate a stereo image showing blood vessels (top and red) and tumor cells (middle and green).

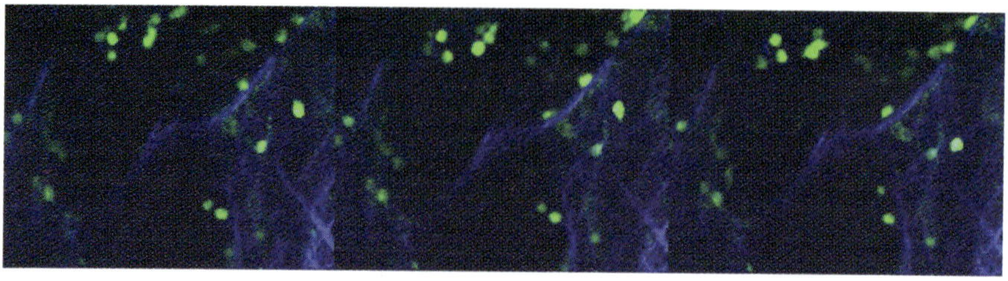

Figure 5. Movement of macrophages in tumors. A knockin of GFP at the lys promoter (courtesy of Dr. T. Graf) (34) was crossed with the MMTV-PyMT strain to generate tumors containing GFP labeled macrophages. The tumors were imaged for GFP (green) and matrix fibers (2nd harmonic, blue). Three frames from a time-lapse sequence are shown, demonstrating the high motility of the GFP labeled cells.

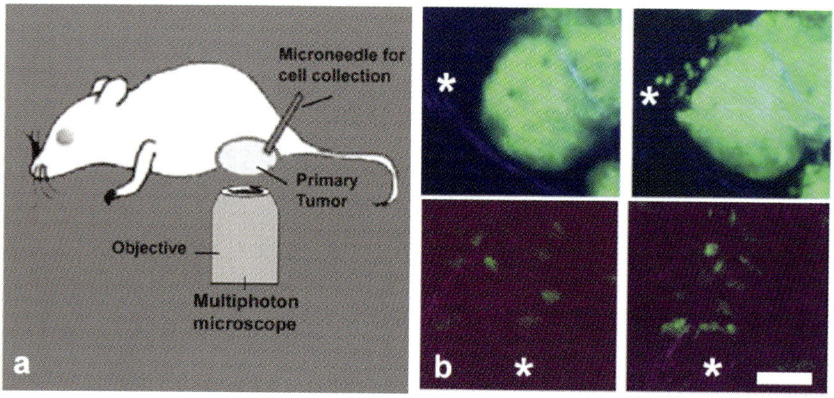

Figure 6. Movement of macrophages and carcinoma cells into collection needles in response to growth factors. A, Model shows how the experiment is performed. The primary tumor in the right number 5 mammary gland is selected to minimize breathing motion, and the position of the collection needle is controlled with a micromanipulator (not shown). B, movement of fluorescent carcinoma cells (top, WAP-Cre/CAG-CAT-EGFP/MMTV-PyMT tumor) and macrophages (bottom, MMTV-PyMT/lys-GFPKi tumor) toward EGF-containing collecting needles. The approximate opening of the collection needle is shown in each field. Each image is a 50-um z-projection and is from a time-lapse series. Images on the right were recorded 90 minutes after images on the left; bar, 25 um. From (36).

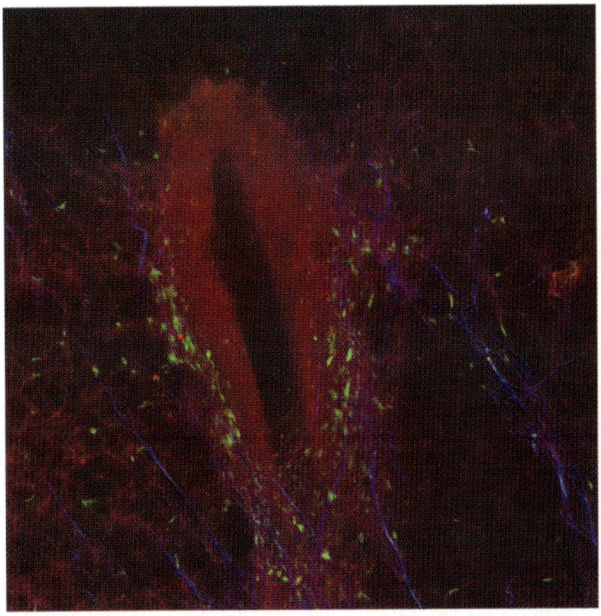

Figure 7. Example of terminal end bud imaging from Mac green female showing green fluorescent macrophages and collagen structures surrounding a terminal end bud (stained with propidium iodide, red). Macrophages (green) and collagen (blue) appear to be co-localized.

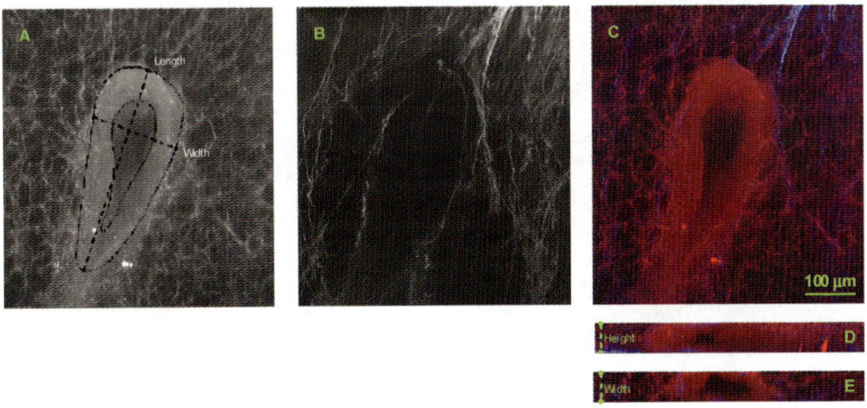

Figure 8. Example of 3-dimensional imaging of terminal end bud in op/+ mice. Fixed whole mounts were imaged with a red filter (A) for propidium iodide staining, and a blue filter (B) for collagen matrix by second harmonic generation, and merged using Image J software (C). These images are 22 um stacks through the largest part of the terminal end bud. The images were resliced in 2 um steps from left to right (D) and top to bottom (E) to produce images showing the length and height, and width and height respectively.

CHAPTER 4

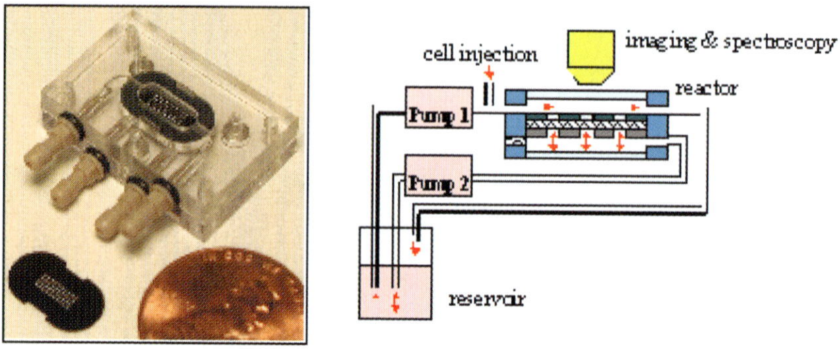

Figure 2. Photo of the silicon-chip scaffold (lower left) and polycarbonate microreactor housing that holds the scaffold and provides four ports for continuous flow of culture medium across the top of the scaffold and through the tissue mass held within the scaffold. An optical window at the top allows in-situ observation of the tissues by two-photon microscopy. Right: Schematic of scaffold and reactor housing shown in cross-section, indicating the connections to recirculation loop and observation window. The main flow of culture medium is across the top of the chip in the upper chamber (pump 1). Flow through the tissue mass in the chip is maintained at a constant rate by either pulling medium through or pumping medium in reverse flow through with the second pump.

CHAPTER 5

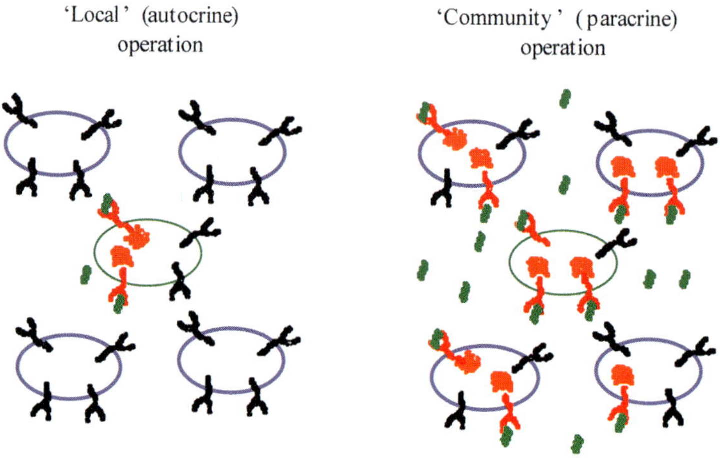

Figure 1. Illustration of autocrine vs. paracrine signaling. The green circle represents a cell that produces a ligand and expresses the receptor for the ligand. The blue circles are neighboring cells that also express the receptor. Red receptors are occupied by ligand, which leads to intracellular signling. (Left) Case of autocrine operation, essentially all of the ligand produced is captured locally. (Right) Case of paracrine operation, with very little ligand captured by the producing cell. (Adapted from DeWitt et al., 2002).

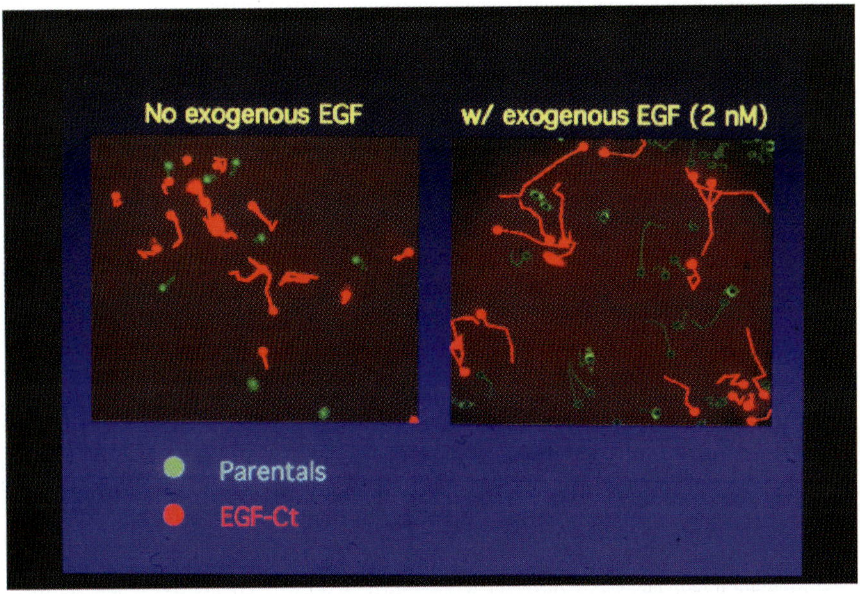

Figure 2. Autocrine producing EGF-Ct cells do not stimulate migration of neighboring parental hMEC cells. The initial location of EGF-Ct (orange) and parental hMEC (green) cells and their subsequent cell paths are shown in the absence (left, Control) and presence (right, +EGF) of 2 nM exogenous EGF. (Adapted from Maheshwari et al., 2001).

INDEX

Cancer Metastasis – Biology and Treatment

1. W.G. Jiang and R.E. Mansel (eds.): *Cancer Metastasis, Molecular and Cellular Mechanisms and Clinical Intervention.* 2000 ISBN 0-7923-6395-7
2. W.G. Jiang, K. Matsumoto and T. Nakamura (eds.): *Growth Factors and their Receptors in Cancer Metastasis.* 2001 ISBN 0-7923-7141-0
3. D.R. Welch (ed.): *Cancer Metastasis-Related Genes.* 2002 ISBN 1-4020-0522-9
4. J.-M. Foidart and R.J. Muschel (eds.): *Proteases and their Inhibitors in Cancer Metastasis.* 2002 ISBN 1-4020-0923-2
5. K. Pantel (ed.): *Micrometastasis.* 2003 ISBN 1-4020-1155-5
6. G. Singh and W. Orr (eds.): *Bone Metastasis and Molecular Mechanisms.* 2004 ISBN 1-4020-1984-X
7. M. Esteller (ed.): *DNA Methylation, Epigenetics and Metastasis.* 2005 ISBN 1-4020-3641-8
8. A Wells (ed.): *Cell Motility in Cancer Invasion and Metastasis.* 2006 ISBN 1-4020-4008-3